International Union of Geological Sciences, Series A, Number 3

Ores in Sediments

VIII. International Sedimentological Congress
Heidelberg, August 31 – September 3, 1971

Edited by
G. C. Amstutz and A. J. Bernard

With 184 Figures

Springer-Verlag Berlin Heidelberg New York 1973

Professor G. C. Amstutz
Mineralogisch-Petrographisches Institut der Universität, Heidelberg/BRD

Prof. A. J. Bernard
Laboratoire de Métallogénie, Ecole Nationale Supérieure de Géologie Appliquée et de
Prospection Minière, Nancy/France

Sponsored by the Society of Geology Applied to Mineral Deposits (SGA) and the International
Association of Sedimentology.

For the cover Fig. 4 on page 230 of this volume have been used.

Number 1 of this series appeared 1969, Number 2 1971 in Schweizerbart'sche
Verlagsbuchhandlung, Stuttgart.

ISBN 3-540-05712-9 Springer-Verlag Berlin Heidelberg New York
ISBN 0-387-05712-9 Springer-Verlag New York Heidelberg Berlin

Preface

In 1963 the first Symposium on "Ores in Sediments" took place as part of an International Sedimentological Congress. At the end of that first Symposium, the group then assembled adopted a resolution printed in the book which resulted from it (AM-STUTZ, 1964, p. 7), and points (3), (4) and (5) read as follows:

(3) The group considers the integration of sedimentology in any study of ore deposits in sediments essential to a correct interpretation. A study of the role of sedimentary processes, including diagenesis, is an important field in pure as well as in applied research on the genesis of mineral deposits.

(4) In particular, the group also considers the knowledge of sedimentary rocks and processes (in regard to both, the fabric and the geochemical detail) a prerequisite for the understanding of subsequent metamorphic processes and their possible role in the deformation and reconstitution of mineral deposits and host rocks.

(5) The group suggests that similar symposia could with advantage be held at future Congresses of the International Association of Sedimentologists.

The Editors wish to thank the International Association of Sedimentology for including another Symposium on ore minerals in its Congress program.

Considerable progress has been made since 1963, as the reader will see from the following pages. A few instances only can be mentioned here. On a small scale, still more details on the diagenetic crystallization differentiation are now available. The recognition of its existence has gained much ground and slowly the simple explanation which it offers for accumulations of certain late diagenetic sulfides (e. g. galena) in late diagenetic spaces such as compaction fissures and intraformational breccias, is accepted and the assumptions of epigenetic mimetic replacements are gradually recognized to be untenable.

Almost all papers of this Symposium refer, in one way or another, to the diagenetic behaviour or role of ore minerals. Most papers also refer to the facies and paleo-geographic relations of ore mineral formation. This facet of the Symposium should be of specific interest to those active in exploration.

An additional type of stratabound deposits has been brought into the limelight and provides an answer to very magny enigmatic deposits at or below erosional unconformities: the karst deposits (BERNARD; PADALINO et al.).

Observations on recent deposits, many still in the process of formation, provide first-hand, direct proof of the formation of ore deposits in sediments. The number of recent deposits known is increasing fast, and more and more ore genetic inter-pretations can be based also on actualistic analogies (HONNOREZ et al.; LEMOAL-LE and DUPONT; PUCHELT; VALETTE).

These are only a few of the new results presented in this Symposium.

If one now looks at the content and tries to classify the papers, various ways of grouping them would appear to be useful. As just mentioned, certain aspects are common to almost all papers. The strongest differences exist with regard to the age of the deposits, the facies and the nature of the ore minerals. The common facets being preponderant, it was decided to arrange the papers alphabetically, except for the introductory paper of the Symposium.

For those interested in a subdivision according to topics, the previous comments plus the following list will help. This list attempts a subdivision of the papers according to the classification given in the first Symposium volume of 1963/64, plus a class of papers on more general principles.

A) Reduzate deposits:

ARNOLD, MAUCHER and SAUPE
BARTHOLOME et al.
BERNARD
BOGDANOV and KUTYREV
BRONDI, CARRARA and POLIZZANO
COLLINS and SMITH
GELDSETZER
HONNOREZ et al.
MONSEUR and PEL
PADALINO et al.
PUCHELT
SAMAMA
SCHADLUN
VALETTE
ZIMMERMANN and AMSTUTZ

B) Oxidate deposits:

BERNARD
BRONDI, CARRARA and POLIZZANO
DOYEN
EARGLE and WEEKS
GERMANN
LEMOALLE and DUPONT
MENGEL
PADALINO et al.
PUCHELT
SAMAMA

C) Sulphate and phosphate deposits:

BERNARD
SAMAMA

D) Detrital deposits (placers, sands etc.

ARNOLD, MAUCHER and SAUPE
MENGEL
MONSEUR and PEL
SESTINI
TOURTELOT and RILEY

E) Papers on general principles
(not pertaining necessarily only to A, B, C or D):

BERNARD (Introduction)
BERNARD
MONSEUR and PEL
POPOV
SAMAMA
VALETTE

Last but not least, I wish to thank the Springer-Verlag for the interest in this book and the effort put into its proper presentation. The informed reader will realize that the fast and simplified offset printing method implies some sacrifice regarding the quality of the figures - which depends largely on the quality of the material received.

Thanks are also due to those staff members in the Mineralogical Institute of Heidelberg who contributed freely of their time for editorial reading, especially Dr. R.A. Zimmermann, Dr. R. Saager and Mr. E. Schot. Mrs. W. Ackermann accomplished not only a masterpiece of typing for offset printing, but also contributed much to the editorial detail work, and Mr. E. Gerike upgraded a good number of imperfect drawings.

Heidelberg, November 1972   G.C. AMSTUTZ

Reference

AMSTUTZ, G.C.: Introduction, In: Developments in Sedimentology, Volume 2: Sedimentology and Ore Genesis (ed. AMSTUTZ, G.C.), p. 1-7. Amsterdam: Elsevier 1963/64.

Contents

A Review of Processes Leading to the Formation of Mineral Deposits in Sediments

A. J. Bernard

A few years ago, eminent scientists still postulated seriously a deep seated, magmatic origin of petroleum. These scientists even maintained that the accumulation of petroleum in sedimentary host rocks after long and intricate migration did not speak against its fundamentally plutonic origin.

This was a typical example of a dogmatic way of thinking or behaviour. It allowed the theorists to cling to their ideas, whereas the prospectors - the practical men - were expected to restrict their interest to oil reservoirs and traps. The factual knowledge of oil deposits consequently still progressed, even though the genetic concept evolved only slowly or not at all. Historically, the work of the prospecting geologist must, therefore, be considered as unstrumental for the present state of our understanding of petroleum as a sediment.

The statistical evaluation of the significance of a certain number of oil-forming environments and the fundamental research on the genesis of sedimentary rocks were efforts which are entirely justified by their economic importance. Thus, it is safe to say that the beginning of sedimentology, or at least its rapid development during the last 25 years, was mainly invoked by the stimulations of research on oil.

Oil-forming environments are characterized by the occurrence of a certain subsidence which not only caused the burial and thus preservation of organic, mostly marine matter, but which transformed them also diagenetically into oil. At the same time, the concept of migration yielded progressively to a concept which assigns more importance to synsedimentary traps, i. e. traps which were at least partially closed by sedimentary processes.

The process of oil formation and accumulation represents thus quite an elementary genetic model. Sedimentation and burial of ultrafine, detritic organic material was followed by a diagenetic evolution in situ under the effects of low pressure and temperature conditions. An additional sine qua non condition which is quite specific for hydrocarbons is the fact that the accumulated oil is fossilized almost in situ together with its connate waters. Furthermore, the fluid nature of oil imposes specific conditions on the existence of the deposits.

Having discussed the economically most important sedimentary "mineral", i. e. oil, whose prospection and genesis had so far-reaching sedimentological implications, I shall proceed with a few thoughts on "solid fuels" which were in earlier times as important as oil today.

The abundance of vegetal remnants which are present in coal saved it from "the honour" of a magmatic origin. This is the reason why paleontological and petrological studies since a long time enabled detailed paleogeographic reconstructions. The environments of epeirogenetic intra-cratonic and coastal basins led to successful regional prospecting. Locally, however, the sequential positions (cyclothems) of coal beds still leave unanswered questions concerning their description and interpretation.

If the paleoclimatology of coal deposits, the biological and the Eh-pH conditions of the ultra-detrital accumulation of vegetal matter, the diagenesis or even the epimeta-morphism of these materials and other questions were answered, coal producing environments could be much better understood and thus much better prospected. On the whole, there is still much to be learned from these questions and many more observations need to be made.

With the exception of the chronology problems and the questions which exist on the processes closing oil traps, the conceptual models of fuel deposits are relatively simple. In contrast, economic geologists presently studying metallic ore deposits are confronted with more involved problems which are in the following discussed in the order of their growing complexity.

The problems posed by the heavy mineral placers are best solved by studying the sedimentation processes of detrital rocks. Hydroclassification (based on similarity of properties) apparently answers most of the questions pertaining to the petrography and especially to the grain-size of these deposits. In fluviatile environments, however, the natural jigging of heavy minerals within their associated alluvium often destroys the expected simple grain-size relationships. In spite of our far-reaching knowledge of heavy mineral placers, it is still a hazardous operation to reconstruct the ancient river systems, to localize rapids, and to assess their evolutions, which are responsible for the formation of paystreaks. Very often a prospector has more confidence in a systematic drilling campaign than in a difficult paleogeographical reconstitution.

Finally, a strange theory, which will be discussed during this congress, maintained that there were no pre-Tertiary placers. This, of course, is wrong, but the odd reason of such a tale needs to be mentioned. Economically most placers are workable only if not completely indurated. Once lithified, their grade is usually too low to warrant mining which, in hard rock, would be too costly. The theory mentioned was born by the fact that most pre-Tertiary placers are lithified, which makes their mining uneconomic. For instance, in the discussions on the genesis of the Precambrian uranium and gold conglomerates of the Witwatersrand (South Africa), the old theory was the reason for a lot of misconceptions. The final argument for the sedimentary explanation of this deposit was brought out by sedimentological studies. It was the determination of reworking of daltaic accumulations by coastal currents and their redistribution in bankets, which gave the key to the distribution of workable reefs. To be more exact, the uranium and gold mineralizations are border facies of basins, located less than 60 km from the deltaic zones. Knowing this crucial observation, the prospectors stopped paying attention to the theories of hydrothermal impregnation which were defended by certain authorities, as was the case with the magmatic origin of petroleum.

As simple as it may look, the sedimentation of heavy detrital minerals still leaves some difficult problems. This is for instance the case, when several phases of reworking occurred or when the off-shore prospecting of placers on present-day shelve is considered.

To explain the oolitic iron ores - at least those of the Lorraine - deltaic and detrital sedimentation processes have to be used. In the case of the Lorraine deposits, the sedimentological studies enabled, almost layer by layer, to reconstruct the paleo-geography of the deltaic estuary which occupied the Gulf of Luxembourg in Upper Toarcian times. The remarkably detailed maps resulting from this work indicate the zones of oolite formation and localize the spreading zones of granular iron oolites amidst coastal muddy grounds. The paper of LEMOALLE and DUPONT corroborates this statement on the basis of a study of recent oolites in the Lake Chad.

This model, however, is rendered more complicated by the question of the source and nature of the extraordinary iron supply which led to the Lorraine deposits. A formation by the emersion and erosion of bituminous and pyrite-bearing shales of the Lower Toarcian offers a particularly elegant explanation. Elsewhere, i.e. in the Peine and Salzgitter deposits of Germany, the erosion of a lateritic soil-cover and the near-shore distribution of the iron concentrations is a similarly elegant explanation for a high metal supply.

The localization of marine, epicontinental manganese deposits along ancient shore-lines, be they oolitic or not, may be explained by similar processes. Therefore, the question can be asked, whether the main stage of concentration did not occur before the detrital sedimentation and whether it consisted of chemical precipitation (or flocculation) of metal-rich terrigenous and fluviatile solutions (or suspensions). The zone of mixing of continental acid waters poor in dissolved salts with basic seawaters of high ionic strength is usually a very efficient geochemical trap.

On the whole, terrigenous sources are responsible for the supply of exceptional amounts of metals, the marine and litoral environments providing only the trapping medium. We are approaching now the important problem concerning the nature of these exceptional supplies of metals. The biological Eh-pH theory (and its heterostasic variant) with the aid of the soil science and of climatology considers only terrigenous aspects; however, the pure marine as well as the exhalative contributions (so spectacular when sub-marine) must also be considered. In the order of growing complexity of the problem, the marine, the exhalative and finally the terrigenous supplies or sources are discussed; the latter leads again back to the difficult and important problem of continental ore genesis.

The Marine Sources

Pure marine sources are best illustrated by phosphate sedimentation which can be explained by the classical theory of upwelling cold streams. Seawater has its highest P_2O_5-content at a depth of 350 to 1000 m. Cold currents rising up to the water surface, for instance along continental slopes, penetrate zones where photo-synthesis takes place. In these zones, large portions of the phosphorus are consumed by plankton which in turn gives rise to a particularly rich "biocenosis". A second phosphorus concentration takes place within living organisms. But even the two concentration effects put together can only form rich phosphate deposits, if the corresponding sedimentary environment is free of or at least depleted in phosphorus-free terrigenous material. Upwelling cold currents, without doubt, can inhibit the terrigenous supply and sedimentation in the shelf zones affected by them. By creating an arid microclimate in the continental hinterland adjacent to the zones of upraise, the presence of upwelling cold currents can perhaps even explain the peculiarities of the terrigenous sediments in phosphatic environments (cherts, Mg-clays).

It is easy to drive the structural and paleogeographic consequences of such a model (epochs of phosphate-genesis, e.g.) which may be used either for regional or local prospecting.

Marine evaporites are a good example for a pure marine source of material. The evaporation of seawater leads to successive precipitations of salts which have been studied in great detail. The explanation of very thick salt accumulations by rhythmical layering led very early to the research of the structures which were responsible for the observed sequences, i. e. coastal lagoons, large closed bays, border areas of epicontinental basins etc. Subsidence rhythms, nature and periodicity of seawater inbreaks, importance of the continental water inlets with respect to the basin size, and climatic conditions were, among others, the main factors used to explain the large mineralogical, petrographic and sequential diversity of old evaporite series Only recently interaction pehnomena were noticed between brines (evolving through periodical additons of saline or fresh waters) and already deposited salts. Similar processes (sometimes called "metamorphism") occur during the diagenesis of evaporites, for instance in the case of potassic salts.

As simple as the case of the evaporites may look, the number and the diversity of the parameters which can influence the deposition is so complex that each study of a deposit is always a difficult task. Being aware of their scarcity in ancient series, I shall just mention the extreme diversity of continental evaporites. The complexity of marine evaporite resulting from the evaporation of a unique and well known solution, i. e. seawater, is in the case of continental evaporite amplified by the variability of the original waters. Their composition depends largely on the hydrogeology of the different evaporating lakes.

Incidentally, the very special chemistry of evaporitic and pre-evaporitic environments leads to warranted models for:
- sedimentary deposits of native sulfur in reducing environments with high anaerobic bacterial activities;
- barite related to cherty strata;
- siderite and magnesite deposits which are well explained by the arrival of iron and calcium rich continental waters in penesaline lagoons possessing a syngenetic dolo mite sedimentation.

However, in the last three cases (barite, siderite and magnesite), the contribution is terrigenous, whereas the chemical trap belongs to the evaporitic environment.

The Exhalative Source

I shall be brief about exhalative-sedimentary sources. The submarine environment only provides the adequate conditions for trapping and fossilizing the concentrations of iron, manganese, pyrite and associated chalcophile metals. Quartz-keratophyric and spilitic volcanism of the pre-orogenic troughs, tholeiitic volcanism of the oceanic ridges, and perhaps alcaline volcanism of the submarine seamounts are potential environments for the exhalation of metal concentrations. In these cases, the problem lies more in the nature of the exhalations and in the mode of their formation than in the sedimentary processes which trap the metals. Incidentally, we may recal the difficulties of mining geologists confronted with the petrological problems of submarine lavas or with the problems of pyroclastic accumulations and their volcano-sedimentary reworking. In many respects this is a sedimentological field which justifies further thorough investigations.

I shall not mention aerial exhalations since they rarely give rise to deposits. They provide, however, together with the associated lavas, the elements of exceptional terrigenous sources which are now discussed in more detail.

The Terrigenous Sources

In addition to the stratabound iron and manganese deposits which were discussed previously, the shelf harbours some further stratabound sulfide concentrations, the genesis of which is still controversial. For some investigators they are the results of telethermal circulations; for others they are the results of syn- or diagenetic sedimentary processes. As in the case of petroleum, it was here, too, the initiative of mining geologists which led to the sedimentological examination of mineralized environments. The results of these studies in turn permitted to define the rules of localization of these deposits.

The main process is the fixation of heavy chalcophile metals in reducing H_2S-generating environments. The "Kupferschiefer" are their best known example, even though perhaps somewhat exceptional, as also the Zechstein evaporites. Despite of a very coherent set of arguments, taking into account the possible diagenetic recrystallizations of primary precipitates, some workers are still in doubt of the syngenetic origin of such sulfide concentrations.

Similar environments exist in areas which were more restricted and thus more active than the one of the Kupferschiefer. The most frequent occurrence is probably that of sea bottom heights influencing the synchronous sedimentation of oscillating shales and carbonate series (those of evaporitic environments included). This reminds us of a trap well known to petroleum geologists, i. e. that of the "buried-hills" which often occurs in a less subsiding and more litoral environment. Reducing environments producing syn- and/or diagenetic H_2S exist selectively on the slopes or on the top of these structures which are well known to sedimentologists (lateral pinching, slumping, "rolls", etc.). Similar to the Kupferschiefer, the mineralizations are pyritic or cupriferous in pelitic host rocks, whereas in carbonate host rocks the mineralizations are lead- and zinc-bearing. The height of the structures is not of importance, be it an epeirogenic horst, a reef bioherm or a diapiric upraise of the bottom, provided that the geometry of the trap creates a slowed down sedimentation of adsorbing material and an H_2S-generating environment at the time of an exceptional terrigenous influx of heavy metals.

Tracing this conspicuous terrigenous metal-influx towards the continent, along the sulfide-bearing horizons which are characterized by a distinct geochemical anomaly ("source beds") one encounters the marine, then lagoonal and finally continental red-bed impregnations. No matter whether they are uranium, lead-zinc or copper mineralizations, it is known that the circulation of ground water bodies, through reducing portions of large piedmont flood-plains, is responsible for the formation of metal impregnations. This circulation occurs obviously during the cementation of detritic sediments, i. e. during diagenesis of the host sandstones. But there, too, it is a prerequisite that the circulating waters bear metals!

With this, we reach the problem of the geochemical behaviour of heavy metals on the continents. The study of paleo-alteration profiles discloses that, depending on the parent rocks and climate, leaching of pedological concentrations may occur. Fast and probably instant reworking of soil concentrations seems to be the most general cause of conspicuous terrigenous sources. Volcanic ashes are, indeed, another possibility for such a source, but a very strong argument against this source is the rather strict correlation which seems to exist between the occurrences of the deposits and paleoclimatic zones. This climatological dependance of terrigenous contributions confirms, well enough, that their nature lies mainly in processes related to soil formation.

Let us have a brief insight into the apparent variability of these pedological processes; it is known that nickel is leached together with magnesium from ferralitic profiles situated on ultrabasic rocks. If the drainage of the altered zone is poor (flat topography), nickel and magnesium are stored and concentrated at the basis of the profiles as montmorillonite or garnierite ... Conclusion: look for ferralites developed on ultrabasic rocks in equatorial paleozones, but look on massives which are levelled by erosion! Should such paleo-surfaces be raised by epeirogenesis, a series of reworking may occur about which I do not want to say more, at least with reference to nickel.

The geology of bauxites provides us with much better known examples of alteration successions and of detrital reworkings of profiles, followed by new alterations on differenciated parts. The simple allitization of a nepheline-syenite is a model which has been complicated by the Fe-Al separation of the ferralites. I shall briefly mention the evolution of a ferralitic material sedimented on karst which easily explains the separation between iron and aluminium, and also the bauxite deposits of the Provence, France; at the same time it introduces the new problem of traps related to the pedological alteration.

We are dealing, of course, with concentrations located below unconformity surfaces. At first, the cementation of sulfide ore bodies demonstrates undoubtedly that the residual concentration of copper and silver occurs per descensum. To what extent will the karstic reworking of stratabound pre-concentrations in a carbonate environment not lead to cross-cutting sulfide bodies below the paleo-emersion surface? To what extent are certain uranium, barite and fluorite veins which are, geometrically speaking, strictly linked to unconformity surfaces not caused by per descensum processes, as proposed up to now? I shall close this short review with these two questions which lead away from purely sedimentary phenomena.

In conclusion I would like to emphasize the following thought: It is logical and "cartesian" in scientific research to go from the known towards the unknown and, in geology, from the surface towards the depth. As a paradox, economic geology is a field which has been dominated in its concept as well as in its methods by a theory related to deep seated and, so to say, to magmatic phenomena. I believe, it is about time in 1971 to go back to Descartes and to ask if many of the deposits which occur in sedimentary rocks are not the results of exogenous processes. Sedimentology gave the start, and economic geology follows its example in a way which now appears irreversible.

Diagenetic Pyrite and Associated Sulphides at the Almadén Mercury Mine, Spain

M. Arnold, A. Maucher, and F. Saupe

Abstract

The pyrite of Almadén provides an excellent record of the early history of the sediments. It is especially important as it appears on a regional scale as a constituent feature of the "Criadero Quartzite". The pyrite crystals are built up of a macromosaic nucleus surrounded by a diagenetic cortex. Framboids of the nucleus display frequently a mutual orientation, a structure which is unusual in "Rogenpyrit". The cortices display different growth types (zonal, fibrous and sectorial).

Cinnabar and pyrrhotite appear only within the nucleus and never in the cortex. Cinnabar thus precipitated under equilibrium conditions between pyrite and pyrrhotite. Mercury was present before the addition of the diagenetic pyrite cortices, since it is trapped in the nucleus.

A - Introduction

1. The Problem

Due to its low mobility and its mechanical strength, pyrite is an excellent recorder of the early history of the sediments, and the contained mineralizations. Cinnabar displays opposite features. Therefore, the first mentioned mineral was selected as the object of a microscopic study of the ore at Almadén.

The Almadén pyrite was first studied by MAUCHER and SAUPE (1967) and was interpreted by them as syngenetic (in the meaning of economic geology and not of sedimentology). The new technique of epitaxic oxidation developed by one of us (ARNOLD, 1969 and in preparation) led us to take the problem up again, in order to delimit the conditions of pyrite formation and to know by these means those of cinnabar. The employed techniques aim to recognize the variation of different parameters within a given crystal. For this reason, phase contrast microscopy, after epitaxic oxidation, and microprobe analysis were used.

2. Regional Geological Setting

Almadén is located in the Paleozoic block forming the Iberian Meseta, exactly between the Sierra Morena to the south and the Montes de Toledo to the north (Fig. 1). Above several thousand meters of schists and greywackes belonging probably to the Upper Precambrian and the Lower Cambrian, lies unconformably (Sardic phase) a Paleozoic sequence, extending from Ordovician to Carboniferous. From the Arenig to Siegenian stages inclusive, this series consists of about 2000 m of sedimentary rocks. It is framed in by two quartzite levels, corresponding to these two stages, and is composed of a monotonous sequence of shales, alternating shales and sand-stones, and two other quartzite levels belonging to the Caradoc and the Llandovery[1]. The latter supports the mercury mineralization and its pyrite is described in the following lines.

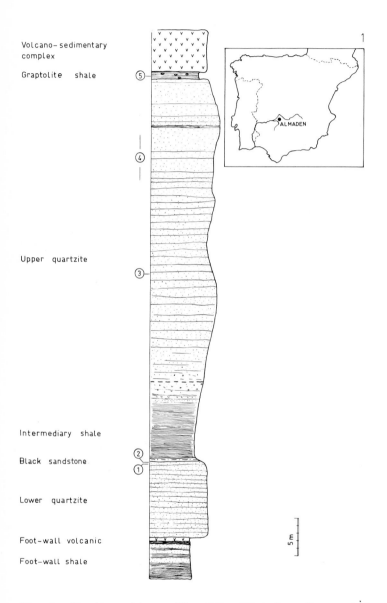

Fig. 1. Stratigraphic profile of the "Criadero Quartzite", with the five sample locations. Location map of Almadén

1) A detailed description with a bibliography is given by SAUPE (1971a).

3. Local Geological Setting

The following stratigraphic summary is based on a recent publication (SAUPE, 1971b).
The "Criadero Quartzite" is built up of two members, framed in and separated by
shales. Its total thickness is about 50 m at Almadén, but it gradually decreases in
all directions. The constituent quartzite is remarkably pure and of diagenetic origin,
the latter being a consequence of the former. The lower member (13 to 15 m) consists at its base of a sandstone, losing progressively upwards its argilo-carbonaceous
cement, thus changing into a quartzite. Its lower and upper limits are sharp; the footwall is formed by the already mentioned shale, and the hanging wall by a black sandstone bed, overlain by black shales. About 10 m higher in the section, these rocks
grade into the second quartzite member which has an approximate thickness of 30 m.
Fig. 1 shows the location of the three ore horizons and the position of the five samples studied in the present paper. Table I contains short descriptions of them (see
also fig. 2 to 5).

Table I - Description of the studied samples

Sample Number	Description of pyrite	Description of host-rock
Top of stratigraphic column		
5	Pyrite corona of 0.5 to 1 cm thickness	Ellipsoidal dolomite nodules con- taining organic matter. Fine and plane internal stratification. Occur within Graptolite shales.
4	Thin pyrite seams (a few mm)	Black quartzite on top of the S. Nicolas ore zone. Fine plane internal stratification.
3	Pyrite nodules, with a faint concentric zoning	Gray quartzite in the footwall of S. Francisco ore zone.
2	Pyrite bed of 2 to 5 cm	Top of a light gray quartzite bed, overlain by black sandstone.
1	Pyrite seam of 0.5 to 1 cm	Upper bed of the S. Pedro ore horizon.
Bottom of the stratigraphic column		

Samples 1, 2 and 5 are well located stratigraphically and their lateral equivalents
may easily be found in the lower levels of the mine. Sample 3 is also stratigraphically well defined, but similar material is rare. Sample 4 is not characteristic for
a given horizon and may be found anywhere in the upper half of the second quartzite
member, though not abundantly. The five selected samples cover all the distinct diagenetic pyrite forms. Remobilized pyrites are encountered in fissures, but are not
described here.

The presence of pyrite in the "Criadero Quartzite" is a regional feature. It was found
by one of us (F. S.) in the town of Almadén, where the quartzite is barren of ore and it
is limonitic in many places in the concession (cubic relicts are frequent). Thus the
pyrite appears as a constituent feature of the "Criadero Quartzite".

4. Epitaxic Oxidation

The following brief description is taken from ARNOLD (1969).

a) Fundamentals

The growth velocity, and at the limit, de dissolution velocity as well, are a function of the crystallographic orientation and, to a smaller degree, of the imperfections contained in, or intersected by the etched surface.

This property is utilized in order to promote the slow formation of a thin transparent sulphate film. This film acts as an interferometer. Thus the crystallographic orientation is expressed by interference colours in the Geffken scale (equivalent of the Newton scale in the case of transmitted light). The imperfections appear through hues of these colours. Dimorphous crystal varieties can also be separated by these means; after epitaxic oxidation they present usually colours of distinct orders.

b) Procedure

The success of the treatment relies mainly upon the slowness of the oxidation but depends also on the state of the polished surface which must be perfect. The polished section is put in a convection current of active oxygen circulating between a source of oxygen and a source of heat.

c) Results

The process may be controlled under the microscope, and under our operating conditions the oxidation velocity is about 3 Å/minute. The range of interference colours which are sought is that which is most sensitive to the eyes and possesses small colour differences. This corresponds to the second half of the first order and the first half of the second order, or to optical retardations of 4350 to 6800 Å. The obtained colours can be perfectly reproduced by a second etching after elimination of a first film of oxidation. The film has a uniform thickness, if its surface is smaller than 30 cm^2 and if both sources (heat and oxygen) are far away from the section.

B - Ore Microscopy[2]

1. Species Present

The following sulphides were noticed: pyrite, cinnabar (rare within the pyrite, though the main ore mineral), chalcopyrite, pyrrhotite, galena and sphalerite. The remaining minerals present, the non-sulphides, are common in normal quartzites. Marca-

---

[2] The authors are indebted to Mr. M. Guyot (C.R.P.G.) whose fine polished sections permitted our observations.

Fig. 2. Sample no. 5: corona of pyrite around a dolomite nodule

Fig. 3. Samples no. 1 and 2 - a: uppermost cinnabar-bearing bed of the S. Pedro ore horizon; b: barren quartzite; c: black sandstone

Fig. 4. Sample no. 4

Fig. 5. Sample no. 3

site, according to previous observations (MAUCHER and SAUPE, 1967), does not occur in the diagenetic pyrite, not even as a relict. However, it was found in small cracks which it cements together with other minerals. Graphite occurs, more or less, as twisted flakes and sometimes as closed figures. No definite organic forms were identified[3].

2. Morphology of Pyrite

The external and thus final morphology of pyrite is constant (MAUCHER and SAUPE, 1967); the cube is the predominating form. Small changes result from the respective importance of the forms {210} and {111} which remain minor in any case. In contrast, the internal morphology of the pyrite is variable from a certain growth stage onwards.

Five structural types were identified after epitaxic etching and investigation under phase contrast microscope. There are, however, transitions between these extreme types. The main interest of such a classification lies in the possibility of linking the associated sulphides (mainly cinnabar and pyrrhotite) to the morphological types. These five types have a common macromosaic nucleus but differ in the structure of their cortices.

a) Macromosaic nuclei

The macromosaic structure of the nuclei is built up of spherules of about 20 μ in diameter which display often an orderly internal texture[4].

The epitaxic oxidation disclosed that the disorientations between these spherules are small. In turn, these spherules are built up by but slightly disoriented grains of less than 1 μ also orderly located. This ordering at two different scales leaves a double set of gaps.

The later evolution takes place according to two manners:
- Polygonalization of the spherules occurs through

---

3) Messrs. J. L. Henry and J. Deunff (Rennes) were kind enough to examine a picture of the most organic-like structure and confirmed its not-organic nature. Their cooperation is herewith acknowledged.
4) Other patterns observed in distinct geological settings are described by ARNOLD (in preparation).

Fig. 6. Structure type I: open structure. Framboids with parallel orientation (Rogenpyrit). The grain limits are emphasized in the lower left corner. At the right border appears, in dark, a set of differently oriented framboids. Epitaxic etching

Fig. 7a. Structure type II: crystals with macromosaic nuclei, achieved by a slight and discontinuous outer zoning. Partially ordered packing. Grain limits graphically emphasized

Fig. 7b. Structure type II: detail of Fig. 7a

Fig. 8. Structure type III: distinctly visible zoning

Fig. 9. Structure type IV: fibers

Fig. 10. Structure type V: crystal with fibers parallel to ⟨111⟩ and zoning parallel to {100}. Epitaxic etching

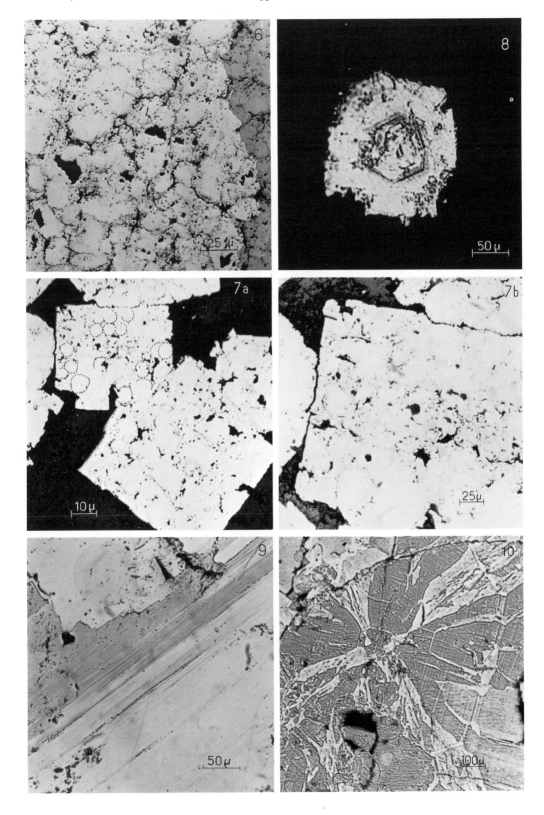

14

1. compaction of an initially plastic material,
2. epitaxic feeding,
until all the available space has been filled up.
- Development of fibers embracing one or several spherules of the same orientation

b) Growth cortices

The following types were distinguished:

Type I (Fig. 6 and 11) - No cortex may appear, and the afore-mentioned structure remains thus open. This is so-called "Rogenpyrit".
This type is the starting point for all the others, since it appears in all as a nucleus.
As the diagenetic evolution is the rule, this type is rarely encountered.

Type II (Fig. 7 and 11) - Almost the complete crystal displays a macromosaic structure, but a slight outer and often incomplete zoning is achieved by growth.

Type III (Fig. 8 and 11) - Growth zoning is well developed and the structure is always a closed one.

Type IV (Fig. 9 and 11) - This type is the most frequent one. For this reason it has been possible to observe it along two growth directions: ⟨100⟩ and ⟨111⟩. The nucleus is still macromosaic, but the cortex is fibrous. The epitaxic oxidation shows that the fibers originated by simultaneous growth of mosaic blocks issued from the central spherules, or more precisely from their outer surface. Their diameter is progressively enlarging outwards, but not all fibers reach the external surface of the crystal In the neighbourhood of the spherules, a selection rapidly takes place so that only crystallites oriented along the ⟨100⟩ directions continue to grow. The achievement of the crystal is therefore only conceivable if the disorientation of the spherules build ing up the nucleus was small . The fibers are contiguous and temporary growth conditions mark all of them (zoning). Thus, this type belongs to the "lineage structure" of BUERGER (1934).

Type V (Fig. 10 and 11) - A lineage structure may be superimposed upon a sectorial structure developed parallel to [111]. This type was seldom encountered.

The crystal nuclei are thus always built up of spherules (cf. Rogenpyrit). The observed diversity lies then in the difference of their later evolution, giving birth to cortices/of different structures. Open structures, as those described as type I or II, occur seldom. Most of them were probably dissolved and their material precipitated on other nuclei which so were preserved. Local conditions, such as movements of the fluids, oversaturation and amount or nature of impurities in suspension, may then be held responsible for the different structures observed. The scheme at the bottom of fig. 11 shows the mutual relations between the five types.

3. Relationship between Pyrite and Subordinate Sulphides

Chalcopyrite and galena: Chalcopyrite occurs exclusively in the gaps between spherules building the nuclei of crystal type II, associated with galena, and crystal type IV associated with cinnabar.

Pyrrhotite: The hexagonal monosulphide, frequently as twinned crystals of 1 to 10μ size, appears only between the spherules of the nuclei of type IV. It was never found in the cortex. An epitaxic relationship between spherules and pyrrhotite inclusions is probable; if a spherule contains several pyrrhotite crystals, they have common extinction under crossed nicols.

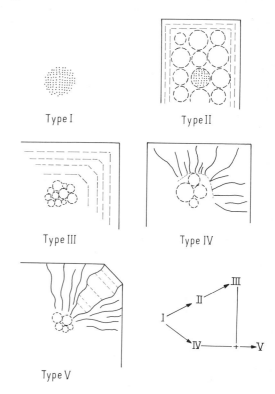

Fig. 11. Schematic representation of
the five growth types and of their
evolutionary trend

Type I   Type II   Type III   Type IV   Type V

Cinnabar: As the economic mineral, it is omnipresent; however, only the inclusions
within the pyrite grains, already mentioned by MAUCHER and SAUPÉ (1967), are of
interest to the present study. These could not have been introduced after fracturing
of the host mineral; cataclasis followed by healing would leave tracks which cannot
escape epitaxic oxidation. Cinnabar appears only in crystals of types II and IV. In
minerals of type II, i.e. of a frequently open structure, cinnabar seldom occurs. As
already mentioned, pyrrhotite was never found in this type. There may be two reasons
for this situation: 1) Pyrrhotite never existed and cinnabar was precipitated normally
(and part of it may have moved away). 2) Cinnabar was introduced later, since the
structure is open.

In type IV cinnabar is frequent, but is always located in the gaps remaining between
the spherules or at the interface nucleus-cortex. Thus there is a strong analogy with
pyrrhotite which appears under the same conditions and in the same type of crystals.
Since cinnabar is rare in type II, this and the other sulphides occur essentially in
type IV.

C - Microprobe Examination

Since the original packing of the framboidal spherules was preserved and is thus
evidence of the conditions prevailing during deposition of the iron sulphides, the
quantitative determination of the contained metals informs us, whether these metals
arrived at the same time as the Fe. In addition, their determination in pyrites of
different generations (framboids, i.e. definitely syngenetic pyrite, as opposed to
the later pyrite of the cortex) would also throw light on later migration.

Keeping this in mind, the work with the microprobe was carried out in three steps[5]:

a) determination of a few elements in small areas (of a few square microns);
b) X-ray diffraction on a given area, in order to explore a given range and to identify all emitted radiations;
c) scanning of zones of special interest.

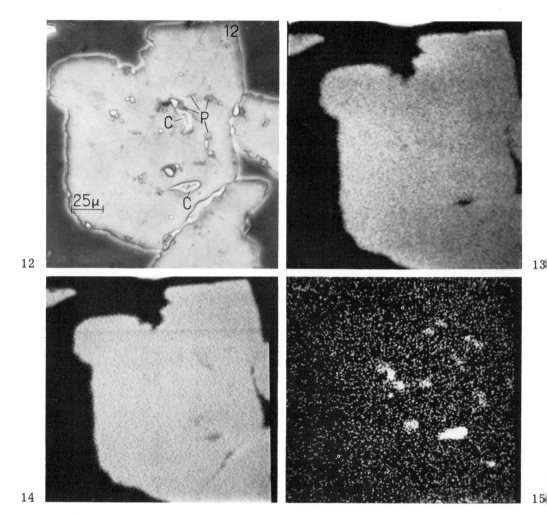

Fig. 12. Pyrite crystal (type IV) with minute cinnabar and pyrrhotite inclusions in the nucleus, and at the interface nucleus-cortex. Phase contrast picture

Fig. 13. Scanning picture of the crystal of Fig. 12: Fe (same scale)

Fig. 14. Scanning picture of the crystal of Fig. 12: S (same scale)

Fig. 15. Scanning picture of the crystal of Fig. 12: Hg (same scale)

5) The first two steps were performed on a Cameca MS 85 microprobe, the last one on a Cameca MS 46. We would like to thank Messrs. Hausser and Ruste of the Metallurgical Laboratory of the E.N.S.M.I.M. (Nancy) for their kind collaboration.

The distribution of Hg is dotty and is probably due to minute cinnabar inclusions which are too small to be determined under the optical microscope. The polish is not perfect and discloses heterogeneities within the pyrite. The search for Hg rich zones is greatly facilitated by a humming device, the frequence of which is proportional to the amount of received impulses. A pyrite crystal could be selected by this means from which Fe, S and Hg exposures were taken.

The Fe picture (Fig. 13) shows mainly the largest pyrrhotite inclusions. The S picture (Fig. 14) discloses more details on the variations of the gradation than the first one: pyrrhotite as well as cinnabar appears. The Hg picture (Fig. 15) shows better than the mere microscopic observation (Fig. 12) that the concentration of Hg is limited to the primitive framboid; the diagenetic cortices do not contain Hg inclusions.

With the microprobe the existence of mercury within the framboids could be checked and thus its early presence confirmed. The other elements were not abundant enough as to make a demonstration of the same possible.

D - Geological Implications and Conclusions

The present mineral assemblage is mainly formed by pyrite with small amounts of pyrrhotite and cinnabar, and occasional traces of galena and chalcopyrite. This assemblage could have originated in different manners:
a) Pyrite may have been transformed partially into pyrrhotite;
b) the first precipitated material was a mixture of various sulphides (mackinawite, greigite) which were transformed into pyrite and pyrrhotite. Cinnabar precipitated about simultaneously;
c) a simultaneous precepitation of the three main sulphides, i. e. pyrite, pyrrhotite and cinnabar, occurred.

Investigation of these possibilities leads us to disregard the first one; indeed, if pyrrhotite should be the consequence of a temperature raise, as is often the case, it should appear in open spaces where the S could move away, e.g. in fractures of the crystals.

At this time it is still impossible to discuss the two latter possibilities. According to observations of present precipitation of Fe sulphides (DOYLE, 1970), pyrite and pyrrhotite are never primary minerals and they are always the result of the transformation of the above mentioned mackinawite and greigite. These latter minerals were never found in sediments other than Tertiary (BERNER, 1970), and no indication of their early presence at Almadén was found. However, these transformations take place rather rapidly.

Even though the exact chemical reactions cannot be stated, it can safely be admitted that the cinnabar precipitated at the same time as the first sulphides, or with a small difference of time. It is known (HARTMANN and NIELSEN, 1969) that the sediments constitute a closed system already a few cm below the interface sea-water - sediment. The precipitation conditions of the system pyrite - pyrrhotite were therefore not very different from those of the earlier iron sulphides, and those of cinnabar were also very similar. The absence of marcasite shows that the pH was >7 (ALLEN et al., 1912)[6]. Since we ignore the activities of the various constituents, it is not possible

6) The so determined Eh-pH conditions are by no means exclusive. The stability diagram of cinnabar shows that it is stable for a pH varying between 0 and 13. In the remobilized pyrite mentioned earlier, where relicts of marcasite appear, some cinnabar is associated. Cinnabar seems therefore to have precipitated in acid conditions in the latter case.

to calculate the exact Eh-pH relation for the equilibrium of pyrite and pyrrhotite. However, assuming that the activity of S would fall between 10^{-1} and 10^{-6}, and neglecting the other constituents which, of course, gives us a rough approximation, it appears (Fig. 16) that the equilibrium is realized in a zone falling close to the stability limit of water: This corresponds to conditions existing in euxinic basins rich in organic matter.

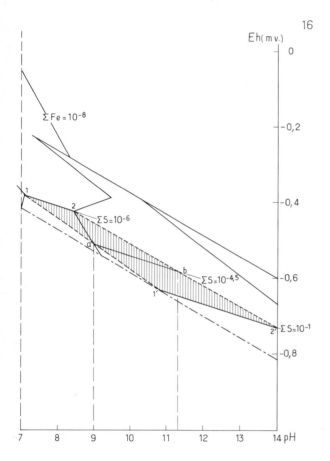

Fig. 16. Equilibrium between pyrite and pyrrhotite reported in a classical Eh-pH diagram. Total dissolved iron activity set at 10^{-8}. Calculation made for total sulfur activity of 10^{-6} (line 1-2), 10^{-1} (line 1'-2') and an intermediate value of $10^{-4.5}$ (line a-b). The hatched zone represents conditions delimited by these extreme values and, thus, the most probable Eh-pH conditions of deposition

The most important findings of the present work are the observations on the two different diagenetic growth patterns of pyrite and the locking of cinnabar by the macromosaic nuclei, as well as its absence in the growth cortices of pyrite, which clearly demonstrate that mercury was present before the diagenetic addition of the pyrite overgrowth cortices.

References

ALLEN, E.T., CRENSHAW, J.L., JOHNSTON, J., LARSEN, E.S.: The mineral sulphides of iron. Am. J. Sc., 33, p. 169-236 (1912).

ARNOLD, M.: L'oxydation épitaxique : une méthode de résolution des structures et micro-structures des bisulfures de fer. Thèse Doct. Spéc. (Géologie appliquée) Nancy, 85 p. (1969).

— Relations entre les structures en framboise et les macrocristaux de pyrite.

BERNER, R.A.: Sedimentary pyrite formation. Am. J. Sc., 268, p. 1-23 (1970).

BUERGER, M.J.: The lineage structure of crystals. Zeitschr. f. Krist., 89, p. 195 -220 (1934).

DOYLE, R.W.: Identification and solubility of iron sulfide in anaerobic lake sediment. Am. J. Sc., 266, p. 980-994 (1968).

GARRELS, R.M., CHRIST, C.L.: Solutions, Minerals and Equilibria. Harper's, New York, 438 p. (1965).

HARTMANN, M., NIELSEN, H.: δ^{34} S-Werte in rezenten Meeressedimenten und ihre Deutung am Beispiel einiger Sedimentprofile aus der westlichen Ostsee. Geol. Rundschau, 58, p. 621-655 (1969).

MAUCHER, A., SAUPE, F.: Sedimentärer Pyrit aus der Zinnoberlagerstätte Almadén (Provinz Ciudad Real, Spanien). Mineral. Deposita, 2, p. 312-317 (1967).

SAUPE, F.: La série ordovicienne et silurienne d'Almadén (Province de Ciudad Real, Espagne). Point des connaissances actuelles. Colloque sur l'Ordovicien et le Silurien, Mém. B.R.G.M., 73, p. 355-265 (1971a).

— Stratigraphie et pétrographie du "Quartzite du Criadero" (= Valentien) à Almadén (Province de Ciudad Real, Espagne). Colloque sur l'Ordovicien et le Silurien, Mém. B.R.G.M., 73, p. 139-147 (1971b).

Addresses of the authors:

M. Arnold   Centre de Recherches Pétrographiques
F. Saupé   et Géochimiques (C.N.R.S.)
                   54 Vandœuvre-lès-Nancy / France

Prof. Dr. A. Maucher   Institut für allgemeine und angewandte
                        Geologie und Mineralogie der Universität
                        8 München 2 / Germany

Diagenetic Ore-forming Processes at Kamoto, Katanga, Republic of the Congo

P. Bartholome, P. Evrard, F. Katekesha, J. Lopez-Ruiz, and M. Ngongo

Abstract

At Kamoto and in other Katangan deposits, copper and cobalt occur as sulfides (mostly digenite, chalcocite, bornite and carrollite) in two stratiform orebodies. The host rock is dolostone, chert and shale. There is no metamorphism. Mineralization was emplaced before major tectonic deformation.

The two stratiform orebodies are separated by a barren interval. The lower one rests a few feet above an erosional surface which developed over consolidated red dolostone beds completely devoid of sulfides. Above the upper orebody, a sulfide fraction is present, consisting mostly of pyrite with some chalcopyrite.

The mineralized sedimentary rocks present some of the features commonly observed in tidal-flat sediments. However, they have undergone a complex sequence of diagenetic transformations. Several authigenic (or partly authigenic) minerals have crystallized: dolomite, perhaps in part from hydromagnesite + $CaCO_3$; magnesite, from the remaining hydromagnesite; quartz; chlorite and other phyllosilicates; colorless tourmaline; pyrite; other sulfides: chalcopyrite, bornite, digenite, chalcocite, carrollite, etc...

New evidence is presented to show that this diagenesis took place at first in an environment devoid of copper and cobalt and then proceeded while the metals were brought in from an outside source.

The features described may have resulted from a reaction between a hypersaline brine (with high pH and high Eh) flowing through the lower part of the Kamoto Dolostone, and a modified connate water (reducing and less alkaline because of abundant organic matter) present in the upper part.

1. Introduction

Southern Katanga (part of the Republic of the Congo) and Nothern Zambia constitute an extremely rich metallogenic province called the Central African Copperbelt. This province is made of two distinct subprovinces: the Zambian subprovince which straddles the Congo-Zambian border from the 12° S parallel southward and includes the wellknown stratiform deposits of Roan Antelope, Chambishi, Chibuluma, Mufulira etc. in Zambia as well as Musoshi, Tshinsenda and others on the Congolese side; the Katangan subprovince which is entirely within Congolese territory and contains many stratiform ore deposits (fig. 1), in which copper and cobalt sulfides are again noteworthy. In this paper, we are concerned only with the Katangan subprovince, and in particular with the Kamoto deposit located near Kolwezi at the western end of the subprovince.

22

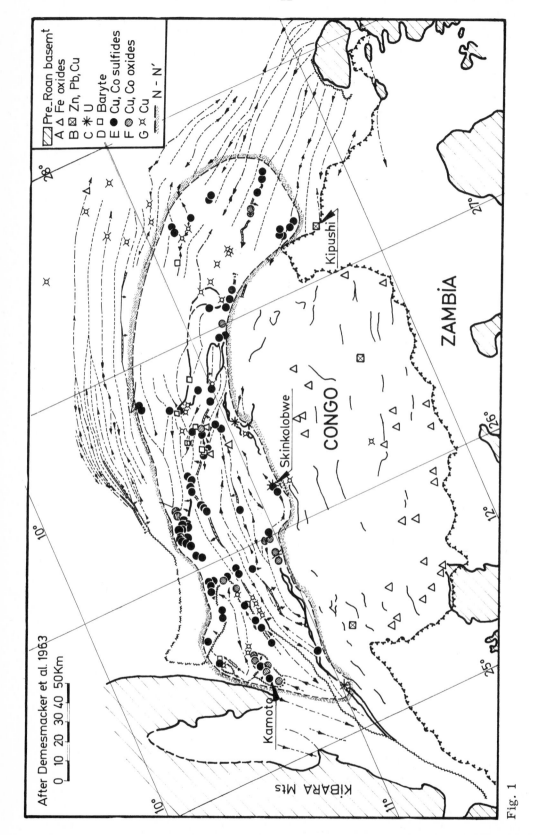

After Demesmacker et al. 1963

0 10 20 30 40 50Km

Pre.Roan basemᵗ
A △ Fe oxides
B ⊠ Zn, Pb, Cu
C ✳ U
D □ Baryte
E ● Cu, Co sulfides
F ◉ Cu, Co oxides
G ⋈ Cu
N – N′

28°

27°

Kipushi

ZAMBIA

CONGO

Skinkolobwe

26°

12°

25°

10°

10°

11°

KIBARA Mts

Kamoto

Fig. 1

For our present purpose, the geology of Southern Katanga may be summarized from
several prior publications (CAHEN, 1954; DEMESMAEKER, FRANÇOIS and OOSTER-
BOSCH, 1963; CAHEN and LEPERSONNE, 1966). A major unconformity truncates
sedimentary units folded during the Kibaran (± 1300 My) or older orogenies. Many
granitic and other intrusives are emplaced in these rocks, which have suffered metamorphism to variable degrees. A thick sedimentary pile was deposited above the unconformity and folded during the Lufilian orogeny, about 650 My ago. The Lufilian
folds follow an arc (fig. 1) north of which the beds are normally flat-lying. The arc
abuts against older rocks observed in the Kibara Mountains at the western end and
the Luina and other domes at the eastern end.

In the eastern part of the arc, the structure is rather simple and consists of a small
number of folds which can be followed over long distances (100 km). In the central
part, folds are more complex and less persistent; some are faulted. In the west,
thrusting has occurred northward over a distance estimated to 60 km; the resultant
thrusted plate, called "Le Lambeau de Kolwezi", contains the Kamoto and neighbouring deposits, and has a rather complex structure illustrated in figure 2. As a rule,
metamorphism has been absent or very slight in the Lufilian arc (1) (however see
BELLIERE, 1951). The region is devoid of intrusives except for minor and sporadic
mafic rocks (JAMOTTE, 1933, and GROOSEMANS, 1934).

From the lack of relations between ore grade and faults, Demesmaeker, François
and Oosterbosch conclude that mineralization was emplaced before tectonic deformation. This conclusion has been widely accepted.

2. Stratigraphy

Within the Lufilian arc and especially within the Kolwezi thrusted plate, the stratigraphy (see OOSTERBOSCH, 1959 and 1962, in addition to previous references) and
the petrology (BARTHOLOME, 1962; KATEKESHA, 1970; NGONGO, 1970) are known
in considerable detail. We now propose the following nomenclature for the main lithostratigraphic units involved (see table 1): the units commonly referred to as the Roan
"System" or "Series" and the Kundelungu "System" above it should be given the rank

(1) No metamorphism is visible in hand specimens from Kamoto. However, under
the microscope, various mineralogical transformations are observed that we
consider as diagenetic. Some of these may be attributed to incipient metamorphism, because the distinction between late diagenesis and incipient metamorphism is arbitrary to a large extent.

Fig. 1. Distribution of ore deposits in Southern Katanga (after Demesmaeker, Fran-
çois and Oosterbosch, 1963).
A: Iron oxide deposits.
B: Base metals vein-type deposits, the most important of which is Kipushi.
C: Uranium vein-type deposits, the most important of which is Shinkolobwe.
D: Baryte vein-type deposits.
E: Stratiform deposits containing sulfides of copper and cobalt with occasional-
ly some uranium, gold and platinum.
F: Stratiform deposits containing oxides of copper. Kamoto and neighbouring
deposits belonging to categories E and F have produced about 240,000 annual
tons of copper in 1969 according to WYLLIE (1970).
G: Other copper occurrences.
N-N': No exposure of the Mines Group is found outside this curve

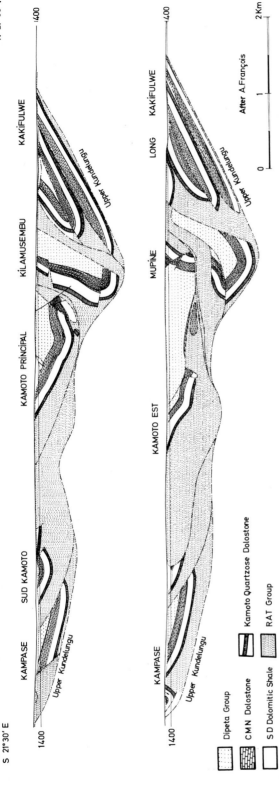

Fig. 2. Two parallel cross-sections through the Kolwezi thrust plate (Lambeau de Kolwezi), showing the geological structure in the Kamoto mine area (after A. François and the Département Géologique of GECOMIN).

The thrust plate rests on younger beds belonging to the Kundelungu Supergroup. Within the thrust plate, the three formations which together constitute the Mines Group have been broken into large blocks. Some of these blocks contain two orebodies; others contain only one (the lower one); others are entirely barren. All of the orebodies are stratiform. The upper ones occur at the base of the S. D. Dolomitic Shale. The lower ones occur in the Laminites-bearing Member of the Kamoto Dolostone

of supergroups. Within the Roan Supergroup, we shall be mostly concerned here with the Mines Group, traditionally called "Serie des Mines" and with the underlying RAT Group.

Table 1. Stratigraphy of part of the Roan Supergroup in the Katanga subprovince of the copperbelt

Groups	Formations	Traditional abbreviations
DIPETA		R. G. S.
MINES	CMN Dolostone	C. M. N.
	SD Dolomitic Shale	S. D.
	Kamoto Quartzose Dolostone	
	- RSC Member	R. S. C.
	- Laminites-bearing Member	⎰ R. S. F. ⎱ D. S. ⎰ Gray RAT
	- (Conglomerate Member ?)	⎱ Red RAT
RAT		

Note: In the Congo, the term Groupe has often been used to designate very large lithostratigraphic units. For example, the Roan Supergroup, together with the overlying Kundelungu Supergroup were included into the "Groupe du Katanga". This unit thus included all the beds deposited after the Kibaran orogeny and before the Lufilian orogeny.

The original meaning of RAT, which is abbreviated from 'roches argilo-talqueuses', is quite inappropriate because the group consists of rocks of various grain sizes made rather uniformly of dolomite and quartz. The abbreviated name is, however, the only term in use to designate this unit. For this reason, we shall keep the designation RAT. Similarly, we shall keep most of the abbreviations coined by the pioneers of Katangan geology, even if the corresponding full name is a misnomer and could be forgotten without loss. Our nomenclature will, therefore, follow the traditional one very closely.

The most striking character of the RAT Group is the uniformly red colour due to disseminated hematite. Detrital quartz, micas and chlorites are abundant in most beds, but some dolomite is always present. There are no sulfides. Bedding is observed, but lamination is quite uncommon. The RAT Group is at least 200 meters thick in the Kolwezi area. Its base is unknown.

We shall subdivide the Mines Group into three formations, in agreement with usage of Katanga geologists (see maps published in DEMESMAEKER, FRANÇOIS and OOS-TERBOSCH, 1963). These formations are from top to bottom:

- The CMN Dolostone, 50 to 125 meters thick, mostly made of beds in which dolomite is mixed with various amounts of detritals. Shale beds are present in the lower part of the formation. Microscopic pyrite is common and carbonaceous material is present especially in the shales.

- The S. D. Dolomitic Shale, 35 to 90 meters thick, mostly an alternation of commonly laminated, locally carbonaceous dolomitic shale and dolomitic silstone beds. Disseminated microscopic pyrite is present in most of the formation, and may be accompanied by chalcopyrite. Carrollite, bornite, chalcocite and other copper sulfides are locally abundant in the lower ten meters: this portion of the formation is the
Upper Orebody at Kamoto and other deposits.

- The Kamoto Quartzose Dolostone, about 30 meters thick, in which dolomite is mixed with very small amounts of detrital material, with considerable amounts of authi
genic quartz and with appreciable local concentrations of sulfides and magnesite. W
shall subdivide this formation into three members:

-- An upper RSC member (15 to 20 meters) which consists almost exclusively of dolomite and authigenic quartz. This member, devoid of sulfides and of carbonaceous matter, usually has no lamination in contrast to the beds above and below
it. Remnants of stromatolitic structures have been found in this member, but
most of the rock is massive and coarse grained, as a result of intense recrystallization. Macroscopic pores are often observed.

-- A lower Laminites-bearing Member which consists of dolostone beds with variable amounts of magnesite in the lower part of the member and increasing amoun
of authigenic quartz in the upper part. Beds consisting almost wholly of magnesit
or authigenic quartz are present. The lowermost bed contains no magnesite, but
much authigenic quartz and chlorite; it is not laminated. Carrollite, bornite and
copper sulfides may be abundant in this Laminites-bearing member, which forms
the Lower Orebody at Kamoto and other mine localities.

According to the traditional Katangan nomenclature, the upper siliceous part of the
Laminites-bearing Member is called R. S. F. The middle magnesite-rich part is called D. S. And the chlorite-rich bed at the base is called RAT Grises (Gray RAT),
which is misleading since, as we shall show in this paper, neither this bed nor the
conglomerate below it should be included in the RAT Group.

Over the 300 km followed by the Lufilian arc, the stratigraphy of the Mines Group
of course displays some variations. All of the units that are given the rank of formations or members do, however, extend over most of the area enclosed by the curve
NN' on figure 1. On the other hand, none of these units has been recognized in the
Zambian subprovince of the Copperbelt. This is the major reason to distinguish between the Katangan and Zambian subprovinces despite of the fact that they both contain stratiform orebodies mineralized with both copper and cobalt sulfides.

3. The Footwall Problem at Kamoto

We are not directly concerned in this paper with the relationship between the Mines
Group and the underlying RAT Group. However, this is a subject of great importance
for those dealing with the genesis of the Kamoto deposit, and we must discuss it
briefly.

It has often been reported (see previous references) that the contact between the two
groups is obscured by tectonic deformation and the breccias that resulted from it.
It seems also that some recent secondary alteration has been taking place along this
contact. For these reasons, the primary nature of this contact has rarely been discussed in the literature. However, the observations that we made of well preserved
strata in two drill cores, as well as in underground workings, do show that, at least
locally at Kamoto, the Mines Group was deposited on an erosional surface truncating

beds belonging to the RAT Group. These underlying beds have already undergone some consolidation and recrystallization when erosion took place. This is witnessed by the following observations:

1) Small fractures observed in the top ten centimeters of the eroded bed are filled with coarse-grained dolomite and do not penetrate into the overlying conglomerate (fig. 3).

2) The conglomerate overlying the erosional surface contains rounded and unflattened pebbles of the same nature as the eroded bed. These pebbles are red because of the finely divided hematitic pigment observed in most of the RAT Group. Between the pebbles, there are coarse detrital grains of hematite similar to that observed in veinlets or recrystallized patches of the eroded bed. Unidentified pebbles, all more or less dolomitic, are also present, but we have observed none derived from the overlying Kamoto Quartzose Dolostone or S. D. Dolomitic Shale.

Fig. 3. Conglomerate below the Lower Orebody
Sample K 286 shows the erosional surface at the base of the conglomerate.
Sample KA 875 shows rounded dolostone pebbles in the conglomerate. Coarse detrital hematite, present between the pebbles, is presumably derived from the recrystallized parts of the underlying beds

In one drill core, the conglomerate is 10 cm thick. Just above it, the red colour fades out, and carrollite porphyroblasts can be found. Stratigraphically above this, there is one meter of poorly bedded and poorly mineralized fine-grained rock, which is grey with some red streaks, followed by the typical chlorite-rich bed which is richly mineralized and forms the base of the Lower Orebody.

Fig. 4

In the other core the conglomerate, the base of which has not been reached, is more than two meters thick. Its top is discoloured and contains small amounts of chalcosine as it grades into the chlorite-rich bed.

No doubt, our observations are too few to afford a conclusive proof of the suggested relationship between the two Groups in all of southern Katanga or even in the Kolwezi district. Yet, pending further observations which will be made when the opportunity arises in underground workings, we suggest as the most probable depositional history for the Kamoto Quartzose Dolostone that sedimentation started on an erosional surface displaying some relief. First, the channels were filled by gravels. Then, when detritals became unavailable, purely chemical sediments started to be deposited on a levelled surface.

With the limited information now available on the extension and thickness of the conglomerate, we cannot be certain that it should be designated as the lowermost member of the Kamoto Quartzose Dolostone. However, we feel that, on a stratigraphic scale, this conglomerate should be included in the same unit as the overlying beds. This is contrary to present usage in which the conglomerate is included together with the underlying beds within the Red RAT because of its red colour (see table 1).

4. General Petrological and Geochemical Features of the Kamoto Ores

The sedimentary beds which constitute the Kamoto orebodies, as well as the barren beds below, above and between them, display many interesting and unusual geochemical and petrological features. Some of these have been described by OOSTERBOSCH (1962) and in a preliminary study by BARTHOLOME (1962). Both of these authors have described the well developed lamination which, as a rule, characterizes the orebodies (fig. 4). By contrast, non-laminated sediments such as the RSC Member and the underlying RAT Group are barren. One exception is the almost massive chlorite-rich bed at the base of the Lower Orebody (the so-called Gray RAT) which is highly mineralized in some areas. Within the orebodies, laminations result from variations in both grain-size and mineralogy. They are closely related to the mineralization (fig. 4 and 5).

Irregular laminations observed in some beds, erosion channels, and the presence of stromatolites in the R.S.C. member suggest tidal flat deposition. We have not found,

Fig. 4. Lamination in the Kamoto orebodies. Tops of photographs correspond to stratigraphic tops of specimens.

Sample K 305 from the lower part of the Lower Orebody consists chiefly of dolomite and magnesite. The latter mineral occurs as white parallel fibers almost normal to the bedding, and as dark crystals which are either isolated in the dolomite matrix or aligned almost continuously in some laminae.

Sample K 302, from the lower part of the Lower Orebody, contains the same dark magnesite crystals in a fine-grained dolomite matrix. An erosion channel is clearly visible.

Sample K 368.1, from the upper part of the Lower Orebody, is a rock in which authigenic quartz prevails over dolomite. The ore grade is much higher than in the samples K 302 and K 305.

Sample K 518 is a typical high-grade dolomitic shale from the Upper Orebody. Sulfides form small nodules in some of the laminae

Fig. 5. Magnesite nodules with sulfides in
a dolostone from the Laminites-bearing
Member

as yet, any gypsum or halite casts. However, magnesite which is abundant in the
lower part of the Laminites-bearing Member suggests a high salinity during deposition. An abundance of authigenic quartz, especially in laminated beds, differentiates
the Kamoto Dolostone from sediments now being deposited in the Persian Gulf and
other modern tidal flats.

Evidence has been presented earlier (BARTHOLOME, 1962) that many post-depositional transformations have taken place in the Kamoto orebodies. Whether or not these
transformations belong to diagenetic processes, is to a certain extent a matter of nomenclature. However, it is well known that progressive metamorphism of dolostones
consists of a sequence of reactions, the first of which are those leading to the formation of talc or tremolite. None of these reactions has taken place at Kamoto (2). This
is in agreement with the fact that fluid inclusions (PIRMOLIN, 1970) indicate maximum temperatures of the order of 200 or 250° C. The inclusions also show that a
fluid phase not generated by metamorphic processes was present at these temperatures.

In addition to high copper and cobalt contents, the Kamoto ores are remarkable for
their low Ni/Co and Na/K ratios. They contain little iron and this iron is almost en-

(2) To our knowledge, tremolite has never been found in beds belonging to the Roan
Supergroup. On the other hand, talc has been reported many times. But talc is
known to form in unmetamorphosed sediments (see for example FÜCHTBAUER
and GOLDSCHMIDT, 1959) in which the fugacity of CO_2 is sufficiently low. The
important point is that, in units of favorable composition such.as the Kamoto
Quartzose Dolostone, the transformation of dolomite + quartz into calcite + talc
has obviously not taken place. However, pseudomorphs after talc or tremolite
may be present very locally at the base of this formation (see paragraph 9).

tirely included in sulfide phases. Microprobe work has shown that in orebodies cobalt occurs only as carrollite Co_2CuS_4. This mineral is apparently in equilibrium with chalcosine, digenite, bornite, chalcopyrite and pyrite.

Microscope examination of several hundred polished sections followed by microprobe work on doubtful minerals has failed to discover a single grain of sphalerite or galena or silver-bearing mineral. Neither is there a discrete nickel mineral, although minor amounts of this element are located in carrollite.

In studying the genesis of the Kamoto orebodies, we are therefore dealing with a highly selective mineralizing process. This very important geochemical feature should be taken into consideration, when one compares Kamoto with the Kupferschiefer, White Pine (BARTHOLOME, 1969) or other possible analogs.

Another remarkable feature of the ore is the low abundance of organic matter. Furthermore, the descriptions presented by DEMESMAEKER, FRANÇOIS and OOSTER-BOSCH (1963) suggest that, within a given stratigraphic unit, an inverse correlation exists between organic carbon content on one hand, and copper-cobalt contents on the other. It seems that beds rich in organic carbon contain pyrite and chalcopyrite, but no minable mineral such as bornite or chalcocite.

5. The Carbonate Fraction

A very pure and well ordered dolomite is a major component of all beds with which we are concerned in this paper. Much of this dolomite is fine-grained (a few tens of microns) and xenomorphic, but the barren RSC Member between the orebodies display large dolomite crystals up to 10 cm across. Medium-grained beds are also found above the Upper Orebody. Hundreds of thin sections stained with Alizarine Red S (FRIEDMAN, 1959) have shown the complete absence of primary calcite (3), even in beds entirely devoid of magnesite.

Some of the dolomite is clearly secondary as shown by recrystallized stromatolitic structures in which no calcite remains. Furthermore, the large dolomite crystals with their many inclusions of authigenic quartz, carrollite and other sulfides must obviously have grown within the sediment after deposition. Some of the dolomite on the other hand, especially in the conglomerate at the base of the Kamoto Dolostone, is clearly detrital. But the origin of most of the fine-grained dolomite remains a problem. From petrological observation, there is nothing to suggest that it has replaced some earlier calcite; some of it may well be a primary chemical precipitate.

In short, dolomite crystals have probably formed by precipitation, by transformation of earlier calcite, and perhaps also by recrystallization, from the time of deposition until the end of diagenesis. Presumably, diagenesis ended at some 200° C in presence of a hypersaline brine as witnessed by fluid inclusions (PIRMOLIN, 1970).

---

(3) In fact, traces of calcite have been found only around pores in the barren RSC Member. Presumably this calcite was deposited during cooling from the solution filling the pores. From the data obtained by ROSENBERG and HOLLAND (1964) and USDOWSKI (1967), it can be seen that a chloride solution saturated with dolomite at high temperature becomes supersaturated with respect to calcite when cooling.

A very pure magnesite occurs with dolomite in many beds belonging to the Laminites bearing Member of the Kamoto Quartzose Dolostone. This mineral is observed as la minae or even thin beds, and as nodules or isolated crystals. Its distribution is obviously controlled by sedimentary features (fig. 5, 6). It seems to be associated with dolomite in all proportions. Magnesite occurs mostly as 0.1 to 1 mm crystals containing inclusions of quartz, dolomite and sulfides, or as fibers perpendicular to bedding (fig. 4). Again, as in the case of large dolomite crystals, the growth must have taken place within the sediment after deposition.

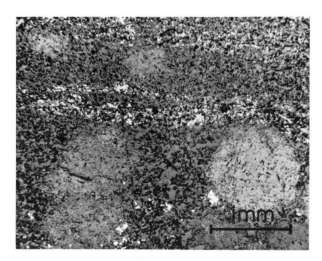

Fig. 6. Magnesite porphyroblasts in a shaly dolostone from the Laminites-bearing Member. Most of the sulfides are found in the detrital laminae (Reflected light)

Petrographic examination affords few clues as to the initial nature of the dolomitic and magnesitic sediments. From analogies with recent occurrences, we infer that the primary precipitates were calcite, aragonite, dolomite and hydromagnesite, perhaps also nesquehonite etc. Post-depositional transformations included the sequence

1. reaction of $CaCO_3$ with hydromagnesite to form dolomite,
2. dehydration of excess hydromagnesite to form magnesite,
3. dolomitization of the excess calcite.

The third stage is required because in many beds the carbonate fraction consists only of ordered dolomite. The complete absence of calcite suggests that, throughout the dolomitization process, a very large excess of magnesium-bearing reagent was available at a moderately elevated temperature.

This makes seepage refluxion (ADAMS and RHODES, 1960) an attractive hypothesis to consider. The facts that the Mines Group rests at least locally upon an erosional surface developed on consolidated terrains and that gravel-filled paleochannels exist at the base of the Mines Group are consistent with this hypothesis. This is why we have insisted on a discussion of the footwall problem at Kamoto.

6. The Quartz Fraction

Authigenic and detrital quartz are readily identified in our rocks. In the Kamoto Quartzose Dolostone, most of the quartz is authigenic as shown by many inclusions of dolomite and sulfides (fig. 7). In fact, the richly mineralized upper beds (RSF) of the Laminites-bearing Member is largely made of such quartz with only minor chalcedony. On the basis of OOSTERBOSCH's mineralogical analyses (1962) and our own, we calculate that at Kamoto within the Kamoto Dolostone alone, the aggregate thickness of authigenic quartz amounts to about 8 meters. The origin of this large amount of silica, which has crystallized in part simultaneously with sulfides, is an important geochemical problem, which is perhaps directly related with the genesis of the orebodies.

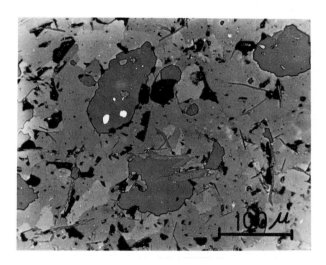

Fig. 7. Authigenic quartz grains with pyrite and dolomite inclusions (Reflected light) from the S. D. Dolomitic Shale. The matrix is dolomite with a small proportion of mica

Some of the laminites observed at Kamoto have the structure usually found in carbonate rocks deposited on tidal flats, although they are mostly made of authigenic quartz. This suggests that silica was brought from some outside source after sediment deposition. It seems unlikely, therefore, that all of the silica was incorporated into the sediment at the depositional interface by settling or chemical precipitation of particles made of some metastable form which later recrystallized as quartz.

In terms of seepage refluxion hypothesis, the source of silica is the high-pH lagoonal water which flowed through the lower part of the Kamoto Quartzose Dolostone. Precipitation of silica occurred during reaction with the overlying connate water maintained at a neutral pH by organic matter decomposition.

7. The Detrital Fraction

In addition to quartz, detrital minerals include abundant muscovite, minor chlorite and accessory tourmaline. On the whole, this detrital fraction has unusually low Na and Fe contents.

In 17 samples of the S. D. Dolomitic Shale, we found that Na is always under 0.1 %, while K varies from 2 to 6 %. These amounts of alkalis are related to the presence of both halite and sylvite as daughter minerals in fluid inclusions (PIRMOLIN, 1970) and are again likely to result from diagenesis. It is well known from experimental work that a chloride brine in equilibrium with Na and K minerals has a low K/Na ratio, which decreases with decreasing temperature. In this study, we are dealing with a brine in equilibrium only with a K mineral: understandably, this brine has a high K/Na ratio. Such a situation may result from two processes: 1) a small amount of a brine of unknown, but ordinary composition equilibrated with a sediment initially poor in sodium; or 2) a large amount of brine initially enriched in K equilibrated with a sediment of unknown, but ordinary composition so as to remove sodium from it. The latter process is in agreement with the idea of seepage refluxion under a hypersaline lagoon.

The Fe content of the detrital fraction, often lower than 0.1 %, results of course from the formation of authigenic sulfides, particularly bornite.

Aggregates of pyrite with a titanium oxide mineral are often observed in the S. D. Dolomitic Shale above the Upper Orebody. Most likely, they result from the transformation of ilmenite. The same titanium oxide is also found within the orebodies, unaccompanied by pyrite. Microprobe examination has shown that this oxide contains little iron (about 1 %). We, therefore, conclude that during diagenesis H_2S was present at a sufficient partial pressure during a sufficient period of time to extract iron almost completely from the detrital components. It is also possible that these detritals were already depleted in iron at the time of deposition.

8. Chlorite and Tourmaline

These minerals, like quartz, are present in both the detrital and the authigenic fractions. Detrital chlorite is green, pleochroic and occurs as more or less twisted flakes lying roughly parallel to the bedding. Authigenic chlorite is colourless and occurs as straight or curved blades with random orientation, and in some cases as inclusions in large sulfide grains (fig. 8). Detrital tourmaline is green and pleochroic; authigenic tourmaline is observed as a colourless overgrowth around it.

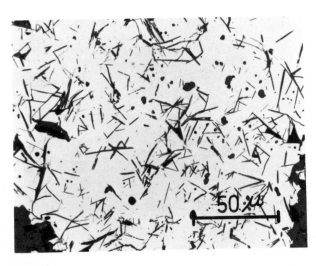

Fig. 8. Authigenic chlorite in chalcocite (Reflected light)

9. Sulfides

Carrollite, bornite, chalcosine and digenite occur in various proportions in the Kamoto orebodies; these minerals give the ore its economic value. Towards the top of the Upper Orebody, chalcopyrite is the major sulfide phase in some beds or laminae. Above the Upper Orebody, pyrite is the major or only sulfide; it is internally zoned with respect to cobalt content (BARTHOLOME, KATEKESHA, LOPEZ-RUIZ, 1971), but otherwise resembles microscopic pyrite such as found in ordinary shales.

The most significant microscopic observations with regard to the diagenetic evolution of the Kamoto orebodies concern pyrite and its relationship to other authigenic minerals.

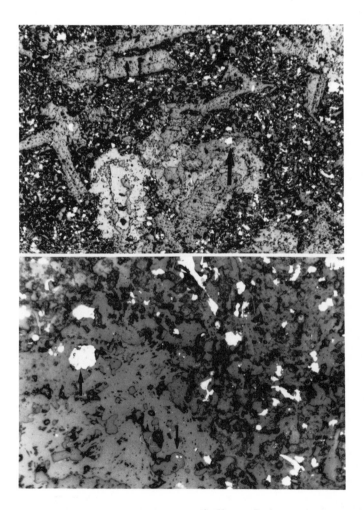

Fig. 9. Pyrite in high-grade ore from the base of the Lower Orebody (Reflected light) This rock consists or large blades consisting of authigenic quartz and dolomite in a fine-grained matrix. The sulfide fraction (chalcocite, digenite and bornite) is almost entirely located in the matrix. A few grains of pyrite are observed: they are found only in the authigenic quartz (see the arrows). The upper picture is enlarged 50 times and the lower one 200 times

This relationship originally observed in a small number of specimens during a preliminary study (BARTHOLOME, 1962), has now been confirmed by microscopic and microprobe examination of many samples.

In summary, our observations show that very small amounts of pyrite are present in most beds within the orebodies. This pyrite occurs as framboids, as isolated grains or as clusters of grains, invariably included within other authigenic grains. In contrast, copper- and cobalt-bearing sulfides are observed mostly as grains of the same size as the major components of the rocks and in contact with all of them.

A few examples will serve to illustrate this distribution of pyrite:

a) Some samples from the chlorite-rich bed at the base of the Lower Orebody consists of a fine-grained matrix in which randomly oriented blades can be seen with the naked eye. The matrix consists of chlorite, bornite, digenite and chalcocite. The blades, which are presumably pseudomorphic after talc or tremolite, consist of a mixture of dolomite and quartz with very minor sulfide content (fig. 9). Pyrite is found only within the quartz portion of these blades.

b) Magnesite nodules are conspicuous in some beds belonging to the Laminites-bearing Member. These nodules, which are coarsly crystallized, are surrounded by a matrix of fine-grained dolomite, quartz, bornite, chalcocite and digenite. Pyrite is observed only in the magnesite crystals and also in the rather larger dolomite crystals formed in the shadow zone on either side of the nodules (fig. 10).

c) In the upper part of the Upper Orebody, some beds contain large, porphyroblastic crystals of carrollite and chalcopyrite in a matrix of dolomite, quartz, bornite and chalcocite. Pyrite is observed only in the porphyroblastic carrollite grains (fig. 1

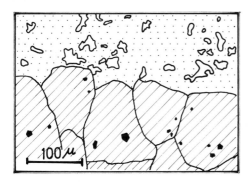

Fig. 10. Pyrite in a magnesite nodule from the Laminites-bearing Member. The matrix of the rock is mostly dolomite (stippled) with chalcocite, digenite and bornite. These minerals consist of xenomorphic grains (white) interstitial to the small dolomite grains of the matrix. A very low amount of pyrite (black) is present in the rock. It is found almost exclusively within large magnesite (shaded) crystals forming the rim of a nodule such as those shown in figure 5

In conclusion, all our observations on the distribution of sulfides are in agreement with the following hypothesis:

1) During early diagenesis, pyrite, the sole sulfide phase, was in equilibrium with the interstitial water. The sediments which were later to become the Kamoto ore

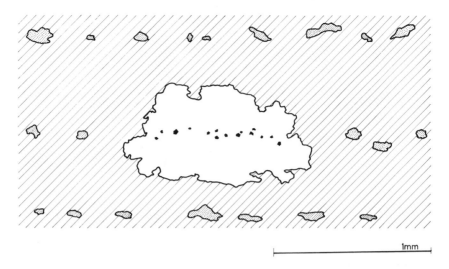

Fig. 11. Relationship between pyrite and other sulfides in and around a carrollite porphyroblast. The drawing shows a carrollite porphyroblast (white) in a dolomite matrix (shaded). This matrix consists of thin laminae rich in bornite blebs (stippled) and of thick barren laminae. Pyrite (black) is found only inside the carrollite porphyroblast at the approximate level of a bornite-rich lamina

bodies were not significantly different from common, unmineralized, tidal flat sediments at this stage. In particular, they were not different from the barren parts of the Kamoto Dolostone and S. D. Dolomitic Shale.

2) During a later stage of diagenesis, the interstitial waters became saturated in cobalt and copper with respect to carrollite, chalcocite, digenite and bornite. Pyrite was, therefore, out of equilibrium and was replaced by copper- or cobalt-bearing sulfides. Some pyrite was protected from complete reaction by being included in some other authigenic mineral grain.

Another independent observation confirms the decreasing availability of iron during stage 2) above: some of the carrollite porphyroblasts display a remarkable internal zoning, with Fe decreasing and Ni increasing from the center toward the periphery.

The picture just presented should be modified slightly in order to take chalcopyrite distributions into account: In most beds, this mineral is found in minor amounts with pyrite as inclusions within other authigenic crystals; in a few beds, it is found in the matrix, usually associated with bornite. Perhaps some chalcopyrite was deposited in equilibrium with interstitial waters during an intermediate stage of diagenesis. Small amounts of chalcopyrite may also have formed simultaneously with pyrite during the first stage; this is suggested by the fact that in the Mines Group outside the orebodies chalcopyrite and pyrite are often found together in the carbonaceous beds.

Conclusions

Evidence has been presented in this paper to show that the Kamoto orebodies consist at least in part of beds initially deposited on tidal flats developed above an erosional

surface. This evidence consists of mineralogical (dolomite and magnesite) and struc
tural (lamination, stromatolites, erosion channels) features. Further research will
include a search for gypsum or halite casts, and for mud cracks (many dolomite-
filled fissures and veinlets are observed in the ore).

As suggested by available evidence, the environment in which the Kamoto Quartzose
Dolostone and the SD Dolomitic Shale were deposited, resembles the one observed
today along the Trucial Coast of the Persian Gulf and along other coasts bordering
arid lands. This environment, which has been described in many recent publications
is remarkable by the fact that, terrigenous input being very low, carbonate sediment
are formed by a variety of supratidal, intertidal and subtidal processes (KINSMAN,
1969; SHINN and others, 1969). The identification of the Kamoto ores as tidal flat se
diments has many implications with regard to the source of the metals, the nature of
the mineralizing phenomena involved, the location, size and shape of the orebodies
and the conditions required for their formation. It is important, therefore, to make
proper interpretation of all possible criteria.

Tidal flat carbonate deposits are conceivably excellent traps for sulfophile metals
from the time of deposition until the end of diagenesis. For, a number of favorable
conditions are fulfilled in them: high content of organic matter, high sulfate content,
low iron content, high initial porosity and assocation with evaporites and chloride-
rich brines.

However, the vast majority of tidal flat carbonates are unmineralized. We have not
reached the point, where we can explain why they are mineralized at Kamoto, let alc
why they contain cipper and cobalt without significant nickel, silver, lead and zinc.
But we have insisted in this paper on several features which characterize the Kamoto
orebodies in contradistinction to modern tidal flat deposits.

These features are, in addition to high copper and cobalt content: the complete absen
of calcite, the abundance of authigenic quartz, the paucity of organic matter and the
lack of porosity. It is tempting to consider them as the result of advanced diagenesis
proceeding under exceptional conditions beyond the stage usually reached in ancient
tidal flat carbonate deposits.

Furthermore, we observe that pyrite and minor chalcopyrite were the first sulfides
to crystallize in many beds. Copper sulfides, bornite and carrollite crystallized late
Outside the orebodies and in particular above the upper orebody, pyrite and minor
chalcopyrite are the only sulfides. We, therefore, suggest that the orebodies result
from a chemical reaction proceeding on a large scale between 1) a stagnant connate
water present in most of the Mines Group (modified by biogenic reduction, by decom
position of organic matter, by reduction of sulfate ion, etc.) and 2) a denser metal-
bearing brine with higher pH and Eh, which flowed through the lower part of the Kamoto Dolostone. This hypothesis is similar to those put forward and elaborated by
D. E. WHITE (1968), W. S. WHITE (1971) and A. C. BROWN (1971) for the mineraliza
tion of the cupriferous Nonesuch Shale at White Pine, Michigan, and by BARTHOLOM
(1963) for other deposits of the Copperbelt.

The source of the brine is open to speculation. Because there is no known intrusive
volcanic activity even remotely related with the orebodies, it seems probable that th
brine originated either by evaporation at the surface of the hydrosphere or by proces
es taking place during diagenesis in the sedimentary pile. The second possibility car
not be tested at the moment. But the first one, implying seepage refluxion from a hy
persaline laggon, seems to be acceptable in view of the evidence presently available

Acknowledgements

We are indebted to Mr. Kandolo, President, and to the Board of Directors of the GECOMIN Company, as well as to their representatives in Kolwezi for the opportunity of visiting and studying the Kamoto mine. Two of us (F. Katekesha and M. Ngongo) have spent several months there in 1969 and again in 1970. They are grateful for the help they received in various circumstances from the GECOMIN's public relations Manager and his staff in Kolwezi.

We wish to extend our thanks especially to Messrs. A. François, Directeur du Département Géologique, R. Oosterbosch, Chevalier, Schayès and Calonne who, directly or indirectly, made this work possible.

We also thank Cl. Monty, G. Monseur and J. Pel, of the University of Liège, for stimulating discussions about tidal flats, carbonate sediments and stratiform sulfide ores. Dr. Monty first suggested a tidal-flat origin for our rocks.

Prof. A. C. Brown has spent much time reading and reviewing this paper. We are grateful to him for the many improvements that he suggested. His patience and his knowledge of copper deposits in sedimentary formations were of great help to us.

We are indebted to Mr. R. Vandenvinne for preparing excellent polished sections, to Mr. M. Gaspar for the microphotographs and to Mrs. J. Weikmans for typing the several successive versions of this manuscript with all the patience required.

Bibliography

ADAMS, J. E., RHODES, M. L.: Dolomitization by seepage refluxion. Bull. Amer. Ass. Petr. Geol. 44, p. 1912-1920 (1960).

BARHOLOME, P.: Les minerais cupro-cobaltifères de Kamoto (Katanga-Ouest) I. Petrographie. Studia Universitatis "Lovanium", Faculté des Sciences (Kinshasa), no. 14, 40 p. (1962).

— Sur la zonalité dans les gisements du Copperbelt de l'Afrique Centrale. Symposium: Problems of postmagmatic ore deposition, V. 1, Prague, p. 317-321 (1963)

— White Pine et Kamoto, deux gisements stratiformes de cuivre. Acad. Royal Sci. Outre-Mer. Bull des Séances (Bruxelles), p. 397-410 (1969).

— KATEKESHA, F., LOPEZ-RUIZ, J.: Cobalt zoning in microscopic pyrite from Kamoto, Republic of the Congo (Kishasa). Mineral. Deposita (Berl.) 6 (1971).

BELLIERE, J.: Manifestations métamorphiques dans la région d'Elisabethville. Publ. Univ. Etat à Elisabethville (Lubumbashi), 1, p. 175-179 (1951).

BROWN, A. C.: Zoning in the White Pine copper deposit, Ontonagon County, Michigan. Econ. Geol., 66, p. 543-573 (1971).

CAHEN, L.: Géologie du Congo Belge. Vaillant-Carmanne, Liège, 577 p. (1954).

CAHEN, L. , LEPERSONNE, J. : The Precambrian of the Congo, Rwanda and Burun‹
   In: RANKAMA, K. , ed. , The Precambrian, 3, Interscience, New York, p. 14‹
   -290 (1966).

DEMESMAEKER, G. , FRANÇOIS, A. , OOSTERBOSCH, R. : La tectonique des gise‹
   ments cuprifères stratiformes du Katanga. In: LOMBARD, J. , NICOLINI, P.
   ed. , Gisements stratiformes de cuivre en Afrique, 2e partie, Association des
   Services Géologiques Africains , Paris, p. 47-115 (1963).

FRIEDMAN, G. M. : Identification of carbonate minerals by staining methods. Jour.
   Sedimentary Petrology, 29, p. 87-97 (1959).

FÜCHTBAUER, H. , GOLDSCHMIDT, H. : Die Tonminerale der Zechsteinformation.
   Beitr. Mineral. Petrog. , 6, p. 320-345 (1959).

GROOSEMANS, P. : Roches basiques de la région de Tenke. Ann. Serv. Mines,
   Comité Spécial du Katanga (Bruxelles), 5, p. 8-13 (1934).

JAMOTTE, A. : Roches basiques et roches métamorphiques connexes de la région
   comprise entre Lufunfu et le Lualaba. Ann. Serv. Mines, Comité Spécial du
   Katanga (Bruxelles), 4, p. 22-55 (1933).

KATEKESHA, F. : Le corps minéralisé supérieur de Kamoto. The University, Liège
   unpubl. thesis, p. 1-60 (1970).

KINSMAN, D. J. J. : Modes of formation, sedimentary associations, and diagnostic
   features of shallow-water and supratidal evaporites. Bull. Amer. Ass. Petr.
   Geol. , 53, p. 830-840 (1969).

NGONGO, R. M. : Le corps minéralisé inférieur de Kamoto. The University, Liège,
   unpubl. thesis, p. 1-62 (1970).

OOSTERBOSCH, R. : La série des mines du Katanga. Bull. de Géologie du Congo
   Belge et Ruanda-Urundi (Lumumbashi), 1, p. 3-7 (1959).

—   Les minéralisations dans le système de Roan au Katanga. In: LOMBARD, J. ,
   NICOLINI, P. , ed. , Gisements stratiformes de cuivre en Afrique, 1e partie,
   Association des Services Géologiques Africains, Paris, p. 71-136 (1962).

PIRMOLIN, J. : Inclusions fluides dans la dolomite du gisement stratiforme de Ka‐
   moto (Katanga occidental). Ann. Soc. Geol. Belgique (Liège), 93, p. 193-202
   (1970).

ROSENBERG, P. E. , HOLLAND, H. D. : Calcite-dolomite-magnesite stability rela‐
   tions in solutions at elevated temperatures. Science, 145, p. 700-701 (1964).

SHINN, E. A. , LLOYD, R. M. , GINSBURG, R. N. : Anatomy of a modern carbonate
   tidal-flat, Andros island, Bahamas. Jour. Sedimentary Petrology, 39,
   p. 1202-1228 (1969).

USDOWSKI, H. E. : Die Genese von Dolomit in Sedimenten. Springer-Verlag, Berlin,
   95 p. (1967).

WHITE, D. E. : Environments of generation of some base-metal ore deposits.
   Econ. Geol. , 63, p. 301-335 (1968).

WHITE, W. S. : A paleohydrologic model for mineralization of the White Pine copper
deposit, northern Michigan. Econ. Geol. , <u>66</u>, p. 1-13 (1971).

WYLLIE, R. J. M. : Katanga copper-Gecomin's open pit mines achieve record production. World Mining (San Francisco), April 1970, p. 36-40 (1970).

Addresses of the authors:

Prof. P. Bartholomé, ⎫   Laboratoire de Géologie,
Dr. P. Evrard   ⎬   Université de Liège
Dr. F. Katekesha   ⎬   Avenue des Tilleuls, 45
Dr. M. Ngongo   ⎭   Liège / Belgique

Dr. J. Lopez-Ruiz   Departamento de Petrologia
Universidad de Madrid
Madrid 3 / Spain

Dr. M. Ngongo   Département Géologique
(present address)   GECOMINES
Kolwezi / Congo-Kinshasa

Metallogenic Processes of Intra-karstic Sedimentation

A. J. Bernard

Abstract

After a brief review of karst weathering processes, the main rock units formed by intra-karstic sedimentation are considered, i. e. detrital and chemical deposits. The behaviour of base metals in traces is then studied in the evolution of the physico-chemical properties of groundwater during its down-trickling course to the water table. Attention is drawn to the trapping effect of the imbibition zone.

Finally, the evolution of both the sulfide deposits and the connate waters during karst fossilisation and further burial is discussed. In particular, the sphalerite-galena-barite association is investigated in order to explain the reasons for the peculiar chemical composition of sulfide fluid inclusions which occur in the so-called "below-unconformity" lead and zinc deposits in carbonate environments.

Introduction

It is well known, at least from text-books, that some cavities of definitely karstic origin may be filled and mineralized by ore-forming solutions.

a) These cavities are definitely karstic since, among many other reasons, they developed exclusively in carbonate environments below surfaces of unconformity or disconformity. Some of these surfaces have been carefully described. The stratigraphic gaps are often represented by hard-pans and selective crustifications (dolomitization, silicification, etc.). Vertical clastic veins, which the surfaces often exhibit, are a strong proof of their erosional and sub-aerial nature.

Thus, the peculiar topographic outcropping patterns of carbonate fromations (sinkholes, "dolines" and poljes, lapies, etc.) could develop more or less intensively according to paleo-climatic activities. Good information on these processes can be derived especially from the so-called residual karstic clays. The preservation of the karstic topographic features, especially "dolines" systems, will then depend on the kind of burial undergone by the surface, i. e. sub-aerial spreading of continental sediments (offlap) or submersion below marine transgressive stratas (overlap).

The karstic nature of the mineralized cavities, the location of which appears strongly related to the vicinity of the unconformity surface, has been recognized as such by outstanding economic geologists, including W. LINDGREN (1933) who wrote about the Tri-State "circle-grounds". More recently, W. H. CALLAHAN (1964) defined all corresponding lead and zinc deposits as strata-bound deposits below unconformities. This type is opposed to the definitely stratiform ore-bodies of the "blanket" or "sheet-ground" deposits, which he called strata-bound deposits above unconformities. Frequently, the two types have been confused and grouped into a single "Mississippi Valley" type (BROWN, 1967). If it is true that both types do occur very often together in the same mineralized area, mining districts exhibiting only one type

44

are also known. For instance, in the mining district of Krakow (Silesia, Poland) only
stratiform deposits above unconformities occur (GRUSZCZYK, 1967), whereas in the
mining district of Sallafossa (Dolomites, Italy) the cross-cutting ore-bodies belong
solely to the below-unconformity type (LAGNY, 1969). The Tri-State circle-grounds
which yielded mainly brecciated lead ores, and the related runs were linked strictly
to the pre-Pennsylvanian unconformity. In constrast, the sheet grounds, which yield
ed disseminated and vuggy sphalerite ores, developed independently of the unconformity surface. In the latter case, it is likely that the ore control is sedimentary as
shown by the ore-bearing chert horizons, which are true sedimentary siliceous mem
bers of the Mississipian limestone and dolomite sedimentation. Thus, it seems illog
ical to continue classifying such different deposits within the same type. Adopting
CALLAHAN's typologic classification one must wonder, whether his distinction is
not at the same time also a genetic one.

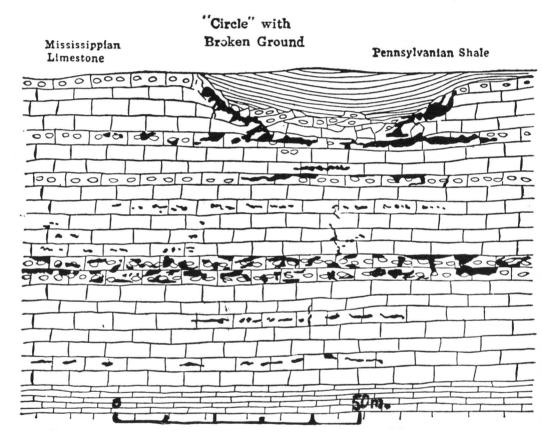

Fig. 1. Well-known diagram of lead-zinc deposits at Joplin (after W. LINDGREN,
1933) showing "broken ground" around "circle" near surface and "sheet-ground"
deposits in Grand Fall chert member below. Black represents ore

From this discussion, it seems advisable to remember that only below-unconformity
deposits show truly cross-cutting relationships with their carbonate country-rocks.
The shapes of the ore-bodies and related runs recall, at least by their size and morphology, the shapes of karstic caves and surrounding fissure vein (or gash-vein) sys
tems.

b) These cavities may be filled and mineralized by ore solutions. "If, as common, the metalliferous deposits are lead and zinc sulfides, then the ore solutions are hydrothermal." Thus could be worded the hydrothermal postulate according to which all sulfides are bound to such a genesis, i. e. to magmatically differentiated juvenile waters (except for true magmatic sulfide melts).

However, as early as in 1897, L. DE LAUNAY demonstrated that the so-called cementation sulfides were derived from supergene processes. The cementation consists of double decomposition reactions between sulfides and sulfates which take place within the sub-aerial part of sulfide ore-bodies. For a long time, the thermodynamic reasons of such double decomposition reactions were ill-understood. Thus the theory of a metallogenic influence of supergene processes has been limited to the cementation of a few chalcophile metals, i. e. Au, Ag, Cu, etc. , according to the SCHURMANN's experimental rule.

It was not until GARRELS's work (1954) that the problem of supergene sulfidization (particularly for lead and zinc) was correctly stated in general terms. Supergene sulfide generation by reduction of base metal sulfates in solutions is quite possible for lead and zinc. Although suspected by some authors (LINDGREN, among others), the economic importance of supergene sulfidization has been underestimated and limited to anecdotic instances, e. g. galena crystals deposited on old iron spikes, newly formed crystals of sphalerite observed when dewatering old mine workings. The implications of galena and sphalerite stalactites discovered in karstlike caves (FERUGLIO, 1969) are quite different.

In fact, the mineralized filling of karstic cavities could be explained in three and only three ways:
1. Hydrothermal ore-bearing solutions opened the cavities and deposited the mineralized filling at the same time. In France this idea led to the definition of "substitution deposits" (RAGUIN, 1961).
2. Cavities are karstic, but the mineralized filling (sulfides and accompanying non-ore minerals) was deposited later by hydrothermal solutions. Originally LINDGREN's idea, this is the telethermal concept which, according to BARNES and CZAMANSKE (1967), explains sulfide precipitation by the reaction of ascending hydrothermal solutions with meteoric waters trickling down through the cave system.
3. The karst-forming agencies caused both opening of the caves and deposition of the mineralized filling.

The first two hypotheses have been presented and supported elsewhere; therefore, only the third will be developed on the following pages.

I - The Karstic Model

Let us consider a whole karst system, as described by J. CVIJIC in 1918. Apparently he exhausted the topic, as studies dealing with the dynamic evolution of karst seem to have been left to the speleologists' care since that time. In fact, most of the published papers about karst subjects during the last decades have been devoted to the superficial zones (LELEU, 1966); the deepest zone, called that "of general imbibition" by CVIJIC, has been left in oblivion. Although normally not accessible for speleologic exploration, this is the zone where most of the ore-bodies occur. Thus, the genesis of these ore-bodies can be understood only by paleo-karst studies.

CVIJIC defined three zones mainly on the basis of the hydrodynamic properties of the flow of meteoric waters which, of course, influence the shape and distribution of caves and related dissolutions fissures (fig. 2). The flow properties also determine the main characteristics of the internal sedimentation within the karst cavity system (SANDER, 1936). Internal sedimentation is a very important process which, unfortunately, has been largely forgotten, even by the most learned authorities in economic geology. Similar to external sedimentation, internal sedimentation yields first detrital and then chemical deposits.

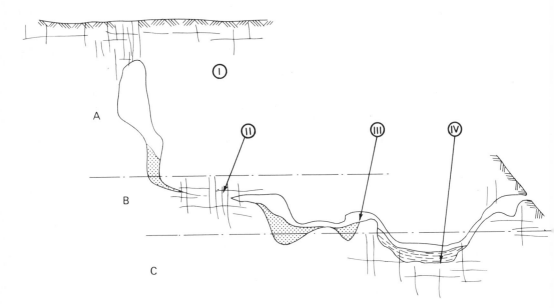

Fig. 2. Simplified section of mature karst system (after J. CVIJIC, 1918)

A - Percolation zone: meteoric waters percolate vertically through carbonate formation, i. e. through fissures which they widen, thus opening caves, sink-holes, gash-veins, etc. Irregular flows, occasionally torrential streams; mechanical erosion largely prevails over chemical leaching. Coarse detrital sedimentation.

B - Permanent circulation zone: water circulation is mostly horizontal when rock jointing allows it; resulting caves exhibit marked lateral extent: galleries. Free or forced flow, always irregular, leads to intense chemical erosion, fine detrital sedimentation.

C - General imbibition zone: comprises the country-rock below the water table. Very slow circulation, if any, i. e. stagnant waters. It is essentially a zone of ultra-detrital and chemical sedimentation.

(Roman numbers within circle refer to chemical composition of corresponding waters, see table I)

a) Detrital sedimentation: Internal hydrodynamics, as well as sub-aerial ones, are able to erode, to transport and to deposit. Internal sediments like collapse breccias (similar to colluvial deposits), sedimentary breccias (deposited from torrential streams), carbonate and clay arenites and lutites, derive from this mechanical erosion. The spreading of these detrital elements depends directly on hydrodynamic properties of the internal flow, and therefore appears closely correlated to the cave morphology. So, it is mainly the basal part of the percolation zone which exhibits breccias and calcarenites deriving either from roof-collapse (sub-outcropping or deep caves) or from torrential reworking of these piedmont-like deposits. The result are hydro-classified or grade-bedded deposits of these clastic elements.

The mean speed of flow in the permanent circulation zone ranges from 3 to 100 m/h. This allows only the transportation of ultra-detritals, carbonate and clay muds, which will unavoidably be deposited in the stagnant imbibition zone.

b) Chemical sedimentation: Agressive meteoric waters become more and more loaded with dissolved salts as their speed of percolation decreases. Indeed the longer the contact between waters and either the carbonate country-rock of the cave or the first internal, coarse or fine, detrital sediments, the more complete is the leaching. At the same time, chemical properties of the solutions change, especially pH and Eh. Enough data have been collected by direct measurements to determine the range of variation in the main physico-chemical factors in a karst system (LELEU, 1966):

$$
\begin{array}{ccccc}
6 & < & pH & < & 8.5 \\
1\ atm. & < & P_{CO_2} & < & 2.10^{-4}\ atm. \\
5\ ^oC & < & T & < & 20\ ^oC \\
-0.4\ v & < & Eh & < & -0.8\ v
\end{array}
$$

Table I - Mean composition of karst waters (see figure 2)

	I	II	III	IV
Ca^{++}	80	110	92	70
Mg^{++}	3.0	26	37	25
Na^+	4.5	71	5.9	415
Cl^-	5.0	164	1.8	583
$SO_4^=$	4.1	46	214	79
HCO_3^-	120	168	207	170
r_s	256	587	587	1342
r_{Mg}^{2+}/r_{Ca}^{2+}	0.027	0.40	0.4	0.60
$r_{SO_4}^{2-}/r_{Cl}^-$	0.71	0.21	59	0.10
pH	7	7.5	7.5	8
μ	0.005	0.01	0.01	0.026

1. Water of the percolation zone: low content in dissolved salts and magnesia; bicarbonated and aggressive.
2. Water of the permanent circulation zone: already neutralised; few sulfates, but chloride content is already noticeable; base exchange index: 0.38.
3. Water of permanent circulation zone: neutralised; high sulfate content (Mississippian limestone, S. Dakota, in: WHITE et al., 1963).
4. Water of imbibition zone: calculated composition achieved by modifying normal properties of clayey shale waters by known trends of stagnation; base exchange index: -0.20; basic pH; Ca and Mg contents decreasing; still sulfatic (in fact sulfidised), this water shows a chloride dominant.

Two waters belonging to the circulation zone have been quoted in order to show the relative composition of these solutions.

48

Similarly, direct sampling of karst waters allows us to sketch the evolution of the mean properties (dissolved salts, pH, Eh, etc.) of the supergene solutions (table I). Given appropriate thermodynamic data, the physical and chemical factors (P, T, ionic strength) appear simple enough to solve any problem of the solution or precipitation of major and minor elements.

Because the dissolved salt content is highest and the solubility product of calcite or dolomite is often exceeded, it appears that chemical sedimentation occurs essentiall in the imbibition zone. Further, the high Ca^{++}-content causes rapid and nearly complete flocculation of clay minerals (CAVAILLE, 1960). Thus, the variation of Ca^{++} activity, which depends upon variations of the karstic flow, e.g. between high and low water periods, determines in turn the clay mineral flocculation rate. Karstic varved marls have been explained very cleverly by this process (TINTANT, 1962).

In addition, the stagnation of solutions in the imbibition zone causes negative Eh values. This zone thus appears as an internal micro-euxinic environment depositing black muds rich in organic matter, the carbonate content of which depends on P_{CO_2} variations. When these conditions steadily prevail, anaerobic bacterial colonies settle down and undertake sulfate reduction producing H_2S, as in any hydromorphic environment.

H_2S formation bears very important consequences on the behaviour of base metals in the karst system; this can be qualitatively visualized from the diagrams of GARRELS and CHRIST (1967) (fig. 3). The solubility of the more stable species of the main metals can be claculated more accurately in each different kind of karst solutions. This has been done for Pb, Zn and Ba (table II), taking into account only the

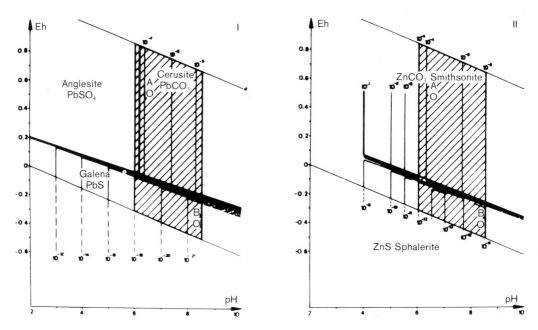

Fig. 3. Diagrams of theoretical stability fields of minerals as functions of pH, Eh and solubility products of various lead (I) and zinc (II) species (after GARRELS, 1953

The range of pH-Eh-variations in karsts is superimposed (shaded area). Travel from A to B represents the presumed mean karst water evolution in mature systems

main species, i. e. sulfide, carbonate and sulfate, the solubility products of which are known either by measurements or from thermodynamic constants. Solubility products of more involved salts (phosphates, arsenates, vanadates, etc.) which are very important for lead, are poorly known, and it did not seem wise to consider them in this model.

Table II shows a drastic change of Pb and Zn solubility at the top of the imbibition zone. Lead and zinc, rather soluble in the three first solutions, become highly insoluble as soon as H_2S evolves significantly. In contrast, Ba^{++} exhibits poor solubility in the first solutions and a rather high one in the reducing waters.

Table II - Solubilities of Pb, Zn and Ba in karst waters (μgr/gr; P_{CO_2} in atm.)

	I	II	III	IV
CO_3Ca	200	270	125	70
Ba_2^+	0.064 (320)	0.060 (220)	0.013 (100)	160
Pb^{2+}	0.060 (300)	0.018 (100)	0.010 (200)	10^{-19}
Zn^{2+}	14	6	3	10^{-16}
P_{CO_2}	$10^{-1.8}$	$10^{-2.1}$	$10^{-2.1}$	$10^{-2.7}$

The dissolved calcite content is qualitative. Figures between brackets represent the dissolved metal contents expressed in μgr/gr with respect to dissolved calcite, i. e. the grade of the assumed parent rock before karst weathering.

The general sulfatation trend of underground circulating waters is likely to provoke an early barium precipitation as barite, which will precipitate more or less in the imbibition zone. So the bulk of metal contained in limestones or dolomites stripped off by karst weathering (i. e. clarke grade or existing pre-concentrations in the form of source beds or true stratiform low grade deposits) is likely to be trapped in this very simple and efficient chemical "throttle". However, before attempting a geochemical balance of the model, let us consider the dynamic evolution of karst systems.

II - Evolution of Karst Systems

CVIJIC (1918) also described the mean dynamic evolution of a karst system in three stages (fig. 4).

This model points out a very important feature of base metal concentration: the movements of the water table depends on both erosional stripping off of rocks and alluvial sedimentation which end with the formation of a peneplain. Former cave and fissure systems of zones I and II (fig. 1) are likely to be buried in zone III. The older a karst system is, the more important is the development of the imbibition zone (III), apparently due to the rising water table in the cavity system. In other words, the thickness of the buried imbibition zone is equal to the variation of the water table during the aging of the system.

In fact, actual karst systems have a more involved history than that outlined by CVI-JIC. Climatic changes and complex water table movements, i. e. epeirogenetic events, marine transgressions or regressions etc., strongly complicate karst evolution. Indeed, rejuvenation of a senile system will lead to a young or mature system much more complex and aggressive than the older one; previous cavities induce the development of new ones (BARBIER, 1960).

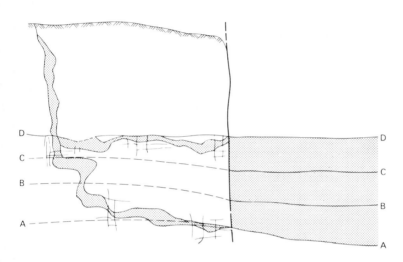

Fig. 4. Diagram showing water table locations during surface levelling after an epeirogenic event. A, B, C are different positions of water tables and the corresponding shape of CVIJIC's zones (A, D). External and internal sediments are shaded

Oscillations of the water table also bear consequences for the internal, intra-karstic sedimentary fillings, and particularly for the mineral distribution. To check this assumption, let us examine the ore filling during a normal, simple cavity evolution from an early to a mature stage. The horizontal caves or galleries where barite is deposited as soon as maturity is reached, generally exhibit early clayey and limonitic coatings, recording cavity formation in the oxidation zone. In other words, after a stage of wall alteration yielding dolomitized or silicified limestones, which are peri-karstic processes similar to the weathering transformations recorded by emersion surfaces, the metalliferous fillings begin with limonitic and/or clayey coatings on which the first barite crystallizes. Most of the time these deposits are banded aggregates of elongated crystals stretching perpendicularly to the walls. This implies a two-dimensional distribution of barite nuclei on free surfaces or, more adequately, on irregularities of free surfaces. In fact, everything happens as if the degree of supersaturation with respect to barite, which allows the precipitation of the first nuclei, was achieved slowly. However, once the nuclei formed, the crystal growth rate suddenly appears very high, each crystal being limited by the growth of the adjacent ones, except in the direction perpendicular to the wall, which remains free. With decreasing $BaSO_4$-concentration in the solution, the growth slackens and then stops. Each band implies such an evolution of the dissolved $BaSO_4$-content in the solution.

When a cavity which has matured to this stage is drowned below the water table, the sudden change in the solubility of $BaSO_4$ generally causes a certain corrosion of the already deposited barite crystals (BARNES and CZAMANSKE, 1967).

While sulfate reduction is gradually increasing, galena will precipitate as it is the less soluble sulfide unless the solutions are exceptionally Zn^{++} rich. Scattered nucleation, two-dimensional and gradually accelerated growth according to a regular increase of the degree of supersaturation will often yield very large crystals, which rest on the last baritic band. Once a certain supersaturation threshold has been passed, precipitation occurs in a three-dimensional nucleation mode yielding xenomorphic, fine grained fabrics. Rather commonly the first sphalerite grains appear at this stage of concentration of the solution. These are generally impure and colloform, i. e. banded "schalenblende" and fibro-radiating ovoids of brown sphalerite.

From this simple model, it is easy to foresee what would happen if such a karstic system was rejuvenated. Introduced into zone I, such karst ores would suffer leaching and oxidation, as occurs in common lead and zinc ore outcrops, e. g. the formation of rusty-colored smithsonite and anglesite. If the former ore is brought into zone II, a new deposition of barite will occur covering sulfides and sulfates of earlier generation.

Detailed studies of both sulfide paragenesis and internal sedimentary fillings thus allow, in simple cases, to comprehend the history of karst-systems before fossilisation and burial.

III - An Attempt at a Geochemical Balance

It is now possible to sketch a balance of the metallogenic activity of karst systems. To give an order of magnitude, let us consider carbonate rocks averaging $100 \mu gr/gr$ Pb. This figure is slightly higher than the lead clarke in carbonate rocks (WEDEPOHL, 1956), but is nevertheless rather common in many carbonate formations exhibiting geochemical anomalies (source-beds) such as the Tri-State Mississippian including the sheet-grounds, the Iglesiente "Metallifero" of Sardinia (Italy), etc. What volume of such a formation is to be stripped off by karst weathering to yield a metal accumulation of 10^5 metric tons of lead, i. e. a small workable lead deposit, at least in Europe? A very simple estimation gives 45 km^2 of beds 10 m thick.

Now, geological studies of several well exposed mining areas show that important concentrations come mostly from the erosion of the lowest 25 m of the now-vanished upper stratigraphic series. A young karst leaches and dissipates all existing preconcentrations. In addition, it can be inferred from previous considerations of the internal sedimentation of karsts, that chemical deposition occurs essentially during the senile stage, when the imbibition zone reaches a maximum extent. Because sulfides behave as chemical sediments they will be concentrated essentially during the latest phases of the evolution of the system.

Taking this fact into account, the figure formerly quoted is changed to a 18 km^2 area, i. e. a square area of 4250 m along one side, of beds 25 m thick. Compared to the extensive surfaces commonly drained by senile karst systems, this new figure appears exceedingly small.

In order to appreciate the duration of such an event, one must adopt a value for the mean rainfall. As a first approximation, the world continental rainfall average can be chosen, 844 mm/cm^2/y. Further, the paleo-climatic conditions can be inferred from the sedimentary facies of carbonate rocks. Magnesian limestones and cherty members correspond to a biostatic period developed through ferralitic weathering processes (ERHART, 1956). This means the climate was equatorial or at least humid and tropical; 1000 mm/cm^2/y would not be an overestimation of the mean rainfall.

It is true that Pennsylvanian climate agencies which were active during the karst weathering of Mississippian formations could be different from the Mississippian climate previously inferred. The works of KELLER et al. (1954) on the Cheltenham Formation, which directly overlies the pre-Pennsylvanian unconformity surface, answer the question. There, continental formations made up of flint clays and diaspore prove that the Pennsylvanian climate was humid and tropical. The leaching activities were very efficient; 1000 mm/cm^2/y is then a minimum value. If one takes the admittedly low solubility rate of 125 μgr/gr for calcite, the formation of the sma deposit previously considered needs 37,500 years. The reader will appreciate that this is an extremely short geological duration.

Thus, the geochemical balance sketched from the drained area, eroded thickness of carbonate rocks, importance of rainfall and the solubility rate of calcite in humid tropical climate may be considered as a very pessimistic estimate of the metallogenic efficiency of karst activities. In many respects this model is certainly one of the most powerful ever known, and the corresponding type of deposit should be much more frequent. Taking into account the tremendous amount of metal contained in the eroded carbonate formations, even when using clarke values, one is compelled to ascribe a very bad yield to the karst model, considered over a whole cycle. As has been emphasized previously, the metallogenetically active stages of the karst cycle are essentially the senile ones. In contrast, the erosional capacity of senile stages appears rather limited. Therefore, it seems necessary to appeal to a delicate epeirogenic regulation in order to provide the erosion of an appreciable thickness of carbonate rocks in a karst at the end of the stage of maturity.

Thus, and this is a very general metallogenic law, a very simple and efficient physico-chemical model could entail workable concentrations only when applied to large volumes of rocks during long periods of optimum efficiency. From this statement one can see that the scarcity and anomalous nature of mineral concentrations result only from the coincidence of favorable geological conditions which, in addition, last over an appreciable length of time.

IV - Later Modifications of Karstic Concentrations

Fossilisation and burial can be distinguished.

a) Fossilisation: Most of the time, fossilisation is achieved by the deposition of an impervious formation above the outcropping head-level of the karst system, although instances are known of karst fossilisation by limestone or sandstone strata. More precisely, karsts are fossilised only when the water table rises above the unconformity surface, for it is well known that karsts continue to work below a permeable cover if the water table lies within the carbonate formation.

Although this fossilisation can occur at any moment of the karst cycle, senile karsts are most frequently overlain.

Continental fossilisation (for instance the Pennsylvanian of the Tri-State) proceeds
by the spreading of fine, muddy, terrigenous sediments in the hydromorphic area of
mature karsts. Internal sediments (including ore deposits) and connate waters may
then be considered as a closed physico-chemical system still able to undergo evolution through burial below an overlying series.

Marine fossilisation is also frequent and sometimes has been invoked in metallogenic
processes. Indeed, the sunken imbibition zone may act as a euxinic trap for base
metals dissolved in sea water. The hydrogeology of the marine transgression will
involve the landward movement of the inclined limit of saline sea water and fresh
karstic waters through the imbibition zone. These movements are generally slow,
and it is likely that the reducing property of the drowned senile karsts will be preserved. In this situation the base metal supply from stagnant bottom sea water could
occur only by cation diffusion through a sub-stationary solvant, thus considerably
limiting the metallogenic range of the phenomenon. One may wonder whether such
marine fossilisation leaves any remarkable evidence except a very conspicuous precipitation of barite.

In any event, the first impervious marine sediments will seal the cave system, which
is now ready for further evolution, in the same manner as for continental fossilisation.

b) Burial: Many transformations resulting from gradually deeper burial may affect
the primary cave fillings. Rises in pressure and temperature, salt enrichment of
connate waters, and related escape of the most volatile gases (H_2S, light hydrocarbons, etc.; cf. MILLS and WELLS, 1919) and the complete halt of bacterial activities above 70 ^{o}C are among the main causes of the changes.

Detailed analysis of these phenomena is beyond the scope of this paper, which is essentially aimed at the sedimentology of karsts. Only the main consequences will be
outlined here.

1. The greater the burial, the greater the changes in connate waters. The general
trend is that of connate waters of oil fields, i.e. chloride enrichment and sulfate depletion. The latter may be explained by bacterial reduction below 70 ^{o}C; up to this
temperature, it is likely that the H_2S-concentration of connate waters increases.
Above 70 ^{o}C the H_2S-concentration certainly decreases through loss of this very volatile gas.

In short, after a 3000 to 4000 m burial, karstic connate waters are likely to closely
resemble oil field connate waters of the Na-Ca-Cl-type. In other words, they will
exhibit a composition very similar to the composition of fluid inclusions of the Tri-
State type (WHITE, 1966; ROEDDER, 1966). This presents a new problem: Will sphalerite recrystallize in such maximum burial conditions (see below) ?

2. The temperature rise and the synchronous evolution of connate waters cause variations in sulfide and barite solubilities. Variations in the solubility of lead and zinc
sulfides below 70 ^{o}C depend on bisulfide complexing, e.g. PbS, H_2S and $Zn(HS)^{3-}$
(BARNES and CZAMANSKE, 1967). However chlorides show an opposite effect. Therefore, the solubility of these sulfides goes through a minimum when the increase of the
salinity of connate waters, and particularly the chloride content overtakes the H_2S-
production (BARNES, 1966).

The solubility of barium sulfate increases with temperature and ionic strength of the
solution, that is to say with the chloride salinity. Obviously, on cooling, a certain
amount of barite will precipitate from the connate waters.

At least, two very simple conclusions may be drawn from these considerations:

a) Solubility variations of sulfides resulting from the antagonistic effect of bisulfide and chloride complexing, cause an increase of crystal size. In fact, with increasing solubility, most of the matter which will dissolve is derived from aggregates having a high specific surface, i. e. the finest crystals in contact with solution, many of which will be completely dissolved. In contrast, for the largest grains, change will be marked only by a surface on which growth stops. Because these surfaces provide existing crystallization centers, they are precisely the ones which will be supplied with newly precipitated sulfides during a later decrease of solubility. Thus, the reasons for the so-called "aging of sulfides" may be easily found in solubility changes during burial. It seems convenient to recall here that burial time is an extremely long period (tens to hundreds of millions years) and that any solubility change, oscillatory or not, will cause an irreversible grain size increase for the two sulfides here considered, galena and sphalerite.

b) The minimum zinc sulfide solubility occurs when chloride complexing overtakes bisulfide complexing. Owing to the relatively high solubility of zinc bisulfide complexes, this is a strong minimum (fig. 5), apparently capable of producing a degree of supersaturation high enough to nucleate new sphalerite crystals. These are light, yellow, translucent, idiomorphic crystals generally assumed to be of late generation, and are particularly sought after and sampled by fluid inclusion

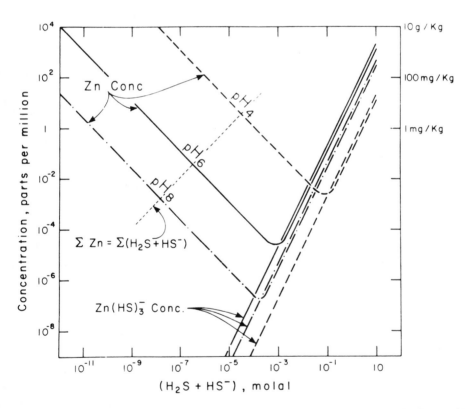

Fig. 5. Total zinc concentration and concentration of zinc bisulfide complex in a solution in equilibrium with sphalerite at 100 $^{\circ}$C, after H. L. BARNES (1966). Ionic strength: 1.0, chloride ion activity: 1.0. This diagram clearly shows the minimum zinc solubility in function of (H_2S + HS^-) concentrations

specialists, for they exhibit beautiful primary inclusions. Therefore, it is not amazing that they register rather high temperature (100 to 150 ⁰C), for they record a very deep, if not the deepest, burial suffered by the deposit.

Conclusions

Among the three genetic hypotheses presented in the introduction, the paleo-karstic one is certainly the easiest to check in view of geological and mineralogical facts. But let us come back to theoretical grounds.

The concept of a hydrothermal solution, capable of dissolving very great amounts of carbonate rock and of depositing at the same time similar quantities of sulfides, meets insurmountable difficulties. Further, the hydrodynamic reasons for carbonate dissolution which occurs selectively below unconformity surfaces (even in the case of a limestone-limestone contact unconformity, as in Dolomites, for instance) remain unexplained by such a hypothesis. Likewise, the ability of an ascending solution to dissolve carbonate rocks and yield a karst mimesis still lacks an explanation. To date nobody, to my knowledge, has dared to explain how an ascending hydrothermal solution can generate within caves an internal sedimentation which is similar in every respect, including fossils, to intra-karstic sedimentation. This argument, I believe, closes the debate.

As to the telethermal hypothesis, the well-known occurrence (CRAWFORD and HOAGLAND, 1968) of reworked sulfides within the internal, detrital, graded-bedding karst sediments leads one to think that, if there has ever been telethermal activity, it interfered with karstification. Therefore, the reason for a systematic time and space coincidence between karsts and hydrothermal activities needs to be clearly and thoroughly argued, which has not yet been done. Finally, one may wonder whether the dilution of telethermal solutions by the karstic flows would not remove most of their proper capacities, including the ability to produce a mineralization.

Chemically, the karst-mineralizing hypothesis appears as plausible as a telethermal theory. Indeed, the karst model is geologically more probable, for it takes into account many more facts. Finally, the karst hypothesis is also much simpler because the time and space coincidence of mineralization and karstic activity is explained by the phenomenon itself.

References

BARBIER, R.: Le rôle des paléokarsts dans la formation des réseaux karstiques actuels et leurs répercussions sur l'hydrogéologie de ces régions. C.R. somm. Soc. Géol. Fr., (Paris), 3, p. 59-60 (1960).

BARNES, H.L.: Sphalerite solubility in ore solutions. Econ. Geol. (Lancaster), monograph 3, p. 326-332 (1966).

BARNES, H.L., CZAMANSKE, G.K.: Solubilities and transport of ore minerals. In: Geochemistry of hydrothermal ore deposits, H.L. BARNES, Ed., New York, p. 334-381 (1967).

BROWN, J.S.: Genesis of stratiform lead-zinc-barite-fluorite deposits in carbonate rocks (the so-called Mississippi Valley type deposits). A symposium, J.S. BROWN, Ed. Econ. Geol. (Lancaster), monograph 3, 443 p. (1967).

CALLAHAN, W. H. : Paleogeographic premises for prospecting for stratabound base metal deposits in carbonate rocks. Cento symposium on Mining Geology, Ankara, p. 191-248 (1964).

CAVAILLE, A. : Les argiles des grottes. Ann. Spéléol. (Paris), XV, 2, p. 383-400 (1960).

CRAWFORD, J. , HOAGLAND, A. H. : The Mascot-Jefferson City Zinc District, Tennessee. In: Ore deposits in the United States 1933/1967, AIME, (New York), p. 242-256 (1968).

CVIJIC, J. : Hydrographie souterraine et évolution morphologique du karst. Revue Trav. Inst. Géogr. alpine, (Grenoble), 6, 4, 56 p. (1918).

ERHART, H. : La genèse des sols en tant que phénomène géologique. Masson et Co. , Paris, 90 p. (1956).

FERUGLIO, G. : Blenda stalattitica nel giacimento dell'Argentiera (Auronzo). Museo Friulano di St. Nat. , (Udine), 13, 28 p. (1969).

GARRELS, R. M. : Mineral species as functions of pH and oxidation - reduction potentials, with special reference to the zone of oxidations and secondary enrichment of sulfide ore deposits. Geochim. Cosmoschim. Acta (Oxf.), 5, 4, p. 153 -168 (1953).

GARRELS, R. M. , CHRIST, C. L. : Equilibres des minéraux et de leurs solutions aqueuses. Gauthier-Villars Ed. , Paris, in-8º, (traduit par R. Wollast), 335 p. (1967).

GRUSZCZYK, H. : The genesis of the Silesian-Cracow Deposits of lead-zinc ores. Econ. Geol. (Lancaster), monograph 3, p. 169-177 (1967).

KELLER, W. D. , WESCOTT, J. F. , BLEDSOE, A. O. : The origin of Missouri fire clays. In: Clays and clay minerals, London, 2nd Nat. Conf. , 1953, p. 7-46 (1954).

LAGNY, Ph. : Minéralisations plombo-zincifères triasiques dans un paléo-karst (gisement de Salafossa, province de Belluno, Italie). C. R. Ac. Sci. , (Paris), 268, D, p. 1178-1181 (1969).

LAUNAY, L. de: Sources thermo-minérales. Recherche, captage et aménagement. Beranger Ed. , Paris, 636 p. (1897).

LELEU, M. G. : Le karst et ses incidences métallogéniques. Sci. Terre, (Nancy), 11, 4, p. 385-413 (1966).

LINDGREN, W. : Mineral deposits. 4th Ed. , Mc Graw-Hill, New York, 930 p. (1933).

MILLS, A. van, WELLS, R. C. : The evaporation and concentration of waters associated with petroleum and natural gas. U. S. Geol. Survey Bull. , (Washington), 693, 104 p. (1919).

RAGUIN, E. : Géologie des gîtes minéraux. Masson Ed. , Paris, 686 p. (1961).

ROEDDER, E. : Environment of deposition of stratiform (Mississippi Valley-type) ore deposits, from studies of fluid inclusions. Econ. Geol. (Lancaster), monograph 3, p. 349-362 (1966).

SANDER, B. : Beiträge zur Kenntnis der Anlagerungsgefüge. Miner. u. petrogr. Mitt. (berlin), 48, p. 27-209 (1936).

TINTANT, M. : Remplissages polycycliques dans le karst de Côte d'Or. S. S. F. , (Paris), 2, p. 59-61 (1962).

WEDEPOHL, K. H. : Untersuchungen zur Geochemie des Bleis. Geochim. Cosmo-
    chim. Acta (Oxf.), 10, p. 69-148 (1956).

WHITE, D. E. , HEM, J. D. , WARING, G. A. : Chemical composition of sub-surface
    waters. Data of Geochemistry. 6th Ed. , Ch. F. , U.S. Geol. Surv. Prof. Paper
    (Washington), 440 F, 67 p. (1963).

WHITE, D. E. : Outline of thermal and mineral waters as related to origin of Missis-
    sippi Valley ore deposits. Econ. Geol. (Lancaster), 13, p. 379-382 (1966).

Address of the author:

Ecole Nationale Supérieure
de Géologie Appliquée
54 Nancy 01 / France

Classification of Stratified Copper and Lead-Zinc Deposits and the Regularities of Their Distribution

Y. V. Bogdanov and E. I. Kutyrev

Abstract

Confined to variegated carbonate-terrigenous formations, sedimentary copper deposits are subdivided into shallow-marine (Mansfeld), lagoon-deltaic (Dzhezkazgan, Udokan) and lacustrine-alluvial (Priuralye, the Colorado Plateau). Enclosed in grey carbonate formations, sedimentary lead-zinc deposits are confined to shallow-marine
carbonate sediments (Mirgalimsay) and coastal-marine terrigenous - carbonate deposits (Sumsar, Uch-Kulach). The transgressive or regressive development of the
paleobasin determines one or other type of the age migration of ore-bearing deposits
and zoning of mineralization. The confinement of sedimentary copper and lead-zinc
deposits to marginal parts of platforms, avlakogens, foredeeps and miogeosynclines
is characteristic.

Igneous-sedimentary copper and lead-zinc are formed due to subfluvial, more rarely
terrestrial acid and basic volcanic activity and in the distance of them.

Copper deposits are formed in connection: 1) with subfluvial acid and basic dacite and
other formations; pyrite deposits of the South Urals, the North Caucasus; 2) with terrestrial basic and acid volcanism (paragenesis of basalt liparite and variegated formations; the Upper Lake and White Pine in the USA, the Minusinsk depression in the
USSR).

Lead-zinc deposits are formed: 1) near the focuses of volcanic activity (liparite, an-
desite - dacite - liparite and other formations; the Ore Altay); 2) in a lateral (or some
chronological) distance of the focuses of volcanic activity (siliceous - terrigenous -
carbonate and other formations ; Zhairem (Atasuy) etc. in Kazakhstan, Filischay in
Aserbaijan, Rammelsberg and Meggen in Western Germany).

- : -

The deposits under discussion are usually confined to (or situated near) certain parts
of the stratigraphic column of thick terrigenous, carbonate-terrigenous and terrige-
nous-volcanogenic sequences of various age - from Precambrian to Neogene inclusive.
The deposits are characterized by the presence of large orebodies of sheet-like or
lenticular shape, which are concordant with the enclosing rocks: cross-cutting ore-
bearing veins also occur in the productive sequences. The commercial value of a deposit depends upon the size of the concordant orebodies.

According to the source of the metal, the deposits are divided into two large groups,
i.e. sedimentary and volcanogenic-sedimentary ones. To consider ore-bearing sediments as parts of the volcanogenic-sedimentary formations is still controversial;

however, the submarine volcanic activity in the basin of sedimentation or terrestria volcanic activity, which took place simultaneously with ore accumulation, presents strong argument in favour of volcanic source of ore elements. It seems that a more detailed classification of sedimentary concentrations should be based on lithological facies pecularities of ore-bearing sediments, and classification of volcanogenic-sed mentary formations on the nature of paleovolcanism.

The stratified sedimentary copper deposits form a single genetic group of ore form- ations, which had originated under similar geological environment. They are charac terized by the association with variegated formations, accumulation of predominantl cupriferous rocks in the source areas resulting from the erosion, and multi-stage genesis. The formation of copper ores took place during the sedimentation stage as well as during the diagenetic, katagenetic, and sometimes metamorphic stages. The main processes of ore formations took place during diagenetic stage (STRAKHOV, 1963; POPOV, 1965).

The stratified copper deposits are characterized by the following features: similar geotectonic position and paleogeographical conditions of copper accumulation; form- ational, stratigraphic, lithological facies and paleotectonic control of mineralizatior specific mineralogical - geochemical composition; zonal nature of mineralization; certain diagenetic, katagenetic and sometimes metamorphic alterations. The magm: tic and tectonic factors, which are very typical for endogenous deposits are repre- sented only by local contact metamorphism of cupriferous rocks in the vicinity of m: matic formations and by displacement of cupriferous horizons along faults.

The proposed classification of stratified sedimentary deposits of copper ores is bas on the association of copper-bearing sediments with various parts of facies profile c variegated sediments.

The volcanogenic-sedimentary copper deposits very rarely occur in variegated form ations. Their only difference from the sedimentary deposits consists in the presenc of ore-bearing volcanites in the productive sequence (Lake Superior, White Pine, th Minusinsk basin, etc.).

There is a marked difference between the geological conditions of formations of vol- canogenic-sedimentary chalopyrite deposits and those of stratified deposits in terri- genous variegated sequences. The former are characterized by the association with eugeosynclinal basaltoid volcanogenic formations (spilite-keratophyre formation etc. The chalcopyrite ore deposits under discussion are often confined to pyroclastic and sedimentary deposits accumulated near active submarine volcanic foci (lavas and ac pyroclastic rocks). Simultaneously with the formation of volcanogenic-sedimentary ore concentration, the hydrothermal copper mineralization took place at different le vels inside the volcanic pipes (SMIRNOV, 1965). The formation of volcanogenic-sedi mentary ore concentrations (like that of sedimentary ore concentrations) was a mult stage process with the diagenesis playing a leading part.

The results of study of the stratified copper ore deposits permit to suggest the fol- lowing classification scheme.

Classification scheme of stratified copper ore deposits:

A - Sedimentary (sedimentary-diagenetic) deposits

    I. The copper slate type (shallow-water marine type): the Mansfeld deposits, Sudets piedmont monocline in Poland.

II. The copper sandstone type:
1. The Dzhezkazgan subtype (lagoon-deltaic subtype):
   Dzhezkazgan, Sary-Oba, Itauz, etc. in Kazakhstan.
2. The Pre-Urals subtype (lacustrine-alluvial subtype):
   Naukat, Varzyk etc. in Central Asia, the Colorado Plateau in the USA,
   the Pre-Urals region.

B - Volcanogenic-sedimentary deposits

I. Related to underwater acid and basic volcanism (spilite-keratophyre, ande-
   site-dacite formations etc.): the uppermost parts of the section of some pyrite deposits in S. Urals, N. Caucasus, the "Kuroko" type deposits in Japan.

II. Related to terrestrial basic and acid volcanism (paragenesis of basaltic, li-
    parite and variegated formations): Lake Superior and White Pine in the USA,
    the Minusinsk basin in the USSR.

C - Metamorphosed deposits

I. Regionally-metamorphosed: Copper sandstone of the Olekmo-Vitim highland
   (Udokan etc.), copper slates and quartzites of the Katanga-Zambia copper
   belt, pyrite deposits of Karelia, fahlbands of Norway and Canada.

II. Contact-metamorphosed: the Krasnoe, Burpalinsk and other deposits in the
    Olekmo-Vitim highland.

Judging from the geochemical combination of elements, shape of deposits and orebodies, association with certain parts of stratigraphic sections of sedimentary and volcanogenic-sedimentary sequences, the stratified lead and zinc deposits form a single
group of ore formations. It was rather recently that the volcanogenic-sedimentary deposits of lead and zinc have been separated into a special group, whereas the similar
group of the chalcopyrite and copper-bearing polymetallic deposits has been distinguished long ago. The results of study of the stratified lead-zinc deposits permit to
suggest the following classification scheme (considering data published both in the
USSR and abroad: STRAKHOV, 1963; POPOV, 1968; SMIRNOV, 1965; AMSTUTZ,
1963; DUNHAM, 1950).

Classification scheme of stratified lead-zinc deposits:

A - Sedimentary (sedimentary-katagenetic deposits)

I. The Mississippi Valley type (shallow-water marine type): the Tri-States
   deposit in the USA, Mirgalimsai deposit in the USSR.

II. The Sumsar type (coastal-marine type): Uch-Kulach, Sumsar, Dzhergalan
    in Central Asia.

B - Volcanogenic-sedimentary deposits (associated with the submarine acid volcanism)

I. The Altai type: near valcanic foci (the liparite, andesite-dacite-liparite, li-
   parite-tuffaceous-shaly formations etc.): Rudny Altai.

II. The Atasu type: laterally (or chronologically) separated from volcanic foci:
    Zhairem, Atasu and other deposits in Central Kazakhstan, Filizchai deposit
    in Azerbaijan, Rammelsberg and Meggen in Germany.

C - Metamorphosed deposits

I. Regionally-metamorphosed: Broken Hill and Mount Iza deposits in Australia;
   Karagaily, Culshad in Kazakhstan.

During recent years, the geologists who specialized in study of economic minerals, have been paying a particular attention to the facts pointing to a directional characte of evolution of ore genesis processes (RUNDQUIST, 1968). Since the formation of stratified ore deposits was synchronous (or nearly synchronous)with the formation of enclosing rocks, the orebodies of this type are particularly convenient for establishing lateral (isochronous) and chronological rows, which both reveal and emphasize the regional regularities of ore distribution (provided the structural patterns of the earth crust are of the same type).

The largest sedimentary deposits of copper, lead and zinc are localized in various structural stages of geosynclinal troughs: in the platform covers only small ore deposits are found. In troughs devoid of any traces of volcanic activity or with a very weak volcanism during early stages, the largest stratified deposits of lead and zinc were formed in lower members of carbonate sequences in favourable facies-paleoge graphic environment. The copper deposits are confined mainly to the troughs filled up with terrigenous rocks, predominantly by paralic members of transgressive-regressive cycles. This control is reflected also in the nature of distribution of coppe and lead-zinc deposits which is characterized by a vertical regional zoning: cupriferous sandstones are confined either to the lower terrigenous variegated geosynclinal formations or to the upper, proper orogenic formations (also variegated ones), whereas the lead-zinc sandstones appear restricted to the middle terrigenous-carbonate formations.

In addition, the character of the metallogenic profile of a trough depends to a great extent upon the "average" composition of slope massifs and troughs basement. If the sections of the slope massif is that of miogeosynclinal type, the conditions are unfavourable for accumulation of copper, but good for accumulation of lead and zinc; the eugeosynclinal type of slope massifs represents conditions favourable for the ferromanganesian mineralization (in siliceous-carbonate sequences) and copper mineralization (in terrigenous sequences). This dependence of the type of metal of a stratified deposit in troughs upon the "average" composition of slopes and basement is eve more conspicuous in the group of volcanogenic-sedimentary ore accumulations. Whe the spilite-keratophyre formation is predominant, the chalcopyrite deposits are forn ed; the predominance of the basalt-andesite-dacite formation is favourable for the ge nesis of copper deposits of the Lake Superior type,and the predominance of the lipa-rite-tuffeceous-shaly formation for that of the lead-zinc deposits (often with iron anc manganese) of the Atasu type.

In general, the regularities of occurrences of various types of deposits in lateral (is chronous) rows of troughs or in their several chronological generations are the sam as in chronological rows of ore-bearing formations of one trough (provided there are corresponding formation rows).

On the whole, relationships between the metallogenic evolutionary rows and the iso-chronous and chronological rows of mineralized troughs are depending upon the following factors: firstly the kind of processes, i. e. exogenous or endogenous, which played the decisive role in the deposits formation, and, secondly, the degree of contrast between the formation making up the slope massifs (basement) and those filling up the geosynclinal troughs. In case of maximum contrast complete rows are formec starting from the border towards the interior of the basin, one may find polymetallic deposits, chalcopyrite and ferro-maganesian ones. This order is reversed with sedi mentary processes: from ferro-manganesian through copper-bearing to polymetallic deposits. In case of minimum contrast, only one of the final members is formed witl or without admixture of facies of intermediate ore formations. For every type of deposits here considered, the contrast of facies or formations i. e. borders of paleoge

graphical facies and of paleotectonic zones, combinations of contrast between basement formations and trough filling rocks, represents the most favourable places for
sedimentation of ores.

Data described above emphasize the need for an evolutionary approach towards the
study of non-ferrous metal deposits and towards the evaluation of potentialities of
areas where formations and tectonic structures controlling ore deposition occur.

Bibliography

AMSTUTZ, G. C. : Bemerkungen zur Genese von kongruenten Blei-Zink-Lagerstätten
in Sedimenten. "Ber. Geol. Ges. ", 8 , Sonderh. I, p. 31-42 (1963).

DUNHAM, K. C. , ed. : The geology, paragenesis and reserves of the ores of lead
and zinc. London (1950).

POPOV, V. M. : Copper. In: "Metally v osadochnykh tolshchakh". Izd-vo "Nauka",
Moscow, p. 3-68 (1965).

—   Stratified lead-zinc deposits of Tien-Shan and Central Kazakhstan. Mater. VII.
Vsesoyuzn. litol. Konf. (1965. g), Izd-vo "Nauka", Moscow, p. 325-343 (1968).

RUNDQUIST, D. V. : On study of phylogeny of economic minerals deposits. Zap.
Vesesoyuzn. mineral. o-va, 97, 2, p. 191-209 (1968).

SMIRNOV, V. I. : Sulphide ore formation in submarine volcanogenic geosynclinal
complexes. In: "Rudonosnost vulkanogenykh formatsii". Izd-vo "Nedra",
p. 30-34 (1965).

STRAKHOV, N. M. : Types of lithogenesis and their evolution in the history of the
Earth. Gosgeoltekhizdat. Moscow (1963).

—   On cognition of submarine volcanogenic-sedimentary rock genesis. In: "Vul-
canogenno-osadochnye formatsii i poleznye iskopaemye". Izd-vo "Nauka",
p. 11-43 (1965).

Address of the authors:

All-Union Geological Research
Institute (VSEGEI)
Leningrad / USSR

Uranium and Heavy Metals in Permian Sandstones Near Bolzano (Northern Italy)

A. Brondi, C. Carrara, and C. Polizzano

Abstract

The discovery of sedimentary galena concentrations in Permian sandstones of Alto Adige (Northern Italy) by means of mineralogical alluvial prospection promoted wide scale exploration in the whole basin of Bolzano. This search confirmed the wide extent and importance of lead mineralization. At the moment, the great extent of the examined basin does not allow us to make any general minerogenetic conclusions, though hypotheses on the formational environment and ore genesis of the first discovered occurrence have been made.

In this work, a general and synthetic description of the first discovery and a short presentation of other mineralizations found both in the Permian sandstones and in the "Tregiovo Formation" are reported.

The authors intend to study the complexity of the mineralizations by means of paleo-geographic criteria in the light of the general sedimentological development of the Permian basin of Alto Adige. The aim of the present work is simply to make some details known and to point out the interest in this type of mineralization in a region which, up to now, has been intensively studied from a minerogenetic point of view.

Introduction

Alluvial prospecting and geological studies for uranium and base metals have been carried out in the Alto Adige Region (Northern Italy) by geologists of the Laboratorio Geominerario - CNEN (Italian Atomic Energy Commission). During the above-mentioned exploration, galena mineralization in the Permian "Arenarie di Val Gardena" was discovered. The alluvial samples with an anomalous content of galena revealed a mineralized belt 14 km long between Lana and Appiano. In this area the most interesting mineralization has been localized at 920 m altitude in the Rio Bavaro valley. The mineralization is interesting from both a minerogenetic and an economic point of view. References to this type of mineralization in that region are few in number and uncertain. Among them, the most significant are the mineralizations of Val Rendena, where galena occurs parageneticaly associated with U and other elements (MITTEM-PERGHER, 1962; DESSAU and PERNA, 1968); the occurrences reported by prospectors of CNEN on Avelengo plateau near Meltina (Relazione CNEN, 1958); those of Castelvecchio and other localities (KLEBELSBERG, 1935). A similar galena concentration has been described by MOSTLER (1966), occurring in "Tregiovo Schists" contained in the underlying Permian volcanics.

Owing to the discovery of galena in Rio Bavaro, a systematic geological search has been carried out over an area of 1600 sq. km around Bolzano (fig. 1). It was possible, therefore, to demonstrate the stratiform character of the mineralization and its overall distribution in that large basin of sedimentation of Permian sandstones. Actually, among 22 localities studied, 19 showed mineralized horizons (fig. 1).

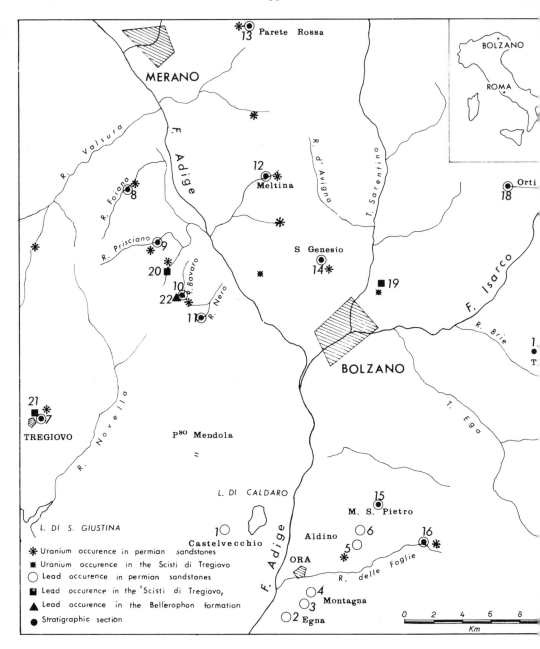

Fig. 1. Map of Permian sedimentary mineralizations

Three occurrences in addition to the above-mentioned localities were found in the "Tregiovo Schists". They present the same formational characteristics as the mineralization in the Permian sandstones. The minerogenetic processes were similar and continuous; therefore, this study includes these also.

67

Geological Situation

The geological sequence of the investigated area belongs to the tectonic unit of the "Southern Alps". From bottom to top the pre-Triassic sequence is composed (fig. 2) of:

- metamorphic basement, Carboniferous in age, represented by epizone metamorphics;
- basal conglomerate ("Verrucano"), lying unconformably on the metamorphic basement, mostly polygenic, with pebbles of quartz and metamorphic rocks with a cherty-sericitic matrix;
- Permian volcanic complex, consisting of three groups, the lower of intertonguing lavas, tuffs and conglomeratic tuffs of dacitic and latit-andesitic composition; the following of ignimbritic rhyodacitic and rhyolitic layers with intercalated sedimentary-volcanic and sedimentary horizons. In the upper part of the rhyodacitic group in the Bolzano area, a sedimentary sequence occurs up to 100 meters thick, stratigraphically corresponding to the "Tregiovo Schists". The same type of sediments is interbedded with the overlying rhyolitic layers. They are made up of marly and calcareous rocks with calcareous and cherty beds, in which various animal and plant remains occur (GIANNOTTI, 1962; MOSTLER, 1966; ULCIGRAI, 1969). The entire thickness ranges up to 2000 meters.

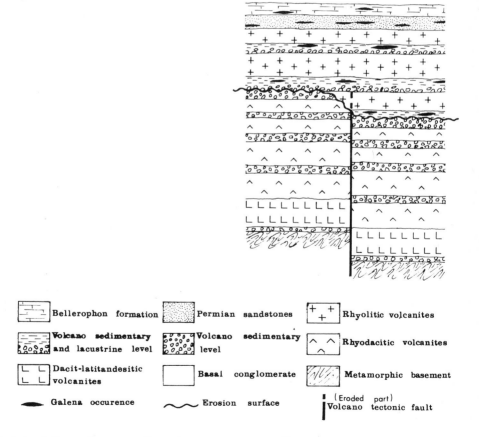

Bellerophon formation — Permian sandstones — Rhyolitic volcanites
Volcano sedimentary and lacustrine level — Volcano sedimentary level — Rhyodacitic volcanites
Dacit-latitandesitic volcanites — Basal conglomerate — Metamorphic basement
Galena occurence — Erosion surface — (Eroded part) Volcano tectonic fault

Fig. 2. Sketch of the volcano-sedimentary series near Bolzano

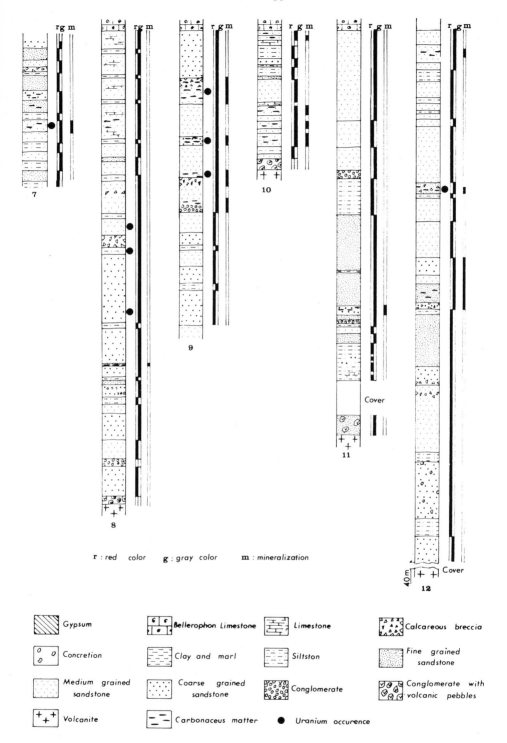

r : red color   g : gray color   m : mineralization

Fig. 3

- Val Gardena Sandstones, made of layers with variable grain sizes, mostly quartzitic, with a siliceous, carbonatic or argillaceous matrix; thickness up to 250 meters;
- Bellerophon Formation, typically evaporitic with calcareous, dolomitic-calcareous
and layers, very rich in fossil remnants; thickness up to 150 meters.

Geological setting and tectonics

There is no unconformity among the above described formations. The only unconformity occurs between the metamorphic basement and the "Verrucano". The last
phase of the Hercynian orogenesis involved uplifting of the basement upon which the
basal conglomerate was deposited. Not much later, the great Permian volcanism
developed along the Hercynian N-E S-W tectonic trends.

Volcano-tectonic collapse took place along the E-W trend. "Tregiovo Schists" and
their equivalent sediments interbedded with ignimbrites were deposited in the volcano-tectonic depression near Bolzano. After the end of the volcanic activity, the
erosion in the investigated area caused the accumulation of the Val Gardena sandstones. Then followed the mixed detrital and chemical sedimentation of the Bellerophon Formation.

Description of the Mineralizations

The purpose of this paper is to point out the extent of the minerogenetic processes
and the economic importance of the galena mineralizations in the Val Gardena sandstones.

Among the explored localities, the most interesting is Rio Bavaro. A previous report on this subject has been printed already (BRONDI et al., 1970). We will refer
to this report describing the main characteristics of other concentrations.

General petrographic features

The thickness of the examined sequences varies from 36 to 250 meters. Nevertheless they can be referred to a single lithologic type that shows monotonous repetitions
in almost all the studied localities. Significant variations are marked by the occurrence of gypsum, limestones and calcareous breccias. They characterize environmental changes during the evolution of the basin (fig. 3 and 4).

In general we can state that, 1) the lower part consists of conglomerates, volcanic
pebbles and coarse-grained sandstones; in places gypsum beds occur; 2) in the central part, thick beds varying from coarse-grained sandstones to siltstones predomi-
nate; 3) in the upper part, typical frequent intercalations of medium to fine-grained
arenaceous beds with mainly carbonate matrices occur, and 4) on the top of the sequence, carbonate beds precede the evaporitic sedimentation of the Bellerophon Formation.

The colours predominating in the sequence are red, gray and whitish-gray. Carbonaceous material occurs mainly in the middle and upper parts. It indicates a break
in the erosional phase with chemical weathering predominating. Similar characteristics are described in the section of Rio Bavaro (fig. 5). The reduction of thickness
in the Rio Bavaro section (34 m), in comparison with the thickness of other sequences

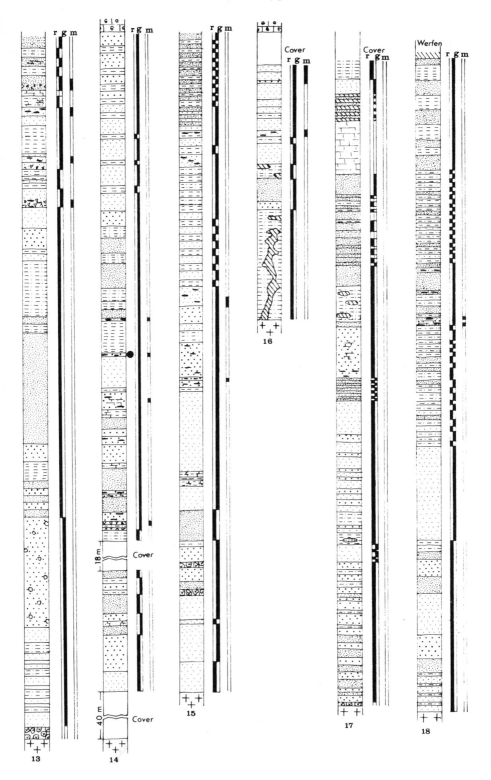

Fig. 4

71

(up to 250 m), could be explained by the occurrence of a morphological high of the underlying volcanic formation.

The presence in the examined sandstones of fluvial cross-bedding, ripple marks, load-casts and the lack of clear signs of both detrital grain sorting and graded bed-

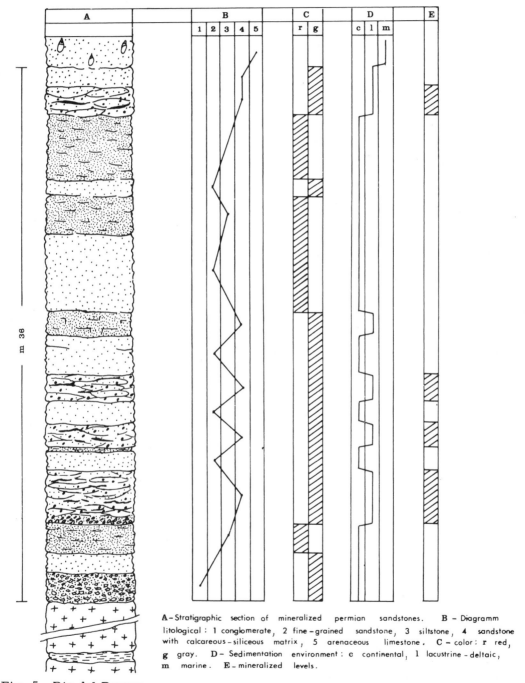

A – Stratigraphic section of mineralized permian sandstones.　B – Diagramm litological : 1 conglomerate, 2 fine-grained sandstone, 3 siltstone, 4 sandstone with calcareous-siliceous matrix, 5 arenaceous limestone.　C – color: r red, g gray.　D – Sedimentation environment : c continental, 1 lacustrine-deltaic, m marine.　E – mineralized levels.

Fig. 5. Rio del Bavaro

ding indicate that sedimentation took place in a fluvial and fluvial-deltaic environment with lacustrine and lagoonal stages. These stages consist of argillaceous inter
calations with a high content of detrital plant material and a carbonatic-siliceous
matrix. Gray colour and occurrence of organic material testify to the stability of
the vegetal cover and, therefore, of a biostatic phase. Later on we will try to cor-
relate the studied mineralizations with this biostatic phase.

Features of mineralized beds

Galena mineralizations discovered in the prospected area show geological character
istics very similar to those of the Rio Bavaro occurrence. The main features are:
a lenticular or stratiform structure within mineralized beds; the host-rocks are always represented by medium-grained gray sandstones containing thin carbonaceous
levels or vegetal remnants; frequent cross-bedding structures and load-casts occur
the matrix is mainly calcitic and siliceous (calcedony, opal); a very simple paragenesis according to the type of mineralization is observed (i. e. disseminated intergranular galena with sizes depending on interspaces; micro-inclusions of sphalerite
within galena; galena rarely associated with copper ores (malachite, azurite); steady
occurrence of pyritic clusters and nodules; sporadic occurrence of uranium ores);
furthermore the galena content of mineralized beds varies from 0. 5 to 3 %; the Pb
content is 10 to 20 times as high as the Zn content; the silver content of Rio Bavaro
galena is 34 ppm; then, the lack of gangue minerals and the fact that the mineralized
beds are mainly localized in the central parts of the sequence with the exception of
Rio Bavaro, where the first three mineralized beds occur in the first ten meters of
the section (fig. 5).

The last point can be explained, as pointed out above, by the occurrence of a paleo-
relief in the Rio Bavaro area. Therefore, while in other localities the phase of coars
detrital filling of the basin was followed by the carbonatic-siliceous phase, in the Rio
Bavaro area the latter began almost immediately. In general, we can state that the
mineralization developed when a phase of rexistasis passed gradually to a phase of
biostasis with the normal plant cover.

This cover led to the sedimentation of organic matter and consequently to the establishment of an environment favourable for the precipitation of metallic ions from
solution. With some reservation, because of the type of environment, an attempt has
been made to identify, by means of previsional diagrams of LOMBARD, one or several preferential phases for the deposition of galena. Studies of this type were applied particularly to the search of stratiform lead deposits by NICOLINI (1964). As
shown in fig. 5, there is a satisfactory agreement between our results and the conclusions of that author.

The above described mineralizations are really localized in the last phases of positive sequences showing recurrences of regressive microphases. This method, clear
ly pointing out the deposition of the mineralization during a determined phase of sedi
mentological evolution of the basin, will be applied to all the other examined sequences.

The minerogenetic process of these metallic concentrations is probably the same as
we pointed out in the case of the Rio Bavaro mineralization. It can be briefly summarized as follows: The waters circulating in the Permian volcanic complex contained metallic ions derived from both the leaching of volcanites and their mineralizations and also from possible exhalative processes. The lead contained in those
waters could precipitate under favourable conditions, i. e. when a geochemical barrier caused by raising the pH at the contact with carbonate rocks or waters occurred

The distribution of the mineralization in the different levels of the volcanic sedimentary sequence (considering not only Val Gardena sandstones but also Tregiovo and
Bellerophon Formations) is therefore closely related to the establishment of these
environments with carbonatic deposition. Owing to these considerations it is possible to state that the galena concentrations belong to the "red-bed" type of ROUTH-
IER (1963) and they represent stratiform "familiar" syngenetic deposits with terrigenous sources (ROUTHIER, 1967).

Other Mineralizations

Besides metal concentrations in Val Gardena sandstones of the investigated area,
other stratiform mineralizations occur in the Bellerophon Formation and in the Tregiovo schists. The first ones form the well-known silver-bearing galena concentrations that have been exploited in the Middle Ages in the area of Trento. A small galena concentration has been found in the Bellerophon Formation of Rio Bavaro just
above the lead occurrence in sandstones. The mineralizations in Tregiovo schists
have been recently studied by MOSTLER (1966) and DESSAU and PERNA (1968). Galena and sphalerite are the most widespread and important ores, whereas copper
ores occur in small local concentrations. According to MOSTLER, the mineralization is synsedimentary and the accumulated metals were derived from weathered
Permian volcanites. The interpretation of the present authors is in agreement with
that of MOSTLER. (Compare also the basic reports by AMSTUTZ, 1962; BERNARD,
SAMAMA, 1970; BOULADON, 1969; and some other authors listed in the bibliography.)

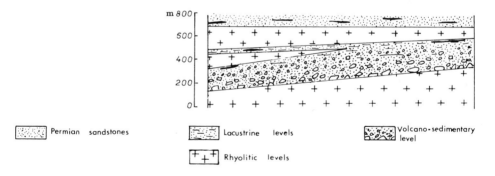

Fig. 6. Stratigraphic position of galena occurrences in the volcano-sedimentary
series near Nalles (Bolzano)

Other galena concentrations have been discovered by the authors in sedimentary
horizons equivalent to the "Tregiovo schists" near Bolzano (fig. 1 and 2). At S. Gia-
como and S. Apollonia (Nalles, in Adige Valley), galena concentrations occur in several beds interbedded with volcanites (fig. 6). Such horizons present fine-grained
sandy matrix with many beds and lenses of chert (fig. 7). A siliceous layer, lacking
in galena occurs with a thickness of 1 meter. PbS and chert precipitation took place
in local lacustrine basins. Very small uranium concentrations occur sparsely at the
top of the sedimentary intravolcanic layers of Adige, Sarentina and Isarco Valleys
(fig. 1).

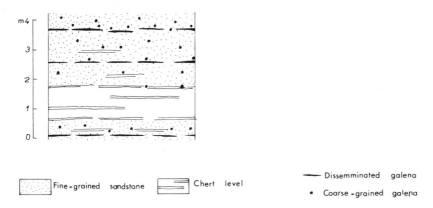

Fig. 7. Galena mineralization in "Scisti di Tregiovo" near Bolzano

Uraniferous Mineralizations

Uranium concentrations occur in the whole area of the Alpine ranges where Permian sandstones outcrop. The most important uranium concentrations occur in the western side of Trento region in Rendena, Daone and Giulis Valleys, not far from the investigated area. Geologically these zones represent a transitional belt between the Alto-Adige plateau and the area of faster subsidence of the Lombard Alps (MITTEMPERGHER, 1970).

The most important uranium concentrations of the investigated area are reported in fig. 1. They have been discovered during a radiometric survey carried out by C.N. E.N. from 1957 to 1960. These mineralizations are quite small and show uranium contents ranging from 500 to 8000 ppm. They are lenticular in shape and concordant with the general setting. The host-rock is always a gray medium-grained sandstone with organic remnants. Association with secondary copper ores and pyrite is frequent. The primary uranium ore is represented by microcrystalline uraninite with spherulitic structure. Local tectonic and diagenetic remobilization caused the transformation of uraninite into pitchblende.

There are two main differences between the uranium concentrations of the Val Rendena group and those of the examined area. The first concentrations are contained in the lower part of the sequences in the form of great lenses. In contrast, the latter are small and localized in the middle or in the upper parts of the sequences. This fact may be explained by the more favourable environment for uranium precipitation of Val Rendena. Here the lower part of the stratigraphic sequence is generally built up with gray conglomerates and coarse sandstones with fluvial features. This coarse grain size indicates the proximity of bed-rock and source-rocks, represented in this case by Permian volcanites. Fresh water circulating in sediments of this type was probably rich in uranium that precipitated in reducing environments due to occurrences of organic material.

The small sizes and the higher position of uranium concentrations in the basin of Bolzano may have been determined by several causes: first of all, the coarse basal levels of the sequences are almost always lacking in organic substances which, in contrast, occur in the higher part; secondly, to a higher position in the stratigraphic sequence generally corresponds a greater distance of the sediments to the source rock. In this case the sediments could be derived from metamorphic rocks of the basement, mostly poor in uranium.

Conclusions

The analogies recognized in the described lead concentrations allow to extend the conclusions drawn for Rio Bavaro to the whole basin; indeed, there is undoubtedly a perfect petrographic standard for the recognition of the most favourable facies for galena occurrence. This standard is sufficient for preliminary exploration purposes. It is, however, inadequate for a study the purpose of which is the confirmation of all the parameters of the minerogenetic processes that led to such lead concentrations. If the characters already described for Rio Bavaro are valid, it is necessary to check the genetic hypothesis for the other mineralizations. The authors are developing a wide program of geologic and geochemical studies supported by adequate petrographic and chemical analyses.

The purpose of this work can be summarized in the following way:

a) Paleogeographic reconstruction of the basin with identification of the physicochemical environment favourable to metal concentrations.

b) Within this reconstruction, detailed research for favourable morphological situations, such as fluvial paleochannels, paleoreliefs, etc. , will be developed. (It must be remembered that the lead concentration of Rio Bavaro is contained in a reduced sequence.)

c) According to other studies carried out on mineralizations belonging to the same type, a search for "zoning" regarding the matrix of the sediments must be developed.

d) The probable source of metallic ions and the paths of transportation and deposition in the general development of the whole basin ought to be recognized by means of analytical data.

e) A general economic evaluation.

The search for uranium is directed towards the leached and bleached formations which are quite characteristic for most of our sequences that can hide "rolls"-type uranium concentrations.

Bibliography

AMIRASLANOV, A. A. : Osnovnie tipi mestorojdienii svinza i zinca. Gosgeoltexisdat, Moscow, p. 211 (1957).

AMSTUTZ, G. C. : L'origine des gîtes minéraux concordants dans les roches sédimentaires. Chron. Mines Rech. Min. no. 308, Paris, p. 115-126 (1962).

BATULIN, S. G. , GOLOVIN, E. A. et al. : Eksogienne epigeneticeskie mestorojdienia urana. Atomisdat, Moscow, p. 322 (1965).

BERNARD, A. , SAMAMA, J. C. : Essai méthodologique sur la prospection des "Red Beds" plombo-zincifères. Science de la Terre, XV, no. 3, Nancy, p. 209-264 (1970).

BOULADON, J. : Contribution à une systématique des gisements de plomb et de zinc. Chron. Mines Rech. Min. no. 385, Paris, p. 215-227 (1969).

BRONDI, A. , POLIZZANO, C. , ANSELMI, B. , BENVEGNU' F. : Rinvenimento di una mineralizzazione a galena nelle arenarie permiane di Nalles (Bolzano). L'Industria Mineraria del Trentino-Alto Adige, vol. III, Trento, p. 171-182 (1970).

BRONDI, A. , GHEZZO, C. , GUASPARRI, G. , RICCI, C. A. , SABATINI, G. : Le vulcaniti paleozoiche nell'area settentrionale del complesso effusivo atesino. Nota I. Atti Soc. Tosc. Sc. Nat. Mem. , Ser. A, 77. Tip. Giardini, Pisa (1970).

CNEN: Ricerche sulle formazioni permiane. Relazione svolta dalla Div. Geomineraria. Rapporto riservato, CNEN, Roma (1958).

DESSAU, G. , PERNA, G. : Le mineralizzazioni a galena e blenda del Trentino Alto-Adige e loro contenuto in elementi accessori. Symp. Int. Giac. Min. Alpi, vol. 3, p. 587-687 (1966).

DOZY, J. J. : Über das Term der Südalpen. Leid. Geol. Meed. , 4, Leiden, p. 123 (1935).

ERHART, H. : La genèse des sols en tant que phénomène géologique. Mason, Paris, 88 p. (1956).

GIANNOTTI, G. P. : Intercalazioni lacustri entro le vulcaniti paleozoiche atesine. Atti Soc. Tosc. Sci. Nat. , vol. LXIX, fasc. 2, p. 598-617 (1962).

GIANNOTTI, G. P. , TEDESCO, C. : Le mineralizzazioni uranifere del Trentino-Alto Adige. L'Industria Mineraria nel Trentino-Alto Adige, Economia Trentina, a. XIII, no. 4-5 (1964).

GÜMBEL, C. W. : Geognostische Mitteilungen aus den Alpen. I: Das Mendel- und Schlerngebirge. Sitz. Akad. Wiss. 1873, München, p. 13-88 (1873).

KLEBELSBERG, R. von: Geologie von Tirol. Bornträger, Berlin, 872 p. (1935).

LOMBARD, A. : Géologie sédimentaire. Les séries marines. Masson, Paris, 724 p (1956).

MITTEMPERGHER, M. : Rilevamento e studio petrografico delle vulcaniti paleozoiche della Val Gardena. Atti Soc. Tosc. Sc. Nat. , Ser. A, fasc. II, Pisa, p. 482 -530 (1962).

— Le mineralizzazioni ad uranio delle Alpi Italiane. Symp. Int. Giac. Min. Alpi vol. 2, p. 319-333 (1966).

— Characteristics of uranium ore genesis in the Permian and Lower Triassic of Italian Alps. In: Uranium Exploration Geology, International Atomic Energy Agency, Wien, Panel Proc. Ser. , p. 253-264 (1970).

MOSTLER, H. : Sedimentäre Blei-Zink-Vererzung in den Mittelpermischen "Schichten von Tregiovo". Mineral. Deposita (Berl.) 1, p. 89-103 (1966).

NICOLINI, P. : L'application des courbes prévisionnelles à la recherche des gisements stratiformes de plomb. Developments in Sedimentology, vol. 2: Sedimentology and ore genesis. Elsevier, Amsterdam, p. 53-64 (1964).

— Gîtologie des concentration minérales stratiformes. Gauthier-Villars, Paris, 792 p. (1970).

ROUTHIER, P. : Les gisements métallifères. Géologie et principes des recherches. Masson, Paris, 1281 p. (1963).

— Le modèle de la genèse. Modèles des théories métallogénétiques. Chron. Min et Rech. Min. , p. 177-190 (1967).

SAMAMA, J. C. : Controle et modèle génétique de minéralisations en galène de type "Red Beds". Mineral. Deposita (Berl.) 3, p. 261-271 (1968).

— Contribution à l'étude des gisements de type "Red Beds". Etude et interprétation de la géochimie et de la métallogénie du plomb en milieu continental. Cas du Trias ardéchois et du gisement de la Largentière. Thèse d'Etat présentée à la Faculté des Sciences de l'Université de Nancy, 450 p. (1969).

ULCIGRAI, F. : Geologia dei dintorni di Tregiovo, Trentino-Alto Adige. Studi trentini di Sc. Nat. Rivista del "Museo Tridentino di Sc.Nat." Sez. A, vol. XLVI, no. 2, Trento, p. 243-300 (1969).

Address of the authors:

Comitato Nazionale Energia Nucleare
Lab. Geominerario C.S.N. Casaccia
00060 S. Maria di Galeria
Roma / Italy

Lithostratigraphic Controls of Some Ordovician Sphalerite

Jon A. Collins and Leigh Smith

Abstract

Carbonate lithofacies analysis of sphalerite-hosting strata in the Western Newfoundland Lower Ordovician St. George Formation was used to ascertain why the mineralization occurs where it does. The host rock, herein referred to as the Lower Limestone, was deposited in two cyclically alternating carbonate bank environments, which produced dolomitic-mottled biointrapelsparite interbedded with dolomite. Diagenesis in the Lower Limestone biointrapelsparite beds consisted of lime mud conversion to micrite, neomorphic growth of microspar, stylolitization and preferential dolomitization of burrows and areas around stylolites. Complete dolomitization of the Lower Limestone dolomite interbeds is attributed to a greater degree of bioturbation.

Dark Grey Dolomite and Cyclic Dolomites overlie the Lower Limestone host and laterally discontinuous solution collapse breccias are developed, especially in the host unit. These breccias are cavern deposits produced during subsidence of the platform after some 100 + m of uplift produced a karst terrane at the end of Lower Ordovician time. Biointrapelsparite beds underwent solution of their calcite portions adjacent to cavern areas during karsting, developing a porous interconnecting network of dolomite mottles. In certain areas, sphalerite grew in colloform fashion about these mottles. Finally, the remaining open spaces were filled with what is now white dolomite. This white-brown mottled rock has been called, inaccurately, a pseudobreccia.

Following cavern filling, the Table Head Limestone was deposited unconformably on the St. George Formation.

Introduction

For the Newfoundland Zinc Mines deposits, situated on the West Coast of Newfoundland (Fig. 1), a study of the sedimentological history and diagenesis of the host rock has been instrumental in unravelling why the mineralization occurs where it does (COLLINS, 1971). Published diamond-drill results indicate about 5.5 million tons grading 7.7 % zinc in several small deposits (Canadian Mines Handbook, 1970-71). The geologic setting is quite similar to that of the Tennessee-type zinc mines (ODER and RICKETS, 1961).

The mineralization consists mainly of sphalerite in a dolomite gangue. To date, various geophysical techniques have been employed but were not successful in delineating direct targets for further exploration. This is not unexpected because disseminated sphalerite is not conductive, has no significant magnetic properties, and is usually not concentrated enough to produce significant gravity anomalies. In short, a thorough understanding of the geology of this type of deposit is apparently the only manner of evolving more efficient exploration techniques for the discovery of additional reserves.

80

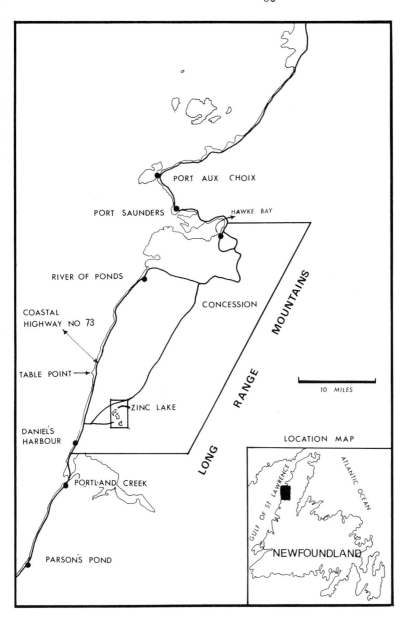

Fig. 1

General Geologic Setting

The region is a level to gently undulating coastal plain averaging ~ 8 km in width
and underlain by a stable platform sequence of Lower Paleozoic carbonate and terrigenous sediments. The Long Range Mountains of Precambrian (Grenville) rocks lie
to the east, and the strata are covered by the Gulf of St. Lawrence on the west (Fig.
The stratigraphy of the area includes the Hawke Bay Formation (Early Cambrian),
St. George Formation (Cambro-Ordovician), Table Head Formation (Middle Ordovician), and the Humber Arm Group (Middle Ordovician). (Fig. 2, SCHUCHERT and
DUNBAR, 1934).

Generalized stratigraphic section (G.S.A. Memoir 1).*

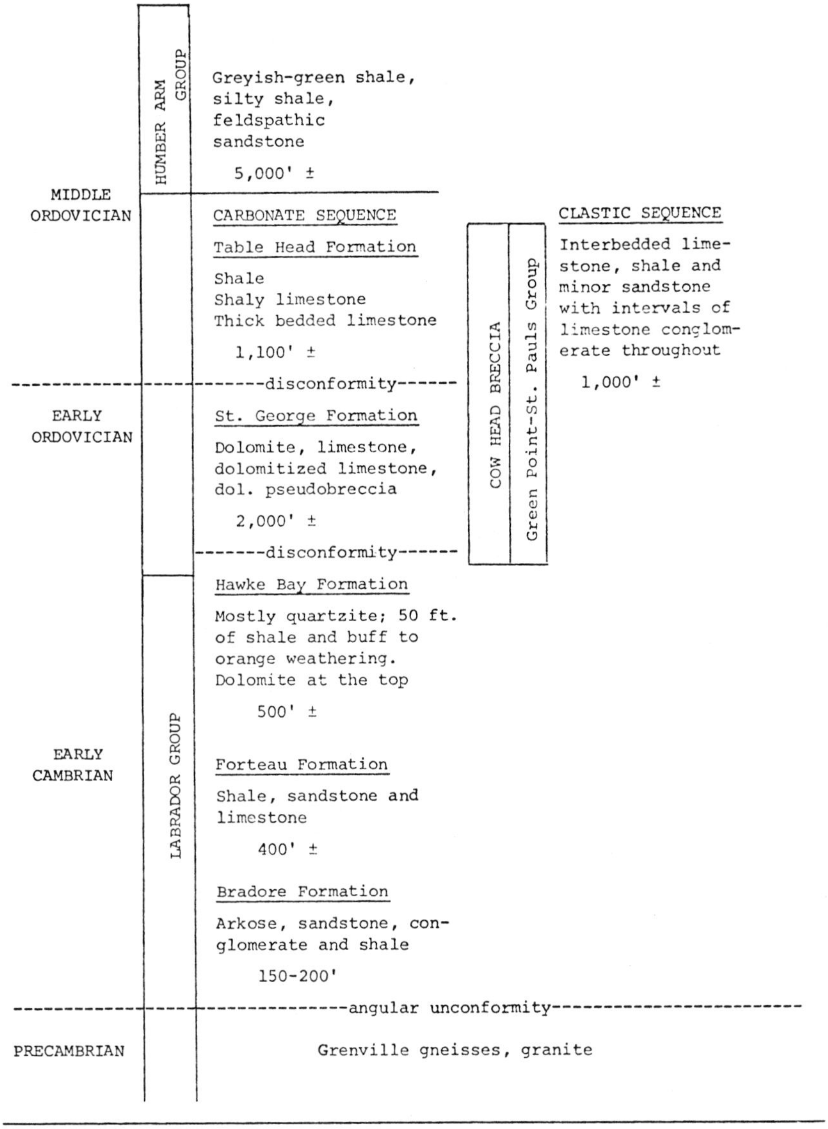

MIDDLE ORDOVICIAN	HUMBER ARM GROUP	Greyish-green shale, silty shale, feldspathic sandstone 5,000' ±		

CARBONATE SEQUENCE   CLASTIC SEQUENCE

		CARBONATE SEQUENCE	COW HEAD BRECCIA / Green Point-St. Pauls Group	CLASTIC SEQUENCE
		Table Head Formation Shale Shaly limestone Thick bedded limestone 1,100' ±		Interbedded limestone, shale and minor sandstone with intervals of limestone conglomerate throughout 1,000' ±

------------------------------------disconformity------

| EARLY ORDOVICIAN | | St. George Formation

Dolomite, limestone, dolomitized limestone, dol. pseudobreccia

2,000' ± | | |

------------disconformity------

| EARLY CAMBRIAN | LABRADOR GROUP | Hawke Bay Formation

Mostly quartzite; 50 ft. of shale and buff to orange weathering. Dolomite at the top

500' ±

Forteau Formation

Shale, sandstone and limestone

400' ±

Bradore Formation

Arkose, sandstone, conglomerate and shale

150-200' | | |

-------------------------angular unconformity-------------------------

| PRECAMBRIAN | | Grenville gneisses, granite | | |

Fig. 2   *After Schuchert and Dunbar (1934).

Folds of low amplitude with axes trending north to northeasterly and normal faults of northeasterly trend and local to regional extent are common.

Lithostratigraphy

The relevant stratigraphic interval is the top 400 feet (125 m) of the Lower Ordovician St. George Formation, beneath the Middle-Lower Ordovician unconformity (Fig. 2). These rocks lend themselves to the following subdivisions - Lower Limestone ore

82

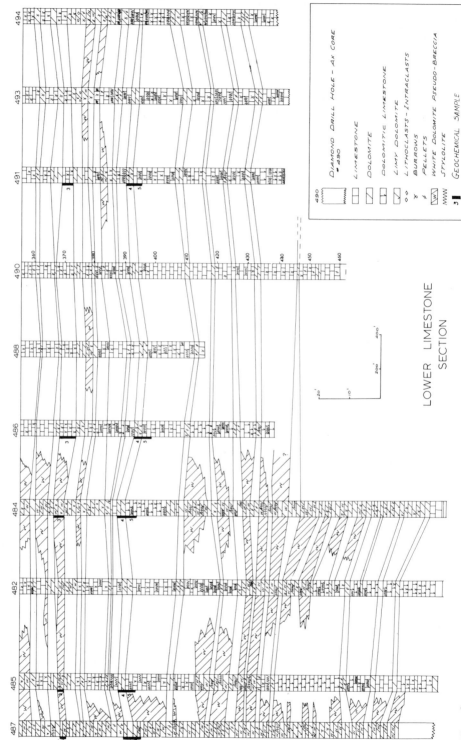

HORIZONTAL DATUM LINE = BASE OF DARK GREY DOLOMITE

LOWER LIMESTONE SECTION

Fig. 3

host (100 - 150+ft. thick, 50 m) (Fig. 3), with Dark Grey Dolomite (50 ft. thick, 15 m) and the Cyclic Dolomites (220 - 230 ft. thick, 68 - 71 m) above.

Method of Analysis

In order to ascertain mineralization controls, the Lower Limestone was traced on a bed-for-bed basis through its various lateral diagenetic changes. Diamond-drill holes on 300 ft. (~90 m) centres were available and hole number 494 was selected for pre-liminary correlation, as it was diagenetically relatively unaltered (Fig. 3). This hole was correlated in detail with 9 other diamond-drill holes in a section line through a mineralized area.

Once correlation had been effected, normal carbonate petrology techniques were used to study the lateral diagenetic changes.

Lower Limestone and Pseudobreccia:

Two basic lithologies are present in the Lower Limestone itself. Two to ten foot-thick units of limestone are interbedded with 6 inch to 4 foot thick beds of homogenous micrite or fine-grained dolomite (Plate 1-C) (Plate 2-A). Pseudobreccia (CUMMING, 1968), the ore host, is a diagenetic facies of the limestone beds in the Lower Lime-stone (Plate 1-D, E, F).

The Limestones are biopelsparite or biointrapelsparite (FOLK, 1962). The pellets are usually less than 1 mm in diameter and are round to ellipsoidal in section. The intraclasts range in size from 1 to 5 mm and consist of pellets, fossil fragments and micrite. They can be quite angular to well-rounded and are usually surrounded by an algal envelope. Fossil fragments rarely comprise more than 5 % of the rock in this unit, and although gastropod fragments (to 1 cm) and bryozoan fragments (to 5 mm) do occur, most biofragments are about 2 mm in size. Ostracods, foramini-fera, gastropods, trilobites, brachiopods, pelecypods, bryozoa and sponges (?) have been identified.

Intraclastic beds up to ~ 15 cm thick are common. These beds consist of 40 % to 50 % intraclasts "floating" in a dark brown micrite matrix. The lithology of the intra-clasts, which are generally fairly well-rounded, consistently appears to be identical to the lithology of the bed immediately underlying them, usually a biopelsparite.

Burrowing has commonly obliterated any of these features that were present, result-ing in a relatively homogenous mass of pelmicrite. Numerous stylolites containing hydrocarbons and pyrite or iron oxide residues intersect the rock at varying angles (Plate 1-A).

Preferential dolomitization of the burrows in the limestone (from 0 to more than 50 %) gives it a mottled appearance (Plate 1-A, B, C, D). This dolomite is microcrystalline and medium to dark grey. Algal envelopes have also become preferentially dolomit-ized locally, and dolomite rhombs are commonly scattered throughout micritic areas.

The medium grey, homogenous, dolomite interbeds have had many of their primary features obliterated by dolomitization. These beds generally consist of fine to medi-um crystalline dolomite (0.2 to 1 mm). Rarely, the beds have not been completely dolomitized, and consist of dark grey, structureless micrite. They usually appear

84

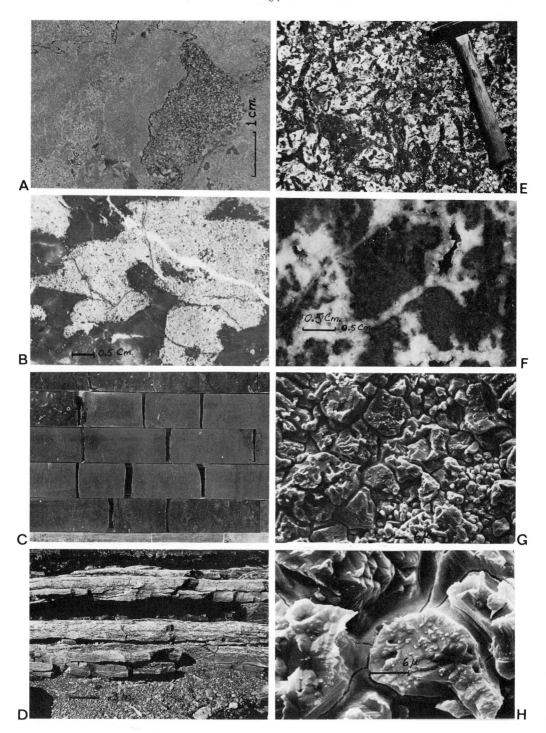

Plate 1

as massive to extensively bioturbated, and rarely display pellets, intraclasts, some algal material, ostracod shells and bedding (Plate 1-C, Plate 2-A).

In outcrop, the limestone appears as relatively massive, medium grey material, some 25 to 60 % of its surface area showing burrows and other features as described. These weather out in relief and usually also have a medium grey coloration. The dolomite interbeds weather recessively relative to the limestone, are massive, and have a light brown-grey colour (Plate 2-A).

The "pseudobreccia" is easy to describe: picture the limestone of the Lower Limestone with all micrite and coarser calcite replaced with very coarse-grained pure white dolomite, leaving only the slightly brownish, dolomite, burrow patches between (Plate 1-D, E, F).

Bedding is a somewhat obscure feature, but particles generally have a preferred orientation of their long axes parallel to it. No terrigenous material, either macro- or microscopic in size, has been noted. Colour variations are attributed to differences in organic content, and to iron oxides and sulfides.

Dark Grey Dolomite:

This unit is comprised mainly of dark grey, pelleted and bioturbated dolomite, which is massive and locally quite well-bedded. It contains a 2-foot thick algal marker unit near its centre and is capped by a light grey to buff, thinly laminated, algal dolomite. Didymograptus c.f. patulus (Hall) and Didymograptus c.f. nitidus (Hall) date the Dark Grey Dolomite as upper (although not necessarily uppermost) Arenigian (A.C. LENZ, personal communication, 1970).

Plate 1 - Descriptions

A - Thinsection under polarized light showing stylolitic and partially dolomitized biointrapelsparite. The dolomite is stratigraphically beneath the stylolite.

B - Polished hand specimen showing light grey dolomite mottle in darker grey biointrapelsparite.

C - Five-foot section of slabbed drill core showing dark grey, dolomite-mottled biointrapelsparite and a light grey dolomite interbed.

D - Pseudobreccia in outcrop at Table Point (Fig. 1). View perpendicular to bedding planes.

E - Bedding-plane surface of pseudobreccia in outcrop. Note how the darker dolomite conforms to a "burrow-trace" shape.

F - Polished hand specimen showing white dolomite incompletely filling open spaces between brown dolomite mottles.

G - Scanning Electron micrograph of micrite pellet (lower right) and microspar cement (upper left). Relief is due to etching. Biointrapelsparite.

H - Scanning Electron micrograph enlargement of microspar grain in G displaying incipient dolomitization and porosity development.

86

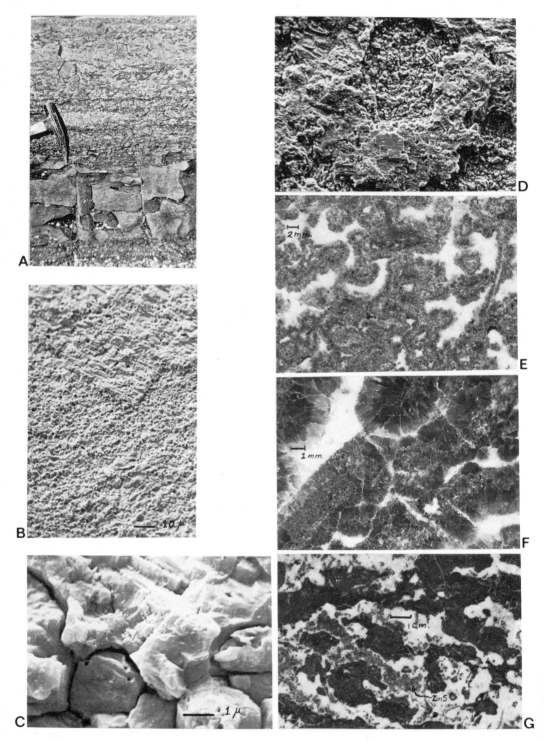

Plate 2

Cyclic Dolomites:

In its simplest form, a Cyclic Dolomite unit consists of a basal conglomerate, six inches to two feet thick, grading upward into dololaminite, argillaminite or argillaceous dololaminite, two to fifteen feet thick. The laminated units grade upward into bioturbites, which vary in thickness from 2 to 20 feet (0.6 to 6 m). This basic cyclic unit is repeated about eight times in the preserved record with some variation in thickness and order.

Solution Collapse Breccias:

These breccias are laterally discontinuous features which are normally best developed in the Lower Limestone interval. Texturally, they consist of angular to subrounded fragments or clasts from a millimeter to several meters in size, in a matrix of mainly fine-grained dolomite and iron oxides or sulfides. Chert pebbles of coarse sand size are common in the matrix. Graded bedding and thin bedding were observed in some of the matrix, and pisolitic and detrital dolomite agglomerations occur near the base of some of the breccia units. The matrix is always much darker in colour than most of the fragments, reflecting a usually much higher organic hydrocarbon and iron oxide content (Plate 3-A, B).

All matrix material found in the collapse breccias was derived from the Lower Limestone, Dark Grey Dolomite and Cyclic Dolomites, while the clasts are fragments of the latter two.

The Unconformable Table Head Limestone

Unconformably overlying these St. George Formation units, particularly the Cyclic Dolomites, is the Lower Middle Ordovician Table Head Formation (Fig. 2). In the

---

Plate 2 - Descriptions

A - Outcrop photograph, perpendicular to bedding, of the Lower Limestone at Table Point. The upper half is limestone showing preferential dolomitization with a biointrapelsparite matrix. The lower part is a dolomite interbed overlying another limestone bed. Scale is 5" hammer head.

B - Scanning Electron micrograph of biointrapelsparite showing dolomite mottle in upper part of photo "growing into" micrite.

C - Detail of B showing intergrain nature of dolomite advance.

D - Scanning Electron micrograph of partially dolomitized pelsparite. Micrite pellet in centre is surrounded by now-dolomitized microspar cement.

E - Mineralized "pseudobreccia". Polished specimen showing colloform sphalerite growth (dark grey rim) around lighter grey dolomite mottles, with final white dolomite infilling.

F - Detailed view of polished section - sphalerite with white and brown dolomite.

G - Polished specimen of "pseudobreccia" showing "second-generation" detrital sphalerite.

88

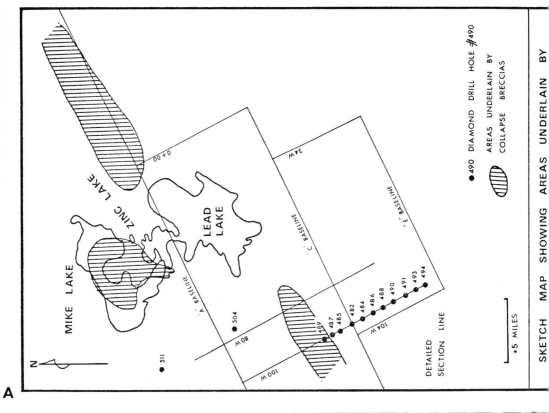

A

SKETCH MAP SHOWING AREAS UNDERLAIN BY

•490 DIAMOND DRILL HOLE #490

AREAS UNDERLAIN BY COLLAPSE BRECCIAS

•5 MILES

B

1 Cm.

area of mineralization, the contact unconformity is a disconformity, with relief commonly 1 to 2 feet vertically over a distance of several hundred feet laterally.

Reconstruction of the Depositional Environment

Phenomena of the limestone beds which must be accounted for in interpreting their
environment of deposition include pellets, intraclasts, burrowed nature, algal affili-
ations, micrite matrix, sporadic intraclastic beds and fauna. The lack of terrigenous
material and the cyclic repetition with dolomite interbeds are also significant.

Micrite content, as well as the poor sorting of pellets and intraclasts, indicate that
the limestone beds were deposited in a low-energy environment, but the ubiquitous
intraclasts confirm that frequent short periods of higher energy did invade this general low-energy regime. Considering these points, the minimal terrigenous content
and the fauna preserved, it is suggested that an open marine platform or bank of regional magnitude and remote from any land mass could produce these sediments at
certain water depths.

The dolomite interbeds signal an environmental change. There is a definite relative
increase in burrowing activity and a much smaller and less diverse faunal content.
The tendency for the dolomite interbeds to be thinner than the limestones probably
indicates a lower depositional rate - i. e. the supply of carbonate was lessened.

It is felt that the alternation of dolomite and limestone beds is the result of a cyclic
interplay of environments due to cyclic shifts in the tectonic regime.

The Dark Grey Dolomite lithologies are all representative of an environment similar
to or slightly deeper than that producing the Lower Limestone dolomite interbeds.

Cyclic Dolomite time was characterized by shallow-water deposition alternating with
periods of subaerial exposure.

Lower Limestone Diagenesis

Lithification of biointrapelsparite interbeds took place within inches of the depositional interface. Intergrain porosity and permeability permitted recrystallization of the
early cement probably also at a quite early stage, into sparry calcite, i. e. microspar
(Plate 1-G) (FOLK, 1965).

Stylolites in the Lower Limestone biointrapelsparite beds are ubiquitous though poorly developed. It is virtually impossible to find an area of more than a square inch
which doesn't have at least some micro-stylolites. Solution on any one stylolite plane
was relatively small. However, the aggregate stylolite solution total, when all stylolites are considered, approaches 40 - 50 % of the Lower Limestone late lithification
bulk. Lateral persistence is limited, individual stylolites often bifurcating or termi-

---

Plate 3 - Descriptions

A - Polished specimen of collapse breccia displaying dolomite clast versus
matrix relationship.

B - Sketch map of small area outlined near Zinc Lake in Figure 1

nating within 5 cm in the case of the better developed ones, and within much shorter distances in the case of the thinner stylolites. These interconnecting network stylolites are usually subparallel to the bedding. Stylolites post-date cementation and are contemporaneous with, or earlier than, dolomitization. Their residues consist of ir oxides, and sulfides with some organic content.

The mottling network of dolomitization is controlled by various other events in the sedimentary history of the rock. Biointrapelsparite beds are incipiently dolomitized with small rhombohedra of dolomite scattered throughout the spar cement areas of the Lower Limestone (Plate 1-H). Scanning Electron Microscope examination shows this dolomite to "attack" micrite areas in an intergrain fashion (Plate 2-B, C). Dolomitization has developed very extensively in areas of stylolitization and dolomite are are commonly bounded by stylolites on top and bottom, though the frequency of stylolites stratigraphically above dolomite patches is higher (Plate 1-A). Considering bot the micro-scale and macro-scale evidence pertaining to the dolomite mottling, burrowing by organisms appears to be the most likely mechanism to prepare the host fo dolomitization, in that it removes organic material (Plate 1-E). This organic material seems to be an inhibiting factor for recrystallization to spar and successively to dolomite, as the greater resistance of pelleted micrite attests. Stylolitization, which concentrates the organic material at the same time as it produces magnesium-rich solutions, both aids and directionally controls the dolomitization process.

Dissolution, Sphalerite Mineralization, and Growth of Coarse-Crystalline White Dolomite

After deposition of the Cyclic Dolomites, some 90 + m of uplift produced a karst terrane on the carbonate platform resulting in cave formation. The calcite dissolved in adjacent beds during this karstification leaving a porous open network of interconnecting dolomite mottles. In these spaces, sphalerite grew in colloform fashion arou the mottles (Plate 2-E, F). The final stage involved filling of all, or nearly all, open spaces by what is now pure white coarse-crystalline dolomite. In some cases, a second period of sphalerite crystallization occurred filling the final open spaces (Plate 2-G). Rare disseminated pyrite and galena occur locally in this mineralized "pseudo breccia".

Thus, biointrapelsparite beds of the Lower Limestone were altered to "pseudobreccia" as their ultimate phase of diagenesis.

Conclusions

A detailed stratigraphic analysis and recognized techniques of carbonate petrology have clearly outlined the processes involved in the mineralization of these Lower Ordovician carbonate rocks.

The depositional environment and early diagenesis of the Lower Limestone determined in advance the type of porosity development during karstification at the end of Lower Ordovician time.

Some control by jointing is suggested for the elongate shape of the cave/collapse breccia areas. As the zinc mineralization occurred preferentially in the porous dolomite network adjacent to these areas, it seems that at some time in the karst cycle, a chemical environment conducive to colloform sphalerite growth occurred.

Although intensely mineralized areas preserve very little evidence useful in unravelling their history, by working from unaltered limestones laterally into areas of mineralization and karstification, significant and useful insight can be gained.

Acknowledgements

The authors wish to thank Cominco Ltd. for permission to publish this paper as well as provision of financial support.

Acknowledgement is made for the use of the scanning electron microscope in the Royal Ontario Museum, provided through a grant from the National Research Council to the Department of Zoology, University of Toronto.

Bibliography

COLLINS, J. A.: Carbonate Lithofacies and Diagenesis Related to Sphalerite Mineralization near Daniel's Harbour, Western Newfoundland. Unpublished M. Sc. thesis, Queen's University, Kingston, Ontario, Canada (1971).

CUMMING, L. M.: St. George - Table Head Disconformity and Zinc Mineralization, Western Newfoundland. Bull. Can. Inst. Min. Metall., Montreal, 1968, p. 721-725 (1968).

FIELDER, F. M., ed.: Canadian Mines Handbook 1970-71. Northern Miner Press Limited Toronto, 247, Canada (1970).

FOLK, R. L.: Spectral Subdivision of Limestone Types. In: Classification of Carbonate Rocks, ed.: W. E. HAM. Amer. Assoc. Pet. Geol., Tulsa, Memoir 1, p. 62-84 (1962).

— Some Aspects of Recrystallization in Ancient Limestones. In: Dolomitization and Limestone Diagenesis, ed.: L. C. PRAY, R. C. MURRAY. Soc. Econ. Paleontol. Mineral., Tulsa, Spec. Publ. 13, p. 14-48 (1965).

ODER, C. R. L., RICKETS, J. E.: Geology of the Mascot-Jefferson City Zinc District, Tennessee. Tennessee Div. Geology Rept. Inv. 12, 29 p. (1961).

SCHUCHERT, C., DUNBAR, C. O.: Stratigraphy of Western Newfoundland. Geol. Soc. America, New York, Memoir 1, 123 p. (1934).

Address of the authors:

Department of Geological Sciences
Queen's University
Kingston, Ontario / Canada

The Manganese Ore Deposit of Kisenge – Kamata (Western Katanga). Mineralogical and Sedimentological Aspects of the Primary Ore

Louis Doyen

Abstract

Criteria are offered for a sedimentary origin of the manganese deposit of Kisenge - Kamata. It is of Precambrian age and has undergone a mesozonal metamorphism. The primary ore belongs to two types, a silicate type and a carbonate type. Both of them always contain graphitized carbon. The carbonate ore frequently contains stromatolitic structures.

Ce dépôt d'âge Précambrien doit être considéré comme étant d'origine sédimentaire. Il a subi un métamorphisme mésozonal. Le minerai primaire appartient à deux types : un type silicaté et un type carbonaté. Tous deux contiennent toujours du carbone graphitisé. Enfin, le minerai carbonaté contient fréquemment des structures stromatolitiques.

Introduction

The important manganese deposit of Kisenge - Kamata, which is worked by the Bécéka - Manganese Company, is situated in Western Katanga (The Republic of Zaire) at

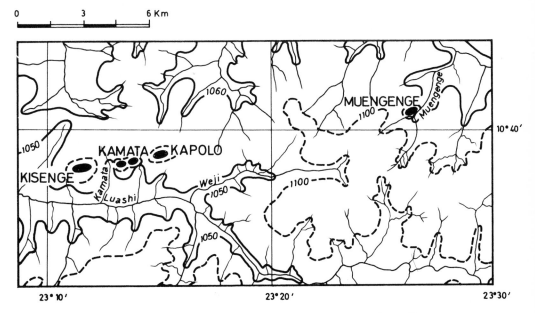

Plate I. Localisation of the Kisenge - Kamata manganese ore deposit

about 25 kilometers south-east of the Tenke - Dilolo railway line about 40 kilometer south of Malonga. The deposit is situated in the High Lulua Basin between the rivers Lukoshi and Luashi. The mineralized zone is located between longitude 23º 10' E. and 23º 30' E. and at latitude 10º 40' (see plate I), (POLINARD, 1932).

This area belongs to a large peneplain that extends to the south as far as the Angolan border. Its elevation above sea level is about 1,100 meters. The ore crops out in th form of elongated hillocks of variable dimensions running in east-west direction. These hillocks are about 20 - 40 meters above the peneplain. They form a chain abo 8 kilometers long. The relief of those mineralized alignments is due to the hardness of the manganese oxides. A lateritic capping covers the greatest part of the region (SCHUILING and GROSEMANS, 1956).

Regional Geology

SEKIRSKY (in POLINARD, 1946, and SEKIRSKY, unpublished) has divided the region of manganese deposits into three zones with distinct petrographic characteristics (se plate II):

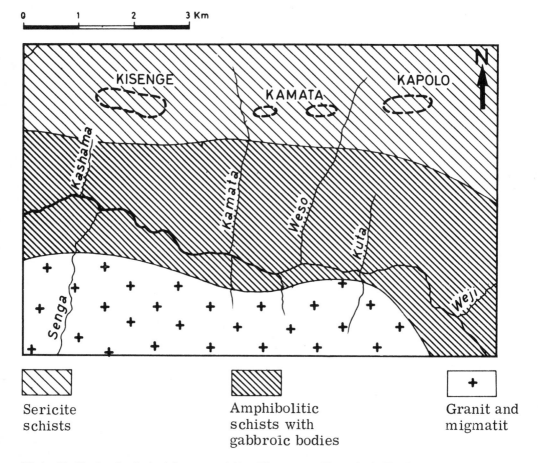

Sericite
schists

Amphibolitic
schists with
gabbroic bodies

Granit and
migmatit

Plate II. Geological sketch-map of the Kisenge - Kamata hillocks region
(after SEKIRSKY)

1) a zone of sericite schists and subordinate quartzites is centred on the watershed Lukoshi - Luashi;
2) a zone of amphibolitic schists following the river Luashi and containing acidic intrusives and gabbroïc bodies;
3) a principally granitic meridional zone.

All the rocks that occur in the Kisenge region belong to a very old series, probably ante-Kibarian and were affected by relatively high grade regional metamorphism. These rocks are well known under the name "Lukoshi Complex".

Pegmatitic intrusions exist both in the mineralized manganese zone and to the south of the mineralized zone. A pegmatite with muscovite crosses the Kisenge ore body.

Two age determinations have been carried out by the Sr/Rb method on a muscovite from a pegmatite that crosses the manganiferous beds in the Musenge region (Loc.: 23^0 25' E. - 10^0 40' S.). The results are as follows:

	Rb	Sr	Sr^{87} rad	$\dfrac{Sr^{87} \text{ rad}}{Sr^{87} \text{ rad } Sr^{87}}$	Age M. years
Sample	ppm	ppm	ppm		
1	4813	1.5	38	99.7	1,900 ⎫
2	4444	3.3	33	99.3	1,790 ⎭ 1,845

The age obtained is 1,845 M. years, it characterizes the orogenesis that has affected the Lukoshi formation or it constitutes a younger limit for this orogenesis (LEDENT, LAY, DELHAL, 1962).

The Deposit

The deposit has been divided into four zones, namely: Kisenge, Principal Kamata, Kamata left side and Kapolo (see plate I). Only the first two are worked.

A) Kisenge

The lithological and mineralogical studies made on the Kisenge 1 drill-hole have shown that, except for its oxidized zone, which contains more than 50 % of manganese dioxide, the Kisenge Deposit was formed by an alternation of garnetiferous and graphitic schists (MARCHANDISE, 1958). The size of the garnets varies from 0.05 to 1.5 mm in diameter. They are nearly always twinned and show subdivisions into segments, which renders the transformation to the more or less hydrated manganese dioxides easier.

The enclosing rocks are composed of highly metamorphosed quartzites containing muscovite, biotite, garnet and tremolite. Below these quartzites, the drill-holes have shown formations richer in mica that grade into quartzitic micaschists with sillimanite, staurolite, tremolite and garnet with increasing depth.

The action of the tropical climate resulted in the weathering of the primary manganese-bearing rocks (Gondite) and the reprecipitation of the manganese in the form of manganese hydroxides and manganese oxides, which led to an enrichment "per descensum" with elimination of silica.

The manganese content is therefore in inverse relation to the silica content and de-creases with depth.

In the deepest drill-holes, it could be established that the primary ore was repre-sented by a gondite of garnet-schist type, the manganese content of which is below 20 %. The garnet constitution varies slightly from bed to bed. Its composition is nearly always the same as that of pure spessartite with a small amount of almandin The refractive index has been determined by POLINARD (1946) on a single sample. He gives a value of n = 1.79 ± 0.005 and a density of 4.10.

B) <u>Kamata</u>

Two drill-holes (Ka 30 and Ka 72) situated on one of the outcrops rich in manganese oxides, have encountered a huge thickness of manganese carbonates in depth, with well-marked stratification. The strike of the beds is east-west and the dip is due south varying from 45 to 70°. These beds are dark grey, sometimes nearly black in colour, crystalline and fine-grained. Practically pure manganese carbonate con-stitutes these rocks. The manganese content varies between 43 % and 46 %. It is worth noting that the theoretical manganese content in a pure rhodochrosite is 47.83 The optical characteristics of the constituents of the rocks are those of rhodochrosit

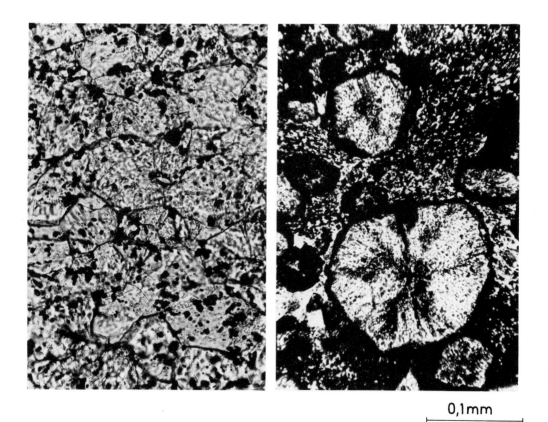

0,1mm

Photo 1. Graphitized matter associated with carbonate
Photo 2. Garnet-grain (spessartite) surrounded by graphitized carbonate

On the other hand, the low contents of Fe (0. 84 %) as well as those of Ca + Mg (1 to 2 %) are incompatible with the presence of ankerite or manganocalcite. The rock, when examined under the microscope, is monomineralic. The carbonate appears as a mosaic of platy crystals of rhodochrosite, the size of the crystals ranging from 50 - 200 microns with 150 as the average.

Graphitic carbonaceous matter, very likely of organic origin, is almost always associated with the carbonate (photo 1). It is localized at the periphery of the rhodochrosite grains and never or rarely as microscopic inclusions. It has frequently been observed at the periphery of the garnets (photo 2). The dimensions of the inclusions vary from 2 to 5 microns and may exceptionally attain 15 microns.

The carbonate filled fissure-veinlets are completely free of carbonaceous matter; they are whitish or pinkish in colour.

The carbonate formations are never garnet-free (photo 3), the garnets showing twinning and a more or less deep alteration into manganese oxides (photo 4).

The carbonates very often contain inclusions of manganese oxides (braunite, bixbyite, manganite), the size of the latter ranging from 5 to 10 microns. Besides, small flakes of nickel and cobalt sulphides coexist with the manganese carbonate and seem to be contemporary.

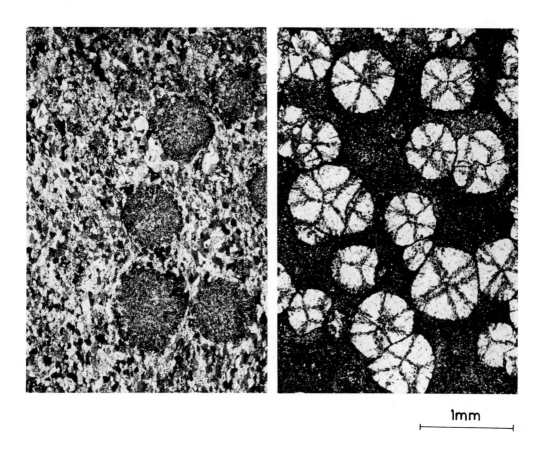

1mm

Photo 3. Relics of garnets in the carbonate mass
Photo 4. Garnets (spessartite) showing twinning

Spectrographic analyses have confirmed the presence of nickel and cobalt and the presence of Cu, Mo, Nb, Ta, Sn, and W in trace amounts.

The carbonaceous matter distributed in even disseminations could be responsible for the presence of low radioactivity (2 to 5 x the B. G.) measured on several samples.

Apparent Succession of the Lithological Units

Below the oxidized zone, the drill-holes have successively encountered:

1. Manganese-carbonate ore, which is finely crystallized, dark-grey or nearly black and crossed by veinlets of pink rhodonite. This formation contains stromatolites at different levels. Those formations nearly always contain garnets, in various proportions. The garnet content of these rocks increases sharply as the contact with the underlying lithological unit is approached.

2. Gondite, very dark, ferruginous, with closely packed garnets, containing traces of manganese carbonate. In depth, these rocks become pure garnet-rocks with a structure resembling that of the quartzites.

3. Graphitic schists: these are highly crushed and strongly graphitic.

4. Gondite: garnet-rock with ferruginous cement, crossed by quartz veinlets.

5. Manganese carbonate ore: grey coloured rock becoming darker and richer in garnets with depth.

6. Transition zone composed of a carbonate mass, very rich in garnets and passing to garnet-schist in depth.

7. Highly graphitic garnet-rocks.

8. Garnet-rock with ferruginous cement, traversed by small quartz veins.

9. Nearly pure graphite with quartz veinlets.

Genetic Discussion

The primary ore deposit of Kisenge must be considered as being of sedimentary origin for the following reasons:

- regular alternation of beds of garnet-rocks, graphitic schists and rhodochrosite;

- presence of manganese oxides as inclusions in the rhodochrosite (braunite, bixbyite, manganite);

- presence of nickel and cobalt sulphides contemporaneous with the crystallization of the manganese carbonate;

- presence of graphitic carbonaceous matter and the existence at the bottom of the formation of a layer of graphitized coal;

- presence of stromatolitic structures in the manganese carbonate ore.

Even though the sedimentary origin of the primary deposit is obvious, some problem concerning mainly the beds of massive and graphitic rhodochrosite exist.

The fact that the rock contains, in the manganese-carbonate mass, biogenic structures resembling COLLENIA may be interpreted as indicating that the manganese carbonate is a product of metasomatic transformations of a pre-existing limestone, a phenomenon comparable to dolomitization. The possibility of a hydrothermal meta-somatism is, however, excluded for obvious reasons.

The formation of rhodochrosite by direct precipitation in marine, lacustrine and lagoon-bottom is, however, well known (VARENTSOV, 1964). Even in the mangani-ferous sediments containing alumino-silicates, rhodochrosite seems to have been the primary manganese-bearing mineral (BERGER, 1965, 1968; BOULADON, 1970).

On the other hand, even moderate metamorphism can form spessartite in such rocks and can also produce structures of the garnet-rocks (ROY, 1966, 1968).

In nearly monomineralic sediments composed of rhodochrosite, a mesozonal meta-morphism can only result in the recrystallization of the carbonates as well as the fromation of spessartite at the expense of the alumino-silicate impurities.

Conclusions

The manganese deposit of Kisenge - Kamata is of sedimentary origin. It was, how-ever, affected by a mesozonal metamorphism.

The primary ore is of two types:
a) the silicate type with a manganese content of about 20 %,
b) the carbonate type with a manganese content of about 40 %.

The Kisenge - Kamata deposit that has produced oxidized manganese ore of excellent quality in the past may furnish in the future a carbonate ore of very good quality.

Bibliography

BERGER, A.: Zur Geochemie und Lagerstättenkunde des Mangans. Gebrüder Born-traeger, Berlin, Stuttgart, 216 p. (1968).

BERGER, P.: Les dépôts sédimentaires de manganèse de la Lienne inférieure. Ann. Soc. Géol. de Belgique, Vol. 88, p. B 245-268 (1965).

BOULADON, J.: Les principaux types de gisements de manganèse et leur importance économique. Revue de l'Industrie Minérale, p. 1-8 (1970).

LEDENT, D., LAY, C., DELHAL, J.: Premières données sur l'âge absolu des formations anciennes du "Socle" du Kasaï (Congo méridional). Bull. Soc. Belge de Géol., T. LXXI, fasc. 2, p. 223-235 (1962).

MARCHANDISE, H.: Le gisement et les minerais de manganèse de Kisenge (Congo Belge). Bull. Soc. Belge de Géol., T. LXII, fasc. 2, p. 187-211 (1958).

POLINARD, E.: Esquisse géologique de la région située au Sud du parallèle de Sandoa-Kafakumba. Ann. Soc. Géol. de Belgique, 54, p. C 100-105 (1932).

— Le minerai de manganèse à polianite et hollandite de la Haute-Lulua. Inst. Roy. Col. Belge, Sc. Nat., Mém. in 8°, T. XVI, fasc. 1, p. 1-41 (1946).

ROY, S.: Syngenetic Manganese Formations of India. Judavpur University, Calcutta, 219 p. (1966).

— Mineralogy of the Different Genetic Types of Manganese Deposits. Econ. Geol. Vol. 63, p. 760-786 (1968).

SCHUILING, H. , GROSEMANS, P. : Les gisements de manganèse du Congo Belge. In: XX Intern. Geol. Congr. Mexico, T. II, p. 131-142 (1956).

SEKIRSKY (Unpublished): Map - in Dossier G. 124. Musée de l'Afrique Centrale, Tervuren (Belgium).

VARENTSOV, I. M. : Sedimentary Manganese Ores. Elsevier, Amsterdam, 119 p. (1964).

Address of the author:

Laboratoire de Géologie Appliquée
Université Libre de Bruxelles
Bruxelles 5 / Belgium

Geologic Relations Among Uranium Deposits, South Texas, Coastal Plain Region, U.S.A. [1]

D. Hoye Eargle and Alice M. D. Weeks

Abstract

Uranium has become an important energy resource in the South Texas Coastal Plain, where it occurs in tuffaceous sandy sedimentary rocks of late Eocene to Pliocene age. The uranium ore contains small amounts of molybdenum and selenium. Deposits range from small, irregular bodies of highly oxidized "yellow ores" near the outcrop to unoxidized "black ores" at depths of from 60 feet (18.3 meters) to at least several hundred feet. The oxidized ore is highly susceptible to leaching accompanied by migration down the dip of permeable sands. The deep unoxidized deposits exhibit well-developed roll-type ore bodies along oxidation-reduction boundaries.

Genetic factors presumed to be common to these deposits are (1) a source of uranium from the diagenetically altered tuffaceous sedimentary rocks, (2) dissolution and transport of uranium by the alkaline carbonate pore-waters of the ash, (3) precipitation of uranium in reducing environments caused principally by hydrogen sulfide and in part by organic matter where waters were retarded by lenticular, clayey or structural barriers, and (4) preservation from leaching by such features as a caliche cap developed during an arid climate.

Four principal geologic environments of uranium ore in South Texas are (1) in near-shore sandstones sandwiched between less permeable beds and overlain unconformably by tuffaceous deposits, (2) in sandstones interfingering with claystones along the sides of major paleo-stream-channel deposits, (3) in sandstone near faults along which natural gas containing hydrogen sulfide has migrated, and (4) in sandstone above the sulfurous caprock of a salt dome.

Introduction

Texas is the most recent State in the United States to become a major producer of uranium. Discovered there in 1954, uranium first came into production in 1960. According to the latest estimates released by the Atomic Energy Commission on reserves by States, Texas ranks third in the Nation with proved reserves of about 6,600,000 tons (6,007,670 metric tons) of 0.16 percent U_3O_8 (U.S.A.E.C.,1971). Deposits of low grade are economic to mine in this region because the ores being mined are at shallow depths, are unconsolidated, and are highly amenable to milling and to extraction of uranium. Although lease prices are high, the costs of exploration, especially of drilling and logging, are said by some operators to be the lowest in the United States.

The uranium areas of Texas are close to oil fields, and several deposits are located along faults that have formed traps for highly productive oil and gas accumulations.

---

1) Publication authorized by the Director, U.S. Geological Survey.

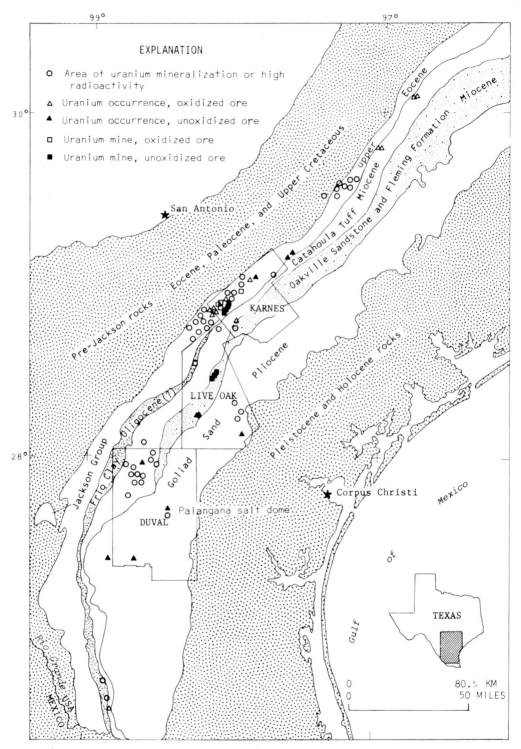

Fig. 1. Geologic map of the South Texas Coastal Plain, showing uranium mines, occurrences, and mineralized areas. Modified from EARGLE, HINDS, and WEEKS (1971, fig. 3)

The strike of formations in the northern part of the region trends northeastward, and the faults nearly parallel the strike. In the southern part of the region the strike is north to north-northwest; the faults, however, trend generally northeastward. The faults are generally downthrown toward the coast, but a few, antithetical to these, are upthrown toward the coast in such a way that grabens are formed. Dips range from more than 100 feet per mile (about 20 meters per kilometer) in the older beds to 30 to 40 feet per mile (6 to 8 meters per kilometer) in the younger beds, reflecting the continual sinking of the Gulf Coast syncline and uplift of the interior.

The part of the Coastal Plain of Texas that receives most attention in uranium prospecting today is the outcrop and shallow subsurface of upper Eocene to Pliocene beds (fig. 1) from east-central Texas to the Rio Grande in southern Texas. The stratigraphy of the producing areas is as follows (an asterisk * indicates a principal or an important uranium-bearing unit):

Age	Formation	Member
Pliocene	Goliad Sand *	
Miocene	Fleming Oakville Sandstone * Catahoula Tuff *	
Oligocene (?)	Frio Clay *	
Eocene	Whitsett (of Jackson Group)	Fashing Clay Tordilla Sandstone * (Karnes County) and Calliham Sandstone (Atascosa County) Dubose Deweesville Sandstone * Conquista Clay Dilworth Sandstone

The uranium region of South Texas comprises three areas named for the counties in which the principal deposits are located: Karnes, Live Oak, and Duval. Deposits in each area are distinctive in stratigraphic position, structure, and type of ore. Figure 2 shows all the mines of the region, both active and inactive, and in both oxidized and unoxidized ores.

Karnes Area

The map of the Karnes area (fig. 3) shows the location of the mines of the deeper downdip ore trend, the outcrop of sands of the Whitsett Formation and the overlying Catahoula Tuff, and the principal faults of the area. The economic deposits are in the upper two of three sandstone members of lagoonal origin in the Whitsett Formation. The Deweesville Sandstone Member, the middle sandstone of the three, contained most of the oxidized ores, now considered generally mined out. The Tordilla, the uppermost of the three sandstones, contains the unoxidized ores now being mined. The uranium deposits lie between the Falls City fault on the northeast, downthrown to the southeast, and the Fashing fault on the southwest, downthrown to the northwest. Note the strong northerly trend of the formations between the ends of the two faults in the western half of the map (fig. 3). Strong induration of rocks locally along fault or fracture zones causes irregularities in the outcrop pattern of some of the sandstones, particularly in the Tordilla Sandstone Member.

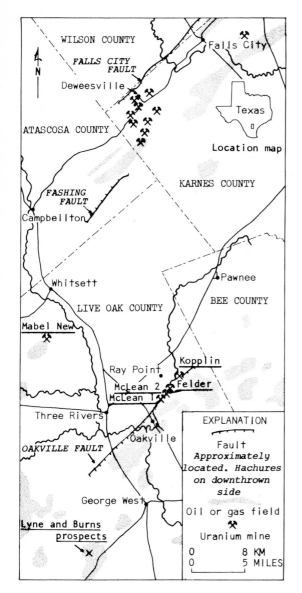

Fig. 2. Map showing the productive mines and oil or gas fields of the South Texas Coastal Plain

Unconformably overlying the truncated host-rock sandstones of the Karnes area is the Catahoula Tuff (Miocene), of continental origin, most of which has been eroded from the areas of the deposits. In the area of current mining, remnants of the Catahoula show that the Catahoula overlaps much of the underlying Whitsett Formation and has gravel-filled channels at its base. Although nonproductive in the Karnes area the tuffs of the Catahoula are considered to have been the principal source of the uranium-mineralizing solutions that formed the deposits of the region.

A cross-section through the Karnes area (fig. 4) shows the three principal sandstone of the lower part of the Whitsett Formation; they are separated by tuffaceous and car

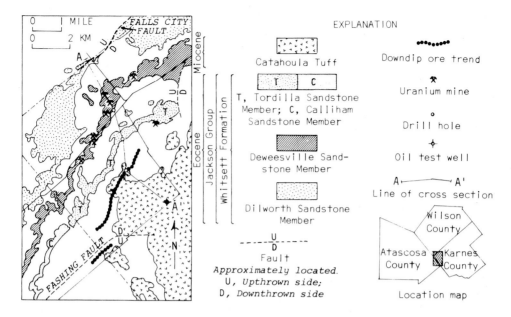

Fig. 3. Geologic map of the Karnes area showing uranium mines and the downdip ore trend. Cross section A-A' is shown on figure 4. Modified from EARGLE and WEEKS (1968, fig. 2)

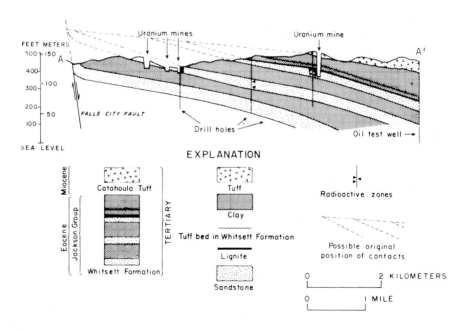

Fig. 4. Cross section A-A' through the Karnes area showing relation of uranium deposits to stratigraphy and structure. Line of section shown on fig. 3. Modified from EARGLE and WEEKS (1968, fig. 3)

bonaceous claystone, siltstone, and lignite. We believe that uranium was leached fr(
the Catahoula and entered each of the underlying sandstones across the unconformity
at the base of the Catahoula. The lowermost sandstone may be barren, because it wa
hidden along the Falls City fault when the uranium deposits were accumulating.

The oxidized ores mined in the Karnes area were small, lenticular, near-surface
bodies or irregular masses of high-grade ore surrounded by low-grade ore that was
generally out of equilibrium in favor of the radiometric assay. These deposits generally were less than 45 ft (14 m) below the surface, in sandstone underlain by claystone. We believe that these deposits are oxidized remnants of formerly unoxidized
roll-front deposits and that erosion has removed most of the overburden that once
covered them.

The downdip unoxidized ore bodies are much larger, more continuous, and more nea
ly in radioactive equilibrium than are the near-surface ores. They occur along a ro)
front, an oxidation-reduction boundary (fig. 5), at or slightly below the present wate
table. The roll front, crescentic in dip section, is parallel to and about half a mile
downdip from the outcrop. Convex downdip, it is bounded on the updip side by partial
ly oxidized, leached, or otherwise altered very pale gray, buff, or yellow barren
sandstone, and on the downdip side by reduced medium- to dark-gray sandstone, the
thickest and richest ore of the deposit. Thickness and tenor of the ore diminish grad
ually downdip from the roll front to an assay cutoff within several hundred feet. The
open-pit mines in the Karnes area are as much as 130 feet (39.6 meters) deep, and
the thickest ore is about 25 feet (7.6 meters) thick in the 25- to 30-foot-thick (7.6 -
to 9.1-meter-thick) host sandstone.

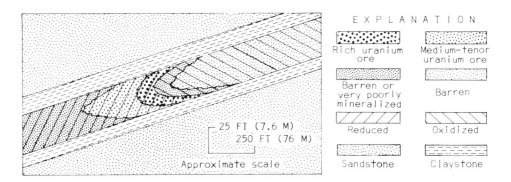

Fig. 5. Idealized cross section showing a roll-type ore body in South Texas

Live Oak Area

Ores in the Live Oak area occur in several types of structures and in several stratigraphic positions.

Oxidized ore of the Mabel New mine (fig. 2), now inactive, was of highly variable
tenor and was derived from uraniferous, pyritic, calcareous sandstone, a few unoxidized masses of which were found in the mine. The ore was found in a sandstone
of the Frio Clay (Oligocene?) near the margins of a sandstone-filled channel as show
in figure 6.

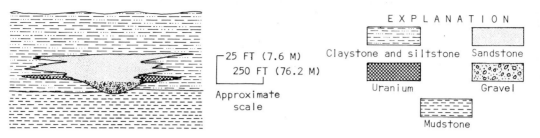

Claystone and siltstone   Sandstone

┌25 FT (7.6 M)
│ 250 FT (76.2 M)
└
Approximate
scale

Uranium   Gravel

Mudstone

Fig. 6. Idealized cross section showing a uranium occurrence in a sandstone-filled channel in South Texas

The Live Oak area has significant downdip deposits of unoxidized ores in which several mines were opened in the late 1960's (fig. 7). The deposits are in the calcareous Oakville Sandstone (Miocene) and are associated with the Oakville and related faults that trend generally northeastward. The Oakville fault is a well-known zone of rather complex faulting, along which oil and gas have been found in sandstones in Miocene and older rocks of the Tertiary.

The ore in the Kopplin mine, the northernmost mine in the area, is in the coarse basal sandstone of the Oakville. This sandstone, about 15 feet (4.6 meters) thick,

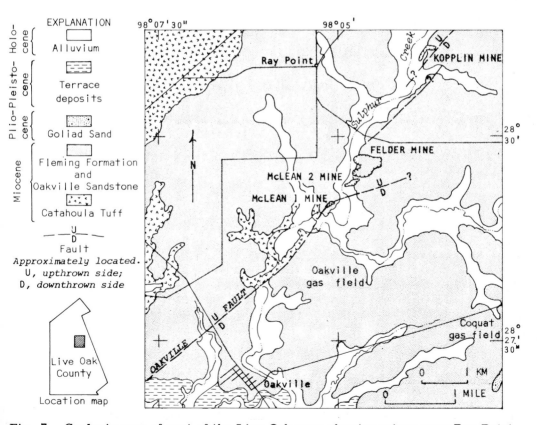

Fig. 7. Geologic map of part of the Live Oak area showing mines near Ray Point and Oakville. Modified from EARGLE, HINDS, and WEEKS (1971, fig. 19)

is overlain by interbedded claystone, siltstone, and sandstone. A fault, with displace
ment of about 100 feet (30 meters), placing younger, more impervious beds against
the basal sandstone, is the downdip limit of the ore. Sulfur-bearing water seeps out
of the ore sandstone updip from the fault.

The McLean 1 mine was opened in a deposit along the Oakville fault itself. On the up
thrown side, ore was found in the basal beds of the Oakville; on the downthrown side,
in higher beds, the ore funnels upward from the fault line (fig. 8). A high concentration of molybdenum minerals and of pyrite occurs along the fault plane in association
with the uranium, but both diminish with distance from the fault.

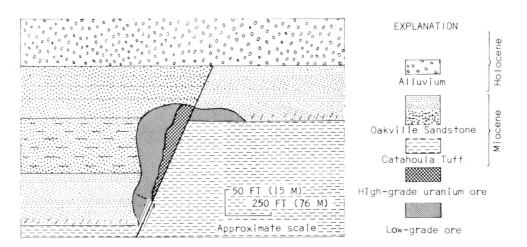

Fig. 8. Idealized cross section showing uranium occurrence along a fault, as typified by the McLean 1 deposit in Live Oak County

The McLean 2 mine, northeast of McLean 1, intercepted the southern edge of an ore
roll that is spectacularly displayed in the adjoining Felder mine. The McLean 2 and
Felder mines contain a conspicuous selenium mineral occurring as a red efflores-
cence in the vicinity of the uranium ore.

The Felder mine, estimated to contain 5 million pounds of U_3O_8 (KLOHN and PICKE
1970), is the largest mine in operation in the region. The ore is in calcareous arkos
sandstone, generally fine to coarse grained, of fluvial origin, and interbedded with
claystone. The deposit lies between two faults, with one of which the McLean ore bo
dies are associated. In cross section down the dip, the ore body of the Felder depos
it has the shape of a crescentic ore roll (fig. 5). The center contains a little ore, but
the greatest concentration of ore is just downdip from the roll front. Selenium is co
centrated near the roll front, and molybdenum occurs as a broad halo around the ura
nium.

KLOHN and PICKENS (1970) pointed out the relation of the position of the Felder deposit and other deposits in Live Oak County to a 300-foot-thick (91-meter-thick) fluvial channel system that trends eastward through the central part of the county (fig.
thence southeastward to the southeastern part of the county. The ore deposits lie not
in the channel itself, but in sandstone masses that finger laterally outward from the
channel. These are probably point-bar as well as sand-splay and other overbank sed

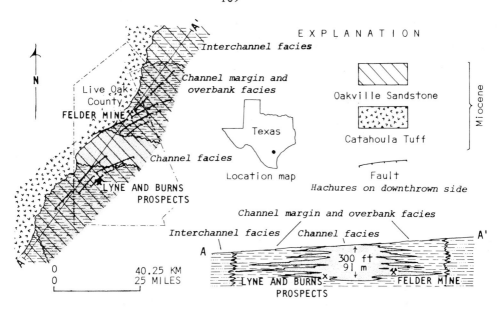

Fig. 9. Generalized geologic map and cross section showing relation of uranium deposits to a fluvial channel system in the Oakville Sandstone (Miocene) in Live Oak County. Modified from KLOHN and PICKENS (1970, fig. 2)

ments that were later covered by finer grained silt and clay. The rocks were later cut by faults, along which many oil and gas fields have been developed throughout Live Oak County (fig. 2). Gas that contains a high percentage of hydrogen sulfide, and even viscous oil, seeps from the faults in this area and permeates the rocks in the vicinity of the mines.

On the west side of the alluvial system described by KLOHN and PICKENS (1970) are the counterparts of the Felder deposit, the Lyne and Burns prospects (figs. 2 and 9). The Lyne prospect lies on both sides of an up-to-the-coast fault that is part of a northeast-trending graben system in southwestern Live Oak County. The Burns prospect is a stratabound deposit that, unlike the Lyne prospect, is not known to be cut by faults.

Duval Area

Many occurrences of anomalous radioactivity and some uranium-mineral localities have been known in the Duval area since 1954. Although the results of much recent exploration are still confidential, drilling is active in that area, and some uranium deposits have been discovered.

The prospects in this area are more varied in geologic setting than are those of the Karnes and Live Oak areas. Prospects have been explored in several formations, ranging in age from Miocene to Pliocene. Most of the prospects have been associated with faults, and several are in grabens. Some prospects have been found in Miocene sandstones whose truncated edges have been overlapped by Pliocene channel sands, as shown in fig. 10. Some are associated with faults that reach the surface. In some places the faults may form the traps for deep petroleum deposits as in the Live Oak area.

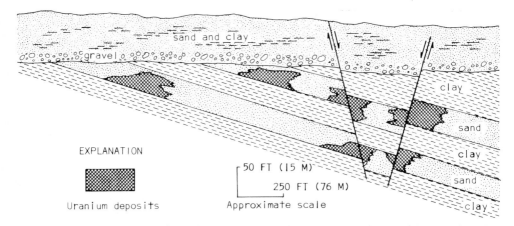

Fig. 10. Idealized cross section showing typical downdip occurrence of uranium deposits associated with unconformities and faults in South Texas

Some radioactivity anomalies and at least one uranium deposit in the Duval area are associated with salt domes. One widely known prospect of this type, believed to be unique, is in the Pliocene sandstone overlying the caprock of Palangana salt dome in east-central Duval County (fig. 1), where, in the late 1950's, a shaft was sunk to a depth of 300 feet (91 meters), reaching sandstones that contain unoxidized ore (fig. 11) There, the deposits are in a reduced environment, strongly affected by accumulation of hydrogen sulfide (WEEKS and EARGLE, 1960).

Origin of the Deposits

The uranium deposits in South Texas are confined to areas where volcanic ash makes up a high percentage of the host rock. These sedimentary rocks average about 50 per

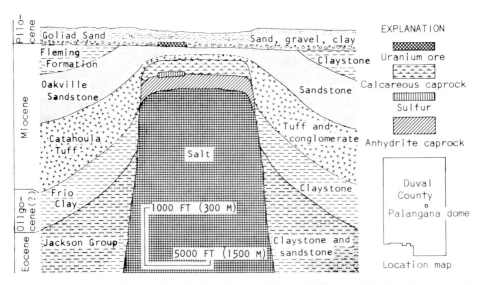

Fig. 11. Idealized cross section through Palangana salt dome, Duval County. Modified from WEEKS and EARGLE (1960, fig. 24.1)

cent ash or diagenetically altered ash; some beds are nearly pure ash. An airborne radioactivity survey (MOXHAM and EARGLE, 1961) of the region showed that the Miocene tuffaceous rocks average 10 to 20 parts per million equivalent uranium (MOXHAM, 1964). Some ash of the older upper Eocene still contains relatively fresh glass shards, although part of the Jackson montmorillonitic clay probably represents ash that was weathered before it was transported to its present location.

Tertiary alkalic igneous rocks from the Big Bend region in western Texas, believed to be in the source area of the South Texas volcanics, contain as much as 45 parts per million uranium (GOTTFRIED and others, 1962). The uranium is disseminated throughout the volcanic sediments, but the total uranium content of the tuffaceous rocks is especially great. GARRELS and others (1957) have emphasized that alkaline carbonate pore water develops in tuffaceous rocks and that such alkaline solution is an excellent agent for leaching and transporting the uranium and other trace elements. The alkaline pore water was important in the diagenetic alteration of the ash as a "built-in" solvent for the uranium.

The uranium deposits also are restricted primarily to the part of South Texas that has been dry enough for the formation and survival of a caliche cap. The caliche is as much as 7 feet (2.5 meters) thick at the Mexican border, but the hard caliche cap is discontinuous at the northeastern end of the region containing uranium deposits in Karnes County. The dry climate helped to concentrate the uranium-bearing solutions and also to protect the deposits from leaching. The tuffaceous rocks have been considerably altered by redistribution of silica as opal and chalcedony and by the formation of zeolite (as clinoptilolite) and montmorillonite.

Uranium is highly soluble as the uranyl ion U^{+6} and will migrate in the ground water until precipitated by reduction. The chief reducing agents are plant fragments (present in some ash beds), hydrocarbons or, in many instances, hydrogen sulfide, which is common both in surficial rocks and near the faults that have trapped the hydrogen sulfide-rich petroleum and natural gas. Precipitation takes place at or near the water table; some uranium, however, may be carried much farther downdip, considerably below the water table, where it forms other deposits.

A study of oxidized uranium deposits in northern Karnes County (MOXHAM, 1964) indicated that considerable loss of uranium by weathering has taken place, in part by stream action and in part by downward migration of ground water. Radiochemical analyses made by ROSHOLT (1959, reported by WEEKS and EARGLE, 1963) showed that some uranium deposited more than 240,000 years ago has been leached, transported along joints and bedding planes, and redeposited in lower beds from shortly after original deposition to only a few thousand years ago.

Thus, the uranium of the deposits seems to have been (1) leached from volcanic ash by alkaline carbonate ground water derived from the diagenesis of the ash, (2) carried downdip to a reducing environment caused principally by hydrogen sulfide and in part by organic matter, (3) precipitated where the waters have been retarded by lenticular, clayey, or structural barriers, and (4) preserved at least in part by an arid climate that allowed relatively little leaching.

Our theories concerning the origin and locality of deposition of the ores of South Texas can be summarized as follows.

1. The principal source of the uranium was the tuff of the nonmarine post-Eocene formations of the region, and the tuff originated in the volcanic regions of northern Mexico and western Texas, whence, together with some erosional debris, it was blown and carried by streams into southern Texas.

2. The uranium, together with other elements, was leached from the volcanic ash by oxygenated and alkaline ground waters in arid climates through weathering and diagenesis.

3. The uranium was carried through stream channels and other permeable rocks to reducing environments below ground-water levels.

4. The uranium was precipitated by reduction with hydrogen sulfide or other gases that permeate the rocks and form seepages along fault lines; also, carbonaceous materials in the host rocks themselves may have furnished some reductants.

5. Deposits were formed along the edges of host-rock strata of good permeability interbedded with claystones or siltstones of poor permeability.

6. Faults that trend generally parallel to the strike were important in localization of some deposits. The role of faults was twofold: they retarded the flow of ground water downdip and thus allowed the reductants opportunity to act on the uranium and other elements in solution, and they permitted the seepage of hydrogen sulfide and other precipitants in solution to penetrate the aquifers.

Acknowledgement

We appreciate the assistance of Beth Ogden Davis in the preparation of the illustrations for this report.

Bibliography

EARGLE, D. H., WEEKS, A. M. D.: Factors in the formation of uranium deposits, Coastal Plain of Texas. South Tex. Geol. Soc. Bull., 9, no. 3, 12 p. (1968).

EARGLE, D. H., HINDS, G. W., WEEKS, A. M. D.: Uranium geology and mines, South Texas. Texas Univ. Bur. Econ. Geology Guidebook no. 12, 59 p. (1971).

GARRELS, R. M., HOSTETLER, P. B., CHRIST, C. L., WEEKS, A. D.: Stability of uranium, vanadium, copper, and molybdenum minerals in natural waters at low temperatures and pressures (abs.). Geol. Soc. America Bull., 68, p. 1732 (1957).

GOTTFRIED, D., MOORE, R., CAEMMERER, A.: Thorium and uranium in some alkalic igneous rocks from Virginia and Texas. In: Geological Survey Research 1962. U. S. Geol. Survey Prof. Paper 450-B, p. B70-B72 (1962).

KLOHN, M. L., PICKENS, W. R.: Geology of the Felder uranium deposit, Live Oak County, Texas. Paper presented at the AIME annual meeting, Denver, Colorado - February 15-19, 1970, Soc. Mining Engineers Preprint no. 70-1-38, New York, Soc. Mining Engineers, 19 p. (1970).

MOXHAM, R. M.: Radioelement dispersion in a sedimentary environment and its effect on uranium exploration. Econ. Geology, 59, p. 309-321 (1964).

MOXHAM, R. M., EARGLE, D. H.: Airborne radioactivity and geologic map of the Coastal Plain area, southeast Texas. U. S. Geol. Survey Geophys. Inv. Map GP-198 (1961).

ROSHOLT, J. N., Jr.: Natural radioactive disequilibrium of the uranium series. U. S. Geol. Survey Bull. 1084-A, 30 p. (1959).

U.S. ATOMIC ENERGY COMMISSION: Statistical data of the uranium industry, January 1, 1971. Grand Junction, Colorado, 54 p. (1971).

WEEKS, A.D. , EARGLE, D.H. : Uranium at Palangana salt dome, Duval County, Texas. In: Short papers in the geological sciences. U.S. Geol. Survey Prof. Paper 400-B, p. B48-B52 (1960).

WEEKS, A.D. , EARGLE, D.H. : Relation of diagenetic alteration and soil-formation processes to the uranium deposits of the southeast Texas Coastal Plain. In: Clays and clay minerals, v. 10 - Natl. Conf. Clays and Clay Minerals, 10th, 1961, Proc. , New York, Macmillan Co. (Internat. Ser. Mons. Earth Sci. , 12), p. 23-41 (1963).

Addresses of the authors:

Mr. D.H. Eargle
U.S. Geological Survey
Austin, Texas / U.S.A.

Dr. A.M.D. Weeks
Temple University
Philadelphia, Pa. / U.S.A.

Syngenetic Dolomitization and Sulfide Mineralization

Helmut Geldsetzer

Abstract

Excellent stratigraphic control and diagenetic changes make it possible to demonstrate the temporal relationship as well as the process of Pb-Zn mineralization in well exposed Precambrian rocks of Helikian age on northern Baffin Island, N.W.T., Canada. Initial deposition of shallow-marine, cratonal sediments was interrupted by regional tilting. Emerging areas were extensively karsted and brecciated, affecting in particular laminated algal carbonates. Subsequent subsidence led to the development of a topographically restricted marine environment and in turn to evaporative conditions. The brecciated carbonate and the extensive subsurface channel system which was formed during the preceding episode of karsting, provided perfect passageways for a refluxing dolomitizing brine. Steady-flow conditions and a continuous supply of seawater resulted in the complete dolomitization of the brecciated and channelled hostrock. Seawater must have contributed most if not all the Mg during dolomitization. Continuous circulation of a seawater brine must have introduced other metal ions as well, which were most likely carried as chloride complexes. These combined with a possibly seasonal supply of sulfide ions generated by sulfate-reducing bacteria. Precipitation of Pb and Zn sulfides took place in open fractures (disseminated occurrences) and in channelways (massive occurrences). Karsting, dolomitization and sulfide mineralization were terminated prior to the deposition of another Helikian sequence of clastics and carbonates which are only partially dolomitized and carry no Pb and Zn sulfides.

Excellent permeability, a proper drainage system for a refluxing brine and a continuous supply of evaporating seawater are the essential criteria that account for this Arctic Pb-Zn deposit. The simplicity of this pattern suggests that syngenetic dolomitization and sulfide mineralization are a common geological event and may be responsible for other Pb-Zn occurrences.

Introduction

A thick Middle Proterozoic carbonate sequence on the Borden Peninsula of northern Baffin Island, North West Territories, Canada (fig. 3), was the object of an extensive exploration program for base metals. During the course of the investigation it became apparent that sulfide mineralization was a temporally very restricted event and genetically associated with epigenetic dolomitization. This conclusion is based upon regional stratigraphic relationships and macroscopic diagenetic changes. Analyses of the distribution of lead, zinc, iron and other critical elements and of the isotopic composition of lead and sulfur are in progress.

General Geology (fig. 1)

A basement complex was intensely metamorphosed during Aphebian time, i. e. durin
the Early Proterozoic (LEMON and BLACKADAR, 1963). Radiometric dates indicate
an age of about 1700 million years (BLACKADAR, 1970). A thick sequence of unmeta
morphosed and almost undeformed basal volcanics and overlying sediments of Helikian age (Middle Proterozoic) were downfaulted into the gneisses of the basement
complex. A thick dolomite unit within the Helikian sequence, the Society Cliffs Form
ation, is the host of massive sulfide mineralization and the focal point of the following discussion. Helikian and older rocks were intruded by gabbroic dikes and sills
of Neo-Helikian, lat Middle Proterozoic age. A profound unconformity separates the
Precambrian rocks from Lower Paleozoic undeformed sediments.

	AGE	LITHOLOGY	TECTONISM
	EARLY PALEOZOIC	SEDIMENTS	UNDEFORMED
	~~	~~~~~~~~~	~~~~~~~~~
PROTEROZOIC	NEO - HELIKIAN 900 m.y.	GABBROIC DIKES	
	- - - - - -	- - - - - -	- - - - - -
	HELIKIAN 900 - 1050 m.y.	SEDIMENTS BASAL VOLCANICS	FAULTED, INTRUDED UNMETAMORPHOSED
	~~	~~~~~~~~~	~~~~~~~~~
	APHEBIAN 1700 m.y.	SEDIMENTS AND VOLCANICS	METAMORPHOSED

Fig. 1

Depositional, Erosional, and Diagenetic History

The stratigraphic and regional restriction of massive sulfides in the Society Cliffs
Formation can best be explained by an analysis of the depositional, erosional, and
diagenetic development of the preserved Helikian rock sequence. A summary of this
analysis is presented on figure 2. The Helikian time interval has been subdivided into depositional and erosional episodes, but for the sake of brevity only those episodes
are considered here which precede and coincide with the time of lead-zinc mineralization in the Society Cliffs Formation.

The Arctic Bay depositional episode

The first depositional episode is characterized by a thick sequence of crossbedded
orthoquartzites which grade upward into black fine-grained clastics. This gradation
takes place by an increasing frequency of fine-grained clastic beds which become the
dominant lithology in the Arctic Bay Formation. Trough crossbedding in the lower

DEPOSITIONAL AND HIATAL EPISODES	DEPOSITIONAL ENVIRONMENT	FACIES NO.	SYMBOL	DESCRIPTION OF LITHOLOGY
STRATHCONA SOUND – ATHOLE POINT DEPOSITIONAL EPISODE	SUBTIDAL	24		SILTSTONE, FINE SANDSTONE AND INTERBEDS OF DENSE FINE CARBONATE UNITS
	FLUVIATILE	23		SANDSTONE
	FLUVIATILE	22		INTERBEDDED SANDSTONE AND CARBONATE CONGLOMERATE
	PALUDAL	21		BLACK FINE CLASTICS
HIATAL EPISODE	1) minor, but locally pronounced uplift 2) localized karsting of algal bioherms and formation of felsenmeers and carbonate boulder conglomerate			
VICTOR BAY DEPOSITIONAL EPISODE	EROSION PRODUCT	20		CARBONATE BOULDER CONGLOMERATE
	SUBTIDAL	19		ALGAL BIOHERM
	SHALLOW SUBTIDAL	18		GREY FLAT PEBBLE CONGLOMERATE
	SHALLOW SUBTIDAL	17		BLACK FINE CLASTICS AND FLAT PEBBLE CONGLOMERATE, REWORKED FACIES 16
	SUBTIDAL	16		BLACK FINE CLASTICS WITH OCCASIONAL DENSE CARBONATE INTERBEDS
	SUBTIDAL	15		THIN BEDDED CALCITIC CARBONATE
	INTERTIDAL TO SUBTIDAL	14		BLACK FINE CLASTICS
HIATAL EPISODE	1) regional uplift 2) extensive and selective karsting to the west 3) development of collapse breccia 4) minor subsidence 5) dolomitization, base metal mineralization and cementation of breccia 6) minor uplift and erosion			
SOCIETY CLIFFS DEPOSITIONAL EPISODE	SHALLOW SUBTIDAL TO SUPRATIDAL	13		VARICOLORED SILTSTONE AND SANDSTONE WITH EVAPORITES
	SHALLOW SUBTIDAL	12		GREY SANDSTONE, MOSTLY QUARTZOSE, LOCALLY ARKOSIC
	SHALLOW SUBTIDAL	11		SAME AS FACIES 10, BUT WITH SANDY MATRIX
	SHALLOW SUBTIDAL	10		GREY FLAT PEBBLE CONGLOMERATE, A REWORKED THIN BEDDED DOLOLUTITE
	SUBTIDAL	9		DOLOLUTITE WITH WHITE DOLOMITE FRAGMENTS
	SUBTIDAL	8		DOLOLUTITE
	SUBTIDAL	7		ALGAL BIOHERMS
	SHALLOW SUBTIDAL	6		FLAT PEBBLE CONGLOMERATE, REWORKED FACIES 4
	SUBTIDAL	5		REGULARLY SPACED ALGAL LAMINAE WITH INTERBEDDED DOLOLUTITE
	SUBTIDAL	4		REGULARLY SPACED ALGAL LAMINAE
	INTERTIDAL	3		NODULAR, IRREGULARLY SPACED ALGAL LAMINAE, DOLOMITIC
HIATAL EPISODE	1) minor uplift 2) erosion increasing to the west			
ARCTIC BAY DEPOSITIONAL EPISODE	SUBTIDAL TO INTERTIDAL	2		BLACK FINE CLASTICS WITH NODULAR ALGAL CARBONATE UNITS & LOCAL SANDSTONE LENSES
	SUBTIDAL	1		BLACK FINE CLASTICS WITH OCCASIONAL DENSE CARBONATE INTERBEDS

Fig. 2

118

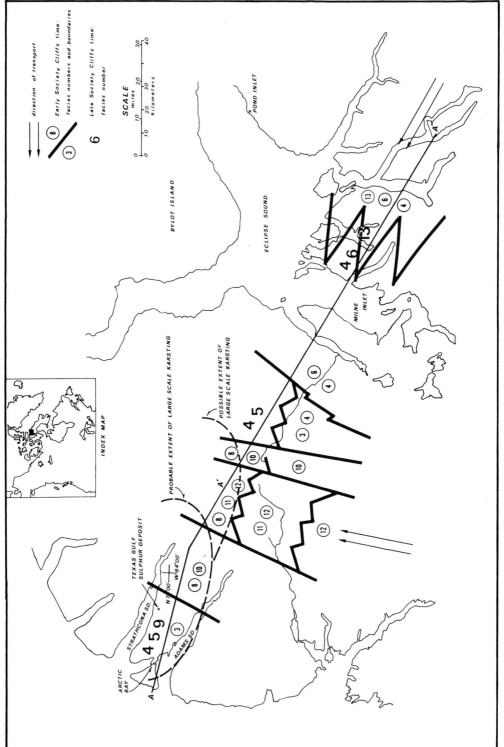

Fig. 3

orthoquartzites suggests a fluviatile environment and a north-westerly direction of transport. The fine-grained clastics are fissile, gypsiferous, frequently mudcracked, marked by a high organic content and disseminated specks of pyrite, and occasionally contain beds of stromatolitic dolomites. The sediments were most likely deposited in a very shallow marine environment.

Wherever the uppermost Arctic Bay Formation was preserved below the unconformity at the base of the overlying Society Cliffs Formation, it is characterized by a considerable increase of stromatolitic dolomites and associated sandstone lenses (fig. 4). This lithologic change reflects a gradually decreasing influx of clastic material from the southeast which appears to be due to a change of the earlier north-westerly gradient to an easterly gradient during Late Arctic Bay time.

During the Arctic Bay depositional episode the area was regionally blanketed by a succession of sediments, the bulk of which was dense and impermeable, forming an effective barrier to the flow of potentially mineralizing solutions.

The post-Arctic Bay hiatal episode

The Arctic Bay depositional episode was terminated by a minor, but regional uplift. The easterly gradient already established during Late Arctic Bay time was reflected by a more pronounced uplift of the western region causing the nearly complete erosion of the strongly dolomitic facies of the Upper Arctic Bay Formation to the west (fig. 4). Thus, the massive dolomite of the overlying Society Cliffs Formation rests directly on black fine-grained clastics over most of the western region. Large shale fragments of the underlying fine-grained clastics are incorporated in the conglomerate at the base of the Society Cliffs Formation indicating that the shale had already been compacted and at least partially lithified during the post-Arctic Bay hiatal episode.

This observation and the lack of mineralization in the basal conglomerate or in the directly overlying dolomites of the Society Cliffs Formation preclude the possibility whereby the metals of the massive sulfides could have been transported upward by connate water discharging from the fine-grained clastics below during compaction, a model favored by JACKSON and BEALES (1967).

The Society Cliffs depositional episode

This depositional episode is discussed in more detail in order to illustrate the effect of lithology and facies distribution on the subsequent mineralization with massive sulfides.

Initial deposition above the erosional surface is characterized by a wide range of lithofacies (figs. 3 and 4) which developed in response to changing rates of subsidence. The easterly gradient which left its imprint during the preceding hiatal episode by a more pronounced uplift and erosion of the western region, remained effective during the Society Cliffs depositional episode as well. Thus, a higher rate of subsidence rapidly established subtidal conditions to the east reflected by a finely laminated, stromatolitic dolomite. The regular spacing and the lateral continuity of the laminae suggest relatively quiet subtidal conditions. Occasional biohermal structures, the laminae of which display the same spacing as the laterally continuous laminae, leave no doubt about the algal origin of the laminated facies. The laminae at the base of the Society Cliffs Formation are locally brecciated, probably indicative of stormy periods during which shallower patches of the subtidal algal mats were reworked.

The western region experienced a much slower rate of subsidence and is characterized by a nodular carbonate which is internally distinguished by irregularly spaced, stromatolitic laminae. The sediment is of typically intertidal origin and strikingly

120

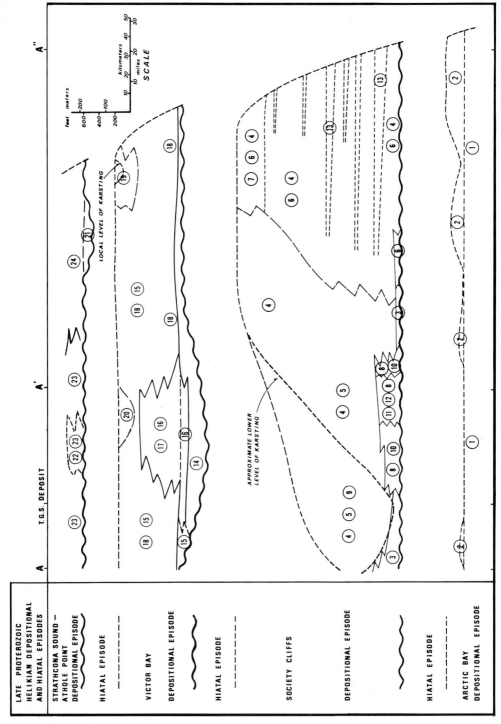

Fig. 4

resembles recent sediments below the Sabkha of the Persian Gulf region (EVANS et al. , 1969). Moulds of gypsum crystals in the core of some nodules and locally developed gypsiferous coatings indicate the originally gypsiferous nature of the now dolomitized nodules.

The intertidal nodular carbonate and the subtidal regularly spaced laminite are separated by a northerly trending belt of sediments which are characterized by sandy flat-pebble conglomerates and coarse-grained, crossbedded, quartzose and arkosic sandstones along the axis of the belt, and by flat-pebble conglomerates and featureless dololutites along the marginal areas of the belt. The distribution, trend and lithic aspect of these facies suggest a northerly dispersal of mostly allochthonous material by longshore currents which paralleled a western shoreline. Space limitations preclude a detailed description of the lithofacies and their spatial relationship. The reader is referred to figures 3 and 4 for further details on the vertical and lateral facies distribution.

The diversity of facies during Early Society Cliffs time gives way to a general predominance of regularly spaced algal laminae reflecting regional subtidal conditions throughout Late Society Cliffs time. This general pattern is interrupted in the easternmost region (figs. 3 and 4) by repeated influxes of clastics from an easterly source. The westward projecting tongues of varicolored crossbedded siltstones and sandstones indicate occasional shallowing reflected by lagoonal and even subaerial deposits.

At the close of the Society Cliffs depositional episode a largely biogenic carbonate sequence, exceeding 2000 feet in thickness to the east, had been deposited across the entire area of investigation. The predominant lithology was an algal carbonate consisting of laminae spaced at regular intervals of about 0. 05 inches. The original carbonate mineral was most likely aragonite or high-magnesium calcite, both unstable and readily attacked under conditions of prolonged subaerial exposure.

The post-Society Cliffs hiatal episode

The carbonate sequence of the Society Cliffs Formation is separated from the overlying Victor Bay Formation by a pronounced regional unconformity. The intervening hiatal episode is reflected by diagenetic changes, mineralization, and erosional fragmentation of the Society Cliffs Formation. A summary of the most critical field observations follows:

1) The laminated facies of the Society Cliffs Formation is almost entirely brecciated in the western region. The depth of brecciation decreases continuously to the east, and the laminated facies is virtually unbrecciated beyond a zone indicated on figures 3 and 4 by the extent and/or level of karsting.

2) Brecciation did not affect the basal nodular carbonate, the flat-pebble conglomerates, the clastic units, and - significantly - the carbonate of the Victor Bay Formation which rests directly on brecciated dolomite in the extreme western area (fig. 4).

3) The brecciated dolomite was thoroughly cemented prior to the deposition of the Victor Bay Formation, because the basal black fine-grained clastics of the Victor Bay Formation have nowhere filtered into the breccia below.

4) Both the breccia and the cement are thoroughly dolomitized whereas the basal carbonates of the Victor Bay Formation directly above are calcareous.

5) Cemented fractures in the breccia contain only specks of sphalerite, galena, and pyrite, whereas channel-like structures at the base of the brecciated section are filles with banded massive sulfides. No massive sulfides or disseminated sphalerite and galena were observed in the formations overlying and underlying the Society Cliffs Formation.

6) The gabbroic dikes contain traces of chalcopyrite, but no sphalerite or galena. Where the dikes cut through the mineralized breccia, the dikes are bordered on both sides by a barren zone which was apparently demineralized during the time of intrusion.
7) The contact between the Society Cliffs Formation and the Victor Bay Formation is always sharp and is mostly documented by a distinct facies change from brecciated or unbrecciated laminite below into black fine-grained clastics above.
8) The overall thickness of the Society Cliffs Formation decreases in a westerly direction.

The above field criteria restrict the time of mineralization to a time interval which postdates brecciation and predates the deposition of the basal limestones and black fine-grained clastics of the Victor Bay Formation.

Discussion

The westerly thinning of the Society Cliffs Formation below the unconformity that separates the algal laminite from the sediments of the Victor Bay Formation, implies that uplift and erosional fragmentation during the post-Society Cliffs hiatal episode was more pronounced in the west reflecting the continued effect of an easterly gradient. The regional and differential uplift above baselevel must have led to an extensive karst development in the topographically higher algal limestones to the west. The corresponding baselevel of karst erosion established of course an easterly gradient. With increasing uplift the groundwater level reached deeper and deeper horizons in the limestone as long as the limestone yielded to the attack by the slightly acidic groundwater. The regularly spaced algal laminite was apparently quite unstable in the vadose zone because this lithofacies was preferentially karsted whereas other lithologies such as the nodular carbonates or the flat-pebble conglomerates were not affected. The instability of the laminite must have been due to an unstable mineralogical composition consisting most likely of aragonite or high-magnesium calcite. Both forms have been shown to be unstable in the vadose zone (FRIEDMAN, 1964; GAVISH and FRIEDMAN, 1969).

The depth of karst erosion was therefore controlled in the west by the thickness of the laminated facies and in the east by the amount of uplift above baselevel (fig. 4). The continued attack by percolating groundwater severely fractured the limestone and ultimately caused extensive collapse and brecciation of the laminated limestone facies. Resistant and impermeable lithologies as well as an interruption of the regional uplift stopped the downward cutting of the lowest level of karst erosion and led to the development of a subsurface channel system such as the one at the base of the brecciated section in the Society Cliffs Formation. Karst erosion considerably increased the porosity and - most important - the permeability of the carbonate sequence.

Continued emergence and karst erosion would have led sooner or later to the complete removal of the laminated facies. The preservation of the breccia indicates that renewed subsidence once again initiated sub-baselevel conditions. The regional distribution of brecciated laminite suggests that the sea readvanced into an area of considerable topographic relief which had been carved out during the period of karst erosion. Any temporary uplifting pulse or eustatic drop of sealevel during this new period of subsidence would have created ideal conditions for the establishment of a restrictive, i.e. evaporative depositional environment. A large number of such uplifting pulses have been documented for the Paleozoic of North America by HAM and

WILSON (1967), and the effect of eustatic sealevel changes upon the generation of evaporative brines has been demonstrated by CRAIG (1969) for the Red Sea brines.

Thus, assuming that slight sealevel changes occurred during the transgression, more or less evaporative conditions were established during times when topographic barriers interrupted or severely restricted the inflow of seawater into shallow marine embayments. Evaporation of seawater increased the salinity and led to the precipitation of aragonite and gypsum. The remaining brine percolated downward into the breccia due to its higher density and, because of its high Mg/Ca ratio, dolomitized the breccia and any existing calcitic cement (fig. 5). The replaced calcium ions combined with free sulfate ions to form additional gypsum. The brine reached the lowest level of karst erosion and followed the easterly gradient of the channel system.

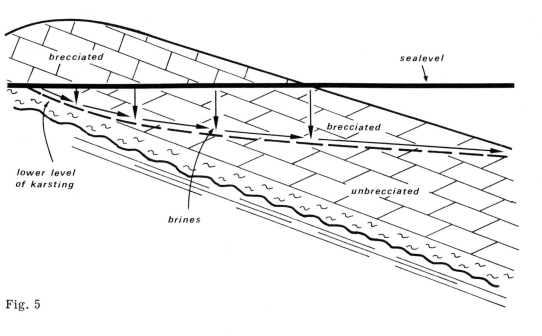

Fig. 5

It is reasonable to assume that the brine was rich in alkalies and chlorine, and poor in magnesium and sulfate ions which were largely lost during dolomitization and gypsum precipitation. The brine must have furthermore contained base metal ions which were probably enriched over the normal concentration in seawater during evaporation at the surface. The base metal ions were probably transported as chloride complexes. The brine most likely carried a normal supply of organic matter and a normal population of bacteria, including sulfate-reducing bacteria. The flow rate of the brine was probably so slow that the oxygen supply was already exhausted in upper levels. The continued influx of organic matter and the availability of gypsum created favorable conditions for sulfate-reducing bacterial activity. This resulted in a small, but continuous generation of sulfide ions which combined with the chloride complexes of zinc, lead, and iron to precipitate sphalerite, galena, and pyrite.

The bulk of the brine must have flowed through the channel system at the base of the brecciated section. Thus, bacterial activity and sulfide generation were most pronounced in the channelways which were gradually filled with massive sulfides. The sulfides are banded and very coarsely crystalline, both good criteria for open void filling. The last spaces just below the roof of cavities were filled by extremely coarsely crystalline dolomite. This suggests that flow conditions were virtually ter- minated when the channelways were blocked by the accumulation of massive sulfides. A lack of nutrient supply discontinued bacterial activity, and the remaining void spa was slowly infilled by dolomite. The paucity of sulfides in fractures may similarly be explained by an inadequate supply of nutrients due to very slow flow conditions which rendered the environment inhospitable for bacterial activity.

It could be argued that the supply of base metals was leached out of the algal laminit fragments by the dolomitizing brine. This would imply, however, that no such leach ing took place during the preceding period of karst erosion. It furthermore would in ply that any epigenetic dolomite is associated with an accumulation of base metals, an association which is certainly not a common geologic feature.

A seawater origin of the base metals is supported by a plot of atomic ratios of coppe lead, and zinc from 114 lead- and/or zinc-bearing deposits (fig. 6) (SANGSTER, 196 The crustal abundance ratio lies considerably outside the fields of major frequencies whereas the seawater ratio (WEDEPOHL, 1969) is surprisingly close to the field of zinc-rich deposits.

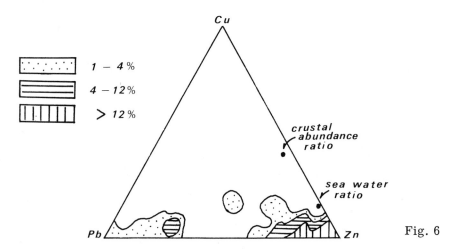

Fig. 6

Conclusions

The preceding discussion illustrates that the formation of the massive sulfides is based upon three fundamental requirements. These are
1) a continuous supply of brine produced by the evaporation of seawater;
2) a proper drainage system;
3) an excellent permeability.

The brine, because of its higher density, establishes the necessary flow condition. The brine must be generated continuously to secure a constant supply of organic nu- trients and sulfate ions for sulfate-reducing bacterial activity.

A proper drainage system must exist to maintain a continuous flow of the brine. A lack of drainage would quickly establish stagnant conditions and terminate bacterial activity due to a rapid depletion of organic nutrients and sulfate ions.

Fig. 7. Regularly spaced algal laminae in the Society Cliffs Formation interrupted by a narrow zone of brecciation

Fig. 8. Nodular dolomite of intertidal origin at the base of the Society Cliffs Formation. Note the person for scale

The permeability must be excellent because a relatively free circulation of the brine is necessary if a large amount of massive sulfides is to be accumulated. The large number of variables in the overall system precludes an exact calculation of the time necessary to create a deposit of economic potential.

The required conditions existed during the post-Society Cliffs hiatal episode on north ern Baffin Island and led to the formation of a Mississippi Valley-type deposit. It is likely that other Mississippi Valley-type deposits share a similar origin.

The concept of a seawater-derived brine as a transporting medium for base metals has been theorized and demonstrated in several publications (see BROWN, 1970, for a comprehensive review). The concept is modified in this report in as much as normal seawater is shown to be the source of the base metals and of the sulfur. The transportation and precipitation of the base metals as well as the generation of sulfide ions and of the brine itself are all part of an uninterrupted process.

Fig. 9. Typical dolomitized breccia in the Society Cliffs Formation. The photo shows a horizontal section of the erosional surface which is overlain at this locality by unbrecciated limestone of the Victor Bay Formation

References

BLACKADAR, R. G.: Precambrian geology, northwestern Baffin Island, District Franklin. Geol. Surv. Canada, Bull. 191, 89 p. (1970).

BROWN, J. S.: Mississippi Valley-type lead-zinc ores. Mineral. Deposita 5, p. 103 -119 (1970).

CRAIG, H. : Geochemistry and origin of the Red Sea brines. in: Hot Brines and Re-
    cent heavy metal deposits in the Red Sea (Degens, E. T. and Ross, D. A. , ed.).
    Springer, New York, p. 208-242 (1969).

EVANS, G. , SCHMIDT, V. , BUSH, P. , NELSON, H. : Stratigraphy and geologic
    history of the Sabkha, Abu Dhabi, Persian Gulf. Sedimentology 12, p. 145-159
    (1969).

FRIEDMAN, G. M. : Diagenesis and lithification in carbonate sediments. Jour. Sed.
    Petrol. 34, p. 777-813 (1964).

GAVISH, E. , FRIEDMAN, G. M. : Progressive diagenesis in Quaternary to Late
    Tertiary carbonate sediments: Sequence and time scale. Jour. Sed. Petrol.
    39, p. 980-1006 (1969).

HAM, W. E. , WILSON, J. L. : Paleozoic epeirogeny and orogeny in the central United
    States. Amer. Jour. Sci. 265, p. 332-407 (1967).

JACKSON, S. A. , BEALES, F. W. : An aspect of sedimentary basin evolution: The
    concentration of Mississippi Valley-type ores during late stages of diagenesis.
    Bulletin Can. Petrol. Geol. 15, p. 384-433 (1967).

LEMON, R. R. H. , BLACKADAR, R. G. : Admiralty Inlet area, Baffin Island, Dis-
    trict of Franklin. Geol. Surv. Canada, Memoir 328, 84 p. (1963).

SANGSTER, D. F. : Some chemical features of lead-zinc deposits in carbonate rocks.
    Geol. Surv. Canada, Paper 68-39, 17 p. (1968).

WEDEPOHL, K. H. , ed. : Handbook of geochemistry, Volume I. Springer, Berlin,
    442 p. (1969).

Address of the author:

Department of Geology,
University of Port Elizabeth
Port Elizabeth / South Africa

Deposition of Manganese and Iron Carbonates and Silicates in Liassic Marls of the Northern Limestone Alps (Kalkalpen)

Klaus Germann

Abstract

Considerable manganese occurrences in the Northern Limestone Alps are restricted to lower and middle Jurassic marls, red limestones and radiolarian cherts. In the red limestones and cherts only minor contents of manganese are concentrated as oxides, forming in the red limestones carbonate-rich manganese nodules, texturally and geochemically comparable to some Recent shallow marine accumulations.

By far the largest quantities of Jurassic manganese are contained in a horizon of thinly laminated upper Liassic marls and shales ("manganese shales") dispersed over more than 250 km distance, from the Allgäu and Lechtal Alps to the Berchtesgaden and Salzburg regions.

Apart from some manganese oxides of a thin weathering crust (e.g. pyrolusite, todorokite) carbonates of the system $CaCO_3$-$MnCO_3$-$FeCO_3$-$MgCO_3$ (Ca-rhodochrosites, and oligonite), and the silicate braunite are the prevailing manganese minerals. They are supposed to be primary precipitates and rarely replacement products of calcitic skeletons. The manganese minerals are characteristically associated with sedimentary iron minerals showing the vertical sequence carbonate - silicate (chamosite) - sulfide, thus demonstrating a lithofacies sequence leading from poorly oxigenated sediments to anaerobic black shale deposits.

A Mn/Fe ratio close to 2 and elevated contents of Cu, Co (up to 0.04 %) and Zn, besides the lack of oolitic fabrics mark the most significant differences between Alpine Jurassic ores and the normal types of sedimentary manganese and iron deposits.

Chemical composition and mineralogy (the occurrence of braunite is known from volcanogenic or metamorphic deposits only) make appear volcanogenic solutions with high contents of Mn and Fe to be the most probable source of manganese and iron in this type of mixed Mn-Fe-deposits. For the first time, in the Northern Limestone Alps volcanic activity during manganese deposition could be proved by layers of celadonite. Celadonite, confirmed by infrared spectroscopy is supposed to represent altered tuff material intercalated in the manganese carbonate ores.

Introduction

The deposition of manganese minerals (oxides, carbonates) in sedimentary sequences, although taking place on the Recent sea floors, is still being discussed regarding fossil ore deposits. Among the major controversial problems are the sources of manganese which either should be continental, hydrothermal-volcanogenic or diagenetic.

The origin of manganese in stratified deposits of oxides and carbonates (HEWETT, 1966) which are supplying the bulk of the world manganese production is of major economic interest: knowledge of the conditions of supply and deposition may serve

to predict distribution and size of economically valuable deposits in fossil sediments
The purpose of this article is to briefly review the mineralogical and chemical characteristics of some Alpine stratified manganese occurrences with respect to their
genetic implications. In particular, influences of previously (CORNELIUS and PLÖ-
CHINGER, 1952; GRUSS, 1958) postulated volcanic activities on the deposition of
these manganese accumulations must be discussed.

Geological Setting

In the Northern Limestone Alps several types of stratified manganese deposits hithe:
to incompletely studied were investigated to clear up their mineralogy, geochemistr
and environment of deposition.
Considerable manganese occurrences in the Northern Limestone Alps are restricted
to lower and middle Jurassic marls, red limestones and radiolarian cherts (Fig. 1).
On the whole, manganese bearing sediments of different facies in the studied area
have been concentrated during a rather short Jurassic time span, their horizontal
distribution, however, being enormous.

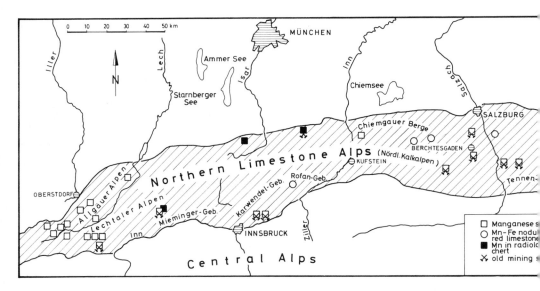

Fig. 1. Location of Jurassic manganese occurrences studied in the Northern Limestone Alps

In the condensed red limestones, and in the radiolarian cherts only minor contents
of manganese are concentrated as oxides.
The ferromanganese nodules and crusts in the red limestones are geochemically and
texturally comparable to some Recent marine accumulations (GERMANN, 1971). In
the fossil nodules the contents of Fe, Mn, and some trace elements as Ni, Co, and
Cu on account of a carbonate content of 60 % are on an average half as high, their
relative abundance, however, being equal to some Recent shallow water occurrences
Close genetic relations of nodule formation to submarine volcanism as demonstrated
by JENKYNS(1970) from Toarcian occurrences of Sicily until now could not be established in the case of the Alpine Jurassic nodules.

131

By far the largest quantities of Jurassic manganese are contained in a horizon of tinly laminated Upper Liassic (Lower Toarcian) marls and shales. The manganese shale ("Manganschiefer") facies of the Northern Limestone Alps is developed over more than 250 km distance from the Allgäu and Lechtal Alps to the Berchtesgaden and Salzburg Mts. (Fig. 1); some comparable deposits are known from the Liassic of Hungaria (DRUBINA-SZÁBO, 1959).

Only presumably workable deposits near Berchtesgaden and Salzburg have been studied in some detail previously (CORNELIUS and PLÖCHINGER, 1952; GRUSS, 1958; GUDDEN, 1969), and a summary on Austrian occurrences has been given by LECHNER and PLÖCHINGER (1956). A comprehensive study on facies, mineralogy and geochemistry of the deposits in the Northern Limestone Alps is given by GERMANN and WALDVOGEL (1971), and GERMANN (1972).

Lithofacies

The manganese shales with a thickness up to 50 meters are composed of a series of dark coloured silty marls and shales (Fig. 2). Predominant rock types are thinly laminated, partly bioclastic siliceous marls with an average carbonate content of

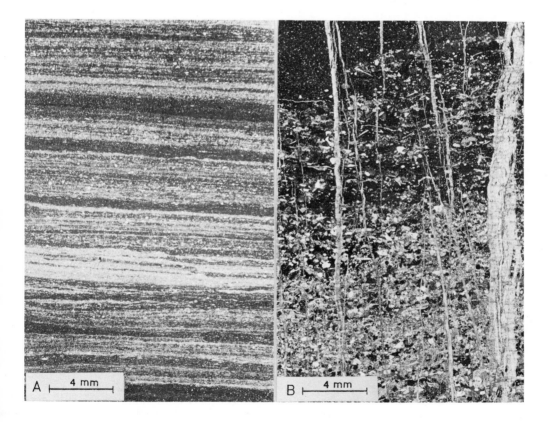

Fig. 2. Ca-rhodochrosite bearing marls of the manganese shale series; thin sections.
A - Laminated silty marl; 17.3 % Mn; Berchtesgaden Alps
B - Echinoidal marl; 10.3 % Mn; Allgäu Alps

50 per cent, their organic carbon content averages at 1 per cent. The varve-like lamination in a millimeter scale is due to the alternation of layers with varying portions of clay or silt material, bituminous substances and pyrite, presumably produced by seasonal sedimentation processes, and absence of burrowing organisms in stagnant waters. The relatively low content of organic carbon, and intercalations of sand and silt layers and bioarenites, however, are indicating conditions of sedimentation not invariably stagnant, presumably produced in shallow shelf seas. Comparable laminated marls and shales, being an ubiquitous facies type of the Lower Toarcian in central and western Europe, after BITTERLI (1963), HALLAM (1967), and KREBS (1969) likewise should have been laid down in relatively shallow water areas.

The manganese shale series is characterized by a lithofacies sequence leading from poorly oxigenated, partly siliceous limestones and marls at the base to sulfide-rich black shales on the top. Both mineralogy and the amount of manganese and iron accumulation are influenced by this facies development.

Mineralogy

Apart from the manganese oxides of a thin weathering crust, among which are pyrolusite and todorokite, only carbonates of the system $CaCO_3$-$MnCO_3$-$FeCO_3$-$MgCO_3$, and small amounts of the silicate braunite have been detected by X-ray diffraction and ore microscopy.
From the frequency distribution of the 104-reflections of the <u>Mn-carbonates</u>, a serie of solid solutions between $CaCO_3$ and $MnCO_3$ with the most frequent members $Ca_{40}Mn$ and $Ca_{10}Mn_{90}$ can be inferred. In no case the contents of $FeCO_3$ and $MgCO_3$ in the Ca-rhodochrosites exceed 15 %. In this series there are gaps near the position of the ordered phases kutnahorite ($Ca_{50}Mn_{50}$) and rhodochrosite which is in contrast to the experimental findings of GOLDSMITH and GRAF (1957) who described a miscibility gap between the $Ca_{50}Mn_{50}$ and the $Ca_{25}Mn_{75}$ carbonate. With increasing manganese content of the samples the carbonates approach to the rhodochrosite composition (Fig. 3), thus demonstrating that the carbonate composition is a function of the total manganese available during carbonate formation.

According to their prevailing crystal sizes in micron dimensions, and the preservation of delicate depositional fabrics (e. g. the millimeter lamination) most of the carbonates seem to be at least early diagenetic or primary precipitates. A coarse-grained variety of manganese carbonates, first described by GRUSS (1958) from the Berchtesgaden Mts. , is exclusively bound to biogenic fragments, thus being a true replacement product. Echinoid fragments entirely or partly consisting of manganese carbonates may show a zonal distribution of calcite and different types of Ca-rhodochrosites resulting from incomplete substitution of calcite by rhodochrosite. The degree of calcite replacement, too, seems to be controlled by the availability of manganese in the ore-forming solutions. Similar conditions of syngenetic replacement of detrital calcium carbonate are suggested by BISCHOFF (1969) for the manganese carbonates in the Red Sea geothermal brine deposits. - Whereas syngenetic precipitation of manganese carbonates in reducing environments is demonstrated on the floor of the oceans (compare HARTMANN, 1964; LYNN and BONATTI, 1965), manganese silicates are lacking in Recent sedimentary accumulations. On the whole, manganese silicates seem of little importance in sedimentary deposits (BORCHERT, 1970), whereas they are formed in large quantities during metamorphism of manganese bearing siliceous sediments. In the Alpine Liassic the silicate <u>braunite</u> ($Mn_7O_8[SiO_4]$) has been detected by X-ray diffraction and ore microscopy in some non-metamorphic deposits of the

Berchtesgaden region, containing hematite and rhodochrosite. Similar to the carbonates, micron-sized braunite crystals are arranged in millimeter laminae. Possibly this type of syngenetic braunite is directly related to warm or hot waters (PARK and MacDIARMID, 1964) with elevated oxidation potential as supposed for the formation of some volcanogenic-sedimentary manganese deposits, in which according to VINCIENNE (1956) and HEWETT (1966) braunite is a common mineral.

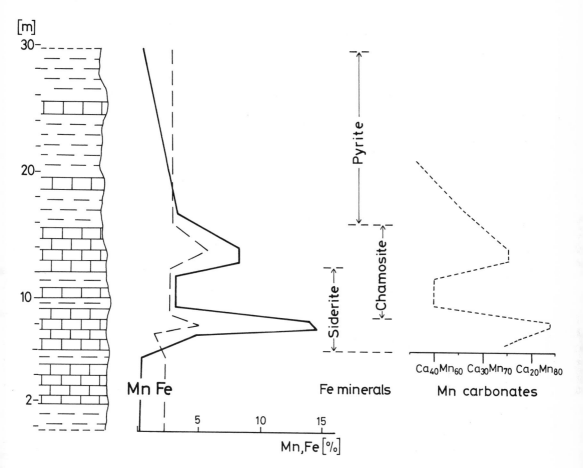

Fig. 3. Idealized section of a typical manganese shale profile of the Allgäu Alps, showing the vertical distribution of metal contents, and prevailing Mn and Fe minerals

The manganese minerals are characteristically associated with a series of iron minerals which, in following the lithologic sequence, from the basis to the top of the manganese shales are arranged in the order (oxide, hydroxide)-carbonate-silicate-sulfide (Fig. 3). Both the carbonate which is a central member ($Fe_{50}Mn_{50}$) of the series rhodochrosite - siderite, and the silicate chamosite are mainly microcrystalline, and demonstrate unaltered depositional fabrics (e.g. detrital chamosite), thus indicating a syngenetic mode of deposition. Oolitic fabrics, known from Recent and fossil sedimentary iron deposits are lacking completely. The sulfides, mainly pyrite and rarely marcasite, are abundant in the manganese-poor black shale facies, where laminated ores of pyrite framboids together with minor contents of chalcopyrite are occurring. Mainly in hematite-goethite bearing ores small amounts of chalcopyrite, sphalerite, and galena were detected in the manganese shales series.

Geochemistry

The manganese contents with an average of 9.9 % are highest in the (hematite) - siderite - chamosite bearing lower parts of the series, where acid insoluble material is less important. With increasing contents of insoluble material the manganese decreases gradually towards the top of the series, and a chamosite - pyrite paragenesi or pyrite alone is prevailing (Fig. 3).

In contrast to the most frequent types of sedimentary iron and manganese deposits the ores of the manganese shales exhibit an average Mn/Fe ratio close to 2 irrespec tive of the prevailing mineral paragenesis. The common separation of Fe and Mn (KRAUSKOPF, 1957) did not take place, the contents of both metals, on the contrary, being correlated (r = 0.21 for 285 samples).

Compared to marine-sedimentary iron and manganese deposits the Alpine ores,besides their high contents of Ca, are characterized by elevated contents of silica. The average composition of manganese-rich Ca-rhodochrosite marls is given in Table 1.

Table 1 - Average composition (%) of Mn-rich marls (10 analyses)

SiO_2	25.67	P_2O_5	0.64
Al_2O_3	3.66	SO_3	0.71
Fe_2O_3 (total Fe)	8.78	CO_2	22.25
MnO	19.34	C_{org}	0.94
CaO	10.26	H_2O	3.28
MgO	3.30	TiO_2	0.19
Na_2O	0.06		
K_2O	0.89		

In constrast to the average composition of comparable black shales (VINE and TOUR-TELOT, 1970), and normal marine Mn and Fe deposits, the trace element contents (Table 2), partly correlated with manganese (Fig. 4), are raised as well.

Table 2 - Average elemental composition (Mn, Fe, Ca, Mg: %; Cu, Ni, Co, Cr, Zn: ppm) and standard deviations (285 analyses by AAS)

	Mn	Fe	Ca	Mg	Sr	Cu	Ni	Co	Cr	Zn
$\bar{x}$	9.9	4.2	10.4	1.6	380	45	82	48	31	53
s	9.9	3.7	8.5	0.86	360	22	39	45	10	27

135

Conclusions

Coexisting Mn and Fe, and elevated trace element contents are hints to a non-detrital, local source of the metals. Considering chemical composition and mineralogy, volcanogenic solutions with high metal contents appear to be the most probable source of Mn and Fe in this type of mixed Mn-Fe deposits, a genetic interpretation which first was given by CORNELIUS and PLÖCHINGER (1952), and GRUSS (1958) for the Berchtesgaden and Salzburg deposits. Marks of Toarcian volcanic activity, however, until now have not been detected in the Northern Limestone Alps. At two shale occurrences (Southern Karwendel Mts. and Berchtesgaden Mts.) which are characterized by average manganese contents higher than 20 %, layers of a greenish celadonite-rhodochrosite rock were found recently. The celadonite, as confirmed by chemical analyses and IR-spectroscopy, is supposed to represent altered tuff material intercalated in the manganese carbonate ores.

In the tuff-bearing manganese ore series which locally reaches a thickness of 3 meters, baryte concretions and high values of cobalt can be conceived as additional signs of volcanic influences. Ni/Co ratios below 0.5 and Mn/Co ratios near 300 in some of these ores should be indicative of volcanogenic origin (DAVIDSON, 1962; ARRHENIUS et al., 1964). Irrespective of the volcanogenic accumulations, Mn, Fe, and Co in general are correlated (Fig. 4). Beyond this, the rhodochrosite - baryte paragenesis, according to BORCHERT (1970), provides important clues to their probable derivation from residual magmatic solutions.

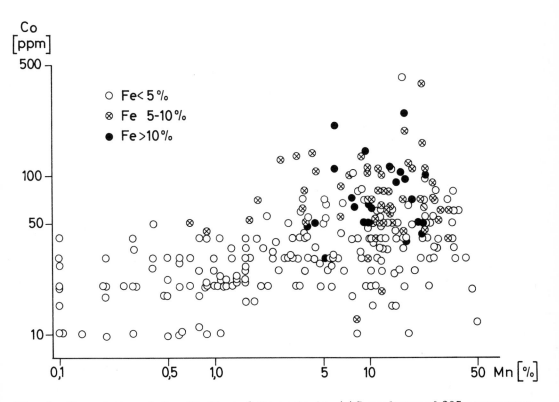

Fig. 4. Correlation of Co with Mn and Fe contents; AAS-analyses of 285 manganese shale samples

Indications of volcanogenic-hydrothermal supply are mostly restricted to mineralogy and chemistry of the ores which additionally are influenced by the particular sedimentary environment. Remains of volcanic sediments being rare exceptions, the ore characteristically are intercalated in sedimentary series without significant changes in facies which should have been developed, however, in the case of terrigenous supply. The strata-bound Jurassic manganese concentrations of the Northern Limestone Alps therefore must be classed with volcanogenic-sedimentary deposits accumulated in temporal or spatial distance from the active centers of volcanism (remote volcanogenic-siliceous group of SHATSKIJ, 1964). Fe, Mn, Ba, and Si are supposed to have been introduced locally into the sea water by volcanic emanations. Similar Recent deposits of Mn and Fe with high contents of Si and Ba precipitated from submarine hydrothermal solutions are known e. g. from the East Pacific Rise (BOSTROM and PETERSON, 1966), and from volcanic islands in the Mediterranean Sea (BONAT et al. , 1971).

Apart from high metal contents of the sea water, enriched by local hydrothermal activities, the development of this Jurassic manganese-iron type of volcanogenic-sedimentary deposits (SOKOLOVA, 1970; SUSLOV, 1970) is favoured by the low oxidation potential of the stagnant sedimentary environment, excluding the fractionation of manganese and iron as oxides. The total amount of metals accumulated indeed being exceedingly high, a great number of local hydrothermal sources on the one hand, and the far reaching distribution of the metals under the prevailing reducing conditions on the other hand, prevented the Liassic manganese and iron occurrences of the Northern Limestone Alps from attaining magnitudes of economic value.

References

ARRHENIUS, G. , MERO, J. , KORKISCH, J. : Origin of oceanic manganese minerals Science 144, p. 170-173 (1964).

BISCHOFF, J. L. : Red Sea geothermal brine deposits: their mineralogy, chemistry and genesis. In: Hot brines and Recent heavy metal deposits in the Red Sea, E. T. DEGENS and D. A. ROSS, Eds. Springer, Berlin-Heidelberg-New York (1969).

BITTERLI, P. : Aspects of the genesis of bituminous rock sequences. Geol. en Mijnbouw 42, p. 183-201 (1963).

BONATTI, E. , HONNOREZ, J. , JOENSUU, O. : Submarine iron deposits from the Mediterranean Sea. Paper presented at the VIII International Sedimentological Congress 1971, Heidelberg (1971).

BORCHERT, H. : On the ore-deposition and geochemistry of manganese. Mineral. Deposita (Berl.) 5, p. 300-314 (1970).

BOSTROM, K. , PETERSON, M. N. A. : Precipitates from hydrothermal exhalations on the East Pacific Rise. Econ. Geol. 61, p. 1258-1265 (1966).

CORNELIUS, H. P. , PLÖCHINGER, B. : Der Tennengebirgs-N-Rand mit seinen Manganerzen und die Berge im Bereich des Lammertales. Jb. geol. Bundesanstalt (Wien) 95, p. 145-225 (1952).

DAVIDSON, C. F. : On the cobalt:nickel ratio in ore deposits. Mining Mag. 106, p. 78-85 (1962).

DRUBINA-SZABO, M. : Manganese deposits of Hungary. Econ. Geol. 54, p. 1078-1094 (1959).

GERMANN, K. : Mangan-Eisen-führende Knollen und Krusten in jurassischen Rot-
kalken der Nördlichen Kalkalpen. N. Jb. Geol. Paläont. Mh. 1971, 3, p. 133-
156 (1971).

— Verbreitung und Entstehung Mangan-reicher Gesteine im Jura der Nördlichen
Kalkalpen. Tschermaks miner. petrogr. Mitt. 17, p. 123-150 (1972).

GERMANN, K. , WALDVOGEL, F. : Mineralparagenesen und Metallgehalte der
"Manganschiefer" (unteres Toarcian) in den Allgäu-Schichten der Allgäuer
und Lechtaler Alpen. N. Jb. Geol. Paläont. Abh. 139, p. 316-345 (1971).

GOLDSMITH, J. R. , GRAF, D. L. : The system CaO-MnO-CO$_2$: solid-solution and
decomposition relations. Geochim. Cosmoschim. Acta 11, p. 310-334 (1957).

GRUSS, H. : Exhalativ-sedimentäre Mangankarbonatlagerstätten mit besonderer Be-
rücksichtigung der liassischen Vorkommen in den Berchtesgadener und Salzburger Alpen. N. Jb. Miner. Abh. 92, p. 47-107 (1958).

GUDDEN, H. : Über Manganerzvorkommen in den Berchtesgadener und Salzburger
Alpen. Erzmetall 22, p. 482-488 (1969).

HALLAM, A. : The depth significance of shales with bituminous laminae. Marine
Geol. 5, p. 481-493 (1967).

HARTMANN, M. : Zur Geochemie von Mangan und Eisen in der Ostsee. Meyniana
14, p. 3-20 (1964).

HEWETT, D. F. : Stratified deposits of the oxides and carbonates of manganese. Econ.
Geol. 61, p. 431-461 (1966).

JENKYNS, H. C. : Fossil manganese nodules from the west Sicilian Jurassic. Eclogae
geol. Helvet. 63, p. 741-774 (1970).

KRAUSKOPF, K. B. : Separation of manganese from iron in sedimentary processes.
Geochim. Cosmochim. Acta 12, p. 61-84 (1957).

KREBS, W. : Über Schwarzschiefer und bituminöse Kalke im mitteleuropäischen
Variscikum. Erdöl u. Kohle 22, p. 2-6, 62-67 (1969).

LECHNER, K. , PLÖCHINGER, B. : Die Manganerzlagerstätten Österreichs. XX. Int.
Geol. Congr. Mexico, Symp. Yac. Manganeso 5, p. 299-313 (1956).

LYNN, D. C. , BONATTI, E. : Mobility of manganese in diagenesis of deep-sea sediments. Marine Geol. 3, p. 457-474 (1965).

PARK, C. F. , MACDIARMID, R. A. : Ore deposits. Freeman, San Francisco, 473 p.
(1964).

SHATSKIJ, N. S. : On manganiferous formations and the metallogeny of manganese.
I. Volcanic-sedimentary manganese formations. Intern. Geol. Rev. 6, p. 1030
-1056 (1964).

SOKOLOVA, E. : Laws governing the distribution of ore concentrations in manganese-
bearing volcanogenic-sedimentary formations. In: Manganese deposits of the
Soviet Union (D. G. SAPOZHNIKOV, Ed.), p. 58-75. Israel Progr. Sci. Transl.,
Jerusalem (1970).

SUSLOV, A. T. : Main features of volcanogenic-sedimentary Fe-Mn-deposits. In:
Manganese deposits of the Soviet Union (D. G. SAPOZHNIKOV, Ed.), p. 58-75.
Israel Progr. Sci. Transl. , Jerusalem (1970).

VINCIENNE, H. : Observations géologiques sur quelques gîtes marocaines de manganèse syngénétique. XX. Int. Geol. Congr. Mexico, Symp. Yac. Manganeso
2, p. 249-268 (1956).

VINE, J.D., TOURTELOT, E.B.: Geochemistry of black shale deposits - a summa report. Econ. Geol. 65, p. 253-272 (1970).

Address of the author:

Freie Universität Berlin
Institut für Angewandte Geologie
1000 Berlin 33 / Germany

Present Day Formation of an Exhalative Sulfide Deposit at Vulcano (Thyrrhenian Sea), Part II: Active Crystallization of Fumarolic Sulfides in the Volcanic Sediments of the Baia di Levante

Abstract

Vulcano is the southermost island of the volcanic Eolian Archipelago. Fumarolic activity is associated with "La Fossa", an active vulcano in the northern part of the island. The trachytic tephra from the last eruption (1890) and the older trachytic tuff, both of which form most of the shore and the bottom of one bay (Baia di Levante) are being mineralized by partly submarine fumaroles. The textures of the mineralized rocks have been examined by reflection microscopy. Pyrite and marcasite, plus minor amounts of sulfur, alunite and opal, cement the grains of the volcanic sand and tuff. Within individual volcaniclastic grains the original Fe-Ti oxides are replaced by iron sulfides, whereas the original silicate minerals (augite, plagioclase and glass) were opalized. Identification of the various phases have been checked by X-ray diffraction and microprobe analyses. Chemical analyses for the major, minor and trace elements in the total mineralized rocks show an enrichment of Fe, Si, S, B and Ba.

The distribution of fumarolic minerals in the surface sand and tuff of the bay and offshore sediments shows the following: 1) The amount of iron sulfides is directly related to the distance from a given emanation site and to its intensity. 2) Within the zone of fumarolic activity the uppermost 30 - 40 cm of sediment of the bay floor are enriched in exhalative sulfides.

We propose a hydrothermal model, based on the hydrogeology of the island, which aims to explain the mechanism of fumarolic activity, and the thermal and geochemical variations, from the time of the last eruption.

Introduction

The formation of sulfide deposits through volcanic exhalations in volcanic-detrital sediments is probably the only extensive hydrothermal process which can be directly studied in nature. Crystallization of FeS_2 has been observed in hot springs in Iceland (BARTH, 1950; ARNORSSON et al., 1967) and in New Britain (STANTON and BAAS-BECKING, 1962), but they are strictly localized phenomena the size of which may not be compared with an ore deposit; moreover, these sulfides are very unstable and do not accumulate unaltered; on the other hand, the formation of sulfides in the hot brines such as those of the Red Sea (DEGENS and ROSS, 1969) cannot be observed directly. In a way, the bay of Porto di Levante of Vulcano gives a perhaps unique opportunity to study "in vivo" the formation of an exhalative ore deposit. After this paper was presented at the VIII International Sedimentological Congress, another case of pyrite deposition by submarine fumaroles was discovered 30 m deep off Punta Banda, Baja California (Mexico) (JAMES, 1970, unpublished).

In a former paper published in Mineralium Deposita, one of us (HONNOREZ, 1969) described the crystallization of sulfides in immerged tuffs of Porto di Levante at Vul cano; the main observations and conclusions of this earlier study will be summarize in the present article. Some of the data published hereafter have been originally pre sented in two theses: a "doctorat du 3^e cycle" (J. VALETTE) and a "Diplomarbeit" (A. WAUSCHKUHN).

Geological setting of the fumarolic activity at Vulcano

The archipelago of the Eolian islands, in the South Tyrrhenian Sea, is an island arc which runs parallel to and 25 km off the northern shore of Siciliy; it appears that it was formed in front of the Sicilian-North African shield during the early Miocene.

Vulcano. the most southern island of the Eolian archipelago, is formed by the follow ing volcanic formations (Fig. 1, after KELLER, 1970 a and b):
1) The oldest basic volcanics of the southern part of island and the younger rhyolitic lavas of the western coast. These two formations have been broken by a caldera where the succeeding volcanic activity took place.
2) The dormant volcano Vulcano itself, the crater of which is called "La Fossa"; its height is 391 m above sea level and 2000 m above the sea floor. The last eruption of this essentially trachytic volcano occurred in 1888-1890.
3) Vulcanello (123 m high) is an adventice volcano mainly formed by leucite tephrite which could have formed in 183 B. C.; its last eruption producing trachytic tephra oc curred in 1550 A. D..
4) The 1 m high isthmus which connects Vulcanello to the northern part of the island it seems to have been completed by a joint eruption of Vulcanello and Vulcano in 155 and by alluvial deposits.
5) The Faraglione, two hills (56 and 27 m high) which appear to be the remnants of another ancient adventice volcano; a major part of its trachyandesitic lavas and tuffs have been strongly altered by fumaroles into sulfates; the thermal springs called "Acqua Calda" at the northern foot of the Faraglione were already used as spas dur ing the XVIII century (TROVATINI, 1786).

Variability of the fumarolic activity at Vulcano

Vulcano has been dormant for the last 80 years and its activity is strictly fumarolic there are moreover many thermal springs debouching at about sea level on the cliffs and beaches around the island. Periodic observations show that this fumarolic acti vity varied considerably during the last century; the extension of the fumaroles, the intensity and temperature and even their products have changes (MATTEUCCI, 1898 DE FIORE, 1914-1915, 1921, 1922, 1924; MORI, 1919; IMBO, 1935; DESSAU, 1934 BERNAUER, 1935a, 1940; HJELMQVIST, 1951; HONNOREZ, 1969; MARTINI and TONANI, 1970).

Before the last eruption, the solfatares of the main crater were so active that sulfur was mined for years. Just after the last eruption, fumaroles were observed in the crater "La Fossa" on its northern and western rims, and on the N, N-NW and S-SE upper slopes of this volcano; there had been eruptions also on "Vulcanello", at the base of "Monte Lentia" around and on the "Faraglione", and in the bay of "Porto di

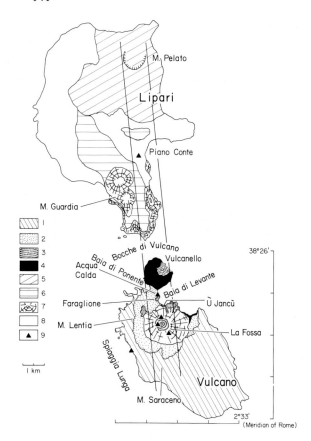

Fig. 1. Geological sketches of Vulcano and Lipari I. (after KELLER, 1970a, b)
and location of the fumarolic activity.
1 = Pre-La Fossa rhyolitic and basic volcanics of east and south Vulcano, respect.
2 = Alluvium of Vulcano
3 = Trachytic and rhyotrachytic lava flows of the young La Fossa, Faraglione and
    Vulcanello
4 = Leucite tephrite lava flows of Vulcanello and early pre-La Fossa
5 = Rhyolitic obsidian flows and pumices from Monte Pelato (North Lipari)
6 = Lower pumice level from Monte Guardia (South Lipari)
7 = Pyroclastics of La Fossa and liparites of the Mte. Guardia group
8 = Older basic volcanics of Lipari
9 = Active fumarolic fields of Vulcano and the extinct one of Lipari

Levante" (Fig. 1). According to BERGEAT (1899) and KELLER (1970a), this fumarolic activity would be localized on a N-S fracture zone on which are also aligned the
volcanoes "La Fossa", the "Faraglione", "Vulcanello", and the recent (300-400 A. D.)
rhyolitic volcano "Monte Pelato" at Lipari (Fig. 1). In 1916, the fumaroles of "Acqua
Calda" invaded the isthmus between Vulcano and Vulcanello killing all plants north
of the Faraglione (DE FIORE, 1920), but they were already reduced in 1933 (BER-
NAUER, 1935a) to the area presently occupied.

At the present time the number of active fumaroles is less than the number 50 years
ago and three fumarolic fields can be arbitrarily defined (Fig. 1):

I. The fumaroles and thermal springs of the "Baia di Levante" extend from the mid-dle Vulcanello-Vulcano isthmus in the north to the northern foot of the "Faraglione" in the south (see Fig. 3) at a place where the spas were called "Acqua Calda"; fuma-roles and thermal springs are spread along the western shore and on the floor of the bay; we set also in this zone the isolated hot spring of "U Jancù" on the southern sho of the bay. None of the fumaroles or thermal springs has a temperature exceeding 100 $^{\circ}$C. Their deposits are the subject of this paper.

II. The fumaroles of the "La Fossa" crater are located on the inside slopes of the crater, on its northern rim and on its outer S-SE upper slope; there is an isolated fumarole on the upper northern slope. This zone includes the hottest fumaroles and their temperatures range presently from 98 up to 210 $^{\circ}$C. They deposit mainly sul-fur, sal ammoniac and boric acid around their vents.

III. The isolated thermal springs of the eastern coast of the island (Punta Nere, Pur ta Lucia and Capo Grillo) and the small group of weak submarine exhalations off "Spiaggia Lunga", on the west coast; the latter coat with iron hydroxides the lower face of the lava pebbles lying at a depth of 0.50 m.

The thermal variations of the fumaroles, particularly those of the crater zone, are quite remarkable. One can distinguish two categories of fumaroles according to thei thermal behavior:

1. The low temperature fumaroles with a temperature corresponding to the boiling point of mineralized water. These are found in all the three zones and their tem-perature appears to have been constant during the last post-eruptive period excep for slight variations due to some atmospheric events.

2. The high temperature fumaroles, comparable in temperature to superheated stea. Their temperature is very unsteady and these variations are obviously not related

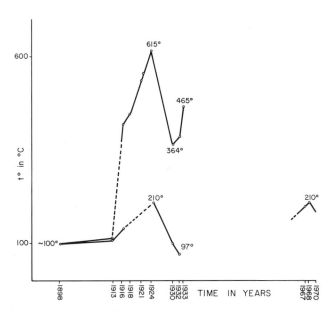

Fig. 2. Time fluctuations of the maximum temperature of the fumaroles in La Foss: crater (upper curve) and on its northern slope (lower curve) (after DE FIORE, 1924; DESSAU, 1934; HONNOREZ, unpublished; IMBO, 1935; MARTINI and TONANI, 1970; MORI, 1919)

to atmospheric events. The temperature variations of the main groups of fumaroles of the volcano "La Fossa" during the last 80 year post-eruptive period are
shown in Figure 2. The 615 °C maximum was reached in 1924 when sulfosalts
were crystallizing.

The mineralogy of the sublimates which coat the vents of the fumaroles and that of
the wall rocks are very sensitive to thermal and chemical variations of the exhalations. The list of minerals found in the fumaroles of "La Fossa" crater has been
published (Table 1, HONNOREZ, 1969).

Table 1 - Chemical analysis of the gases in a shallow submarine fumarole of the
Baia di Levante. The temperature during sampling was about 98-99 °C
and the partial tension of H_2O corresponds to its vapor pressure
(ELSKENS, unpublished).

$$CO_2 + H_2O + SO_2 + H_2S = 97.5\,\%$$
$$O_2 + N_2 + H_2 + CH_4 + etc. = 2.5\,\%$$

Dry gases (= ±88 %)

$$CO_2 = 73\ \%$$
$$H_2S = 0.04\,\%$$
$$O_2 = 2.65\,\%$$
$$N_2 = 9.5\ \%$$
$$CH_4 = 0.2\ \%$$
$$CO = 1.63\,\%$$
$$Ar = 0.6\ \%$$

Acid gases and $H_2 + He + n.d.$ = ±10 %

Some minerals have been previously seen crystallizing in fumaroles, but they have
not been found since. For instance, cannizzarite-galenobismutite and bismuthinite
have been observed crystallizing in 1924 (ZAMBONINI et al. , 1925) and in 1933
(BERNAUER, 1935a; DESSAU, 1934), but they have not been found since; they in-
crusted the altered rocks in the deepest part of some fumaroles of "La Fossa" crater with temperatures of 550 and 615 °C. They were associated with sulfur, sal ammoniac and realgar which was not found later. An iron sulfide with the possible for-
mula FeS formed in small ponds between reefs north of the "Faraglione" in 1916
(DE FIORE, 1921), but has never been found again.

In another paper in preparation we propose a hydrothermal system which tends to
explain the existence at Vulcano of two categories of fumaroles and their variations
during the last century.

The studies of BERNAUER (1933, 1935 a and b, 1936, 1940) on the formation of sulfides and sulfates in the "Fossa" crater and on the beach of the Baia di Levante are
very well known and they will not be recalled here. Striking observations made in
1933 by DESSAU (1934) are the only subsurface information that we have on the deposit of exhalative sulfides at Vulcano. This author had 14 drillings made into the
isthmus between Faraglione and Vulcanello and around the hot springs of Acqua Calda.
The rock sampling could be made only down to 3.5 m because of violent eruptions of

CO_2 first, and steam and water afterwards which occurred every time that a depth of about 3 m was reached. The layers drilled from top to bottom are as follows (DES SAU, 1934):

- few decimenters of volcanic sand hardened by sulfates and sulfur deposited through evaporation of the ground water;

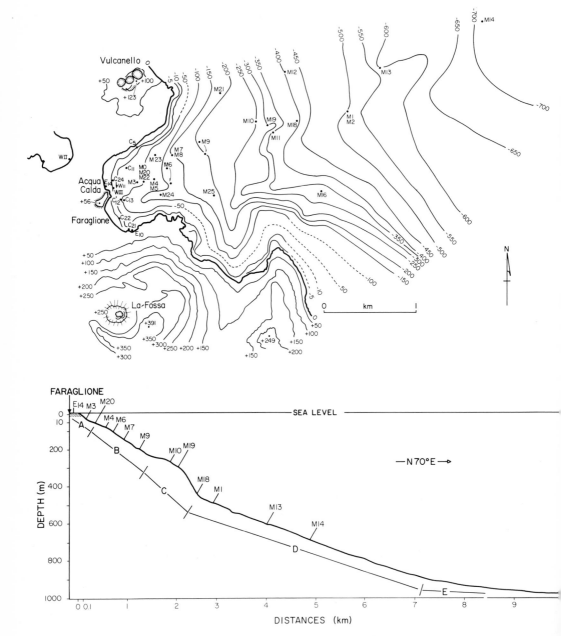

Fig. 3. Original bathymetric map of the Baia di Levante with locations of the samples; bathymetric profile of the Bay along a N 70°E axis, 3.7 x vertical exaggeration (M = samples dredged from the "Mechelen"; E = samples dredged during diving surveys; C and W = samples cored during diving surveys)

- few decimenters of loose volcanic sand and mud soaked by water corresponding to the water table;
- the remaining meters were drilled up to 7 m deep through dry volcanic sand cemented by iron sulfides.

Preliminary S analyses did not show any coherent variations relative to depth (DESSAU, 1934).

The submarine fumarolic field of the Baia di Levante

The morphology of the sea floor in the Baia di Levante results from the conjunction of three adjoining volcanoes: the slopes of "La Fossa" in the south and those of "Vulcanello" in the NW, and the debris of the "Faraglione" between them in the west. The bathymetric map of the Bay is shown in Fig. 3. The submarine slopes are very steep and the continental shelf is thus very restricted or even sometimes missing. The widest sublittoral zone (100 m wide) coincides with the location of the isthmus between Vulcano and Vulcanello. The relief is very uneven and lava peaks and blocks are scattered on the surface; according to the few seismic reflection recordings we have made with a "Boomer", the sediment cover appears to be only a few meters thick. Fig. 3 shows also a median topographic profile in the Bay along a N 70° E azimuth originating from the Faraglione; down to -50 m depth the slope changes from 14 to 40° through 2 or 3 steps (portion A), from -50 to -900 m deep (portions B and D), the mean slope is 8-9° with a sudden steep portion (25-30°) between -290 m and -450 m deep (portion C); deeper than -910 m, the slope is only a few degrees (portion E). It is worth noticing the furrow at a depth of about -100 m. This furrow was also observed on SONAR recordings more in the NE and the SE in the Bay; it appears to be a general topographic feature of the sea floor and it could be a major tectonic line of the caldera.

Thousands of fumaroles can be seen bubbling in the littoral and sublittoral zones of the Baia di Levante, down to a depth of 15 m (10 m, according to SCHULZ, 1970). Except for a few submarine fumaroles scattered in the southern portion of the bay, the fumarolic activity is mainly concentrated opposite and north of the Faraglione in a a 500 m x 100 m area elongated N-S in a direction parallel to the shore line (Fig. 4).

One of us (HONNOREZ, 1969) attributed the absence of fumarolic manifestation beneath the -15 m level to condensation of steam and solution of CO_2 by the hydrostatic pressure of the overlying head of sea water which was mistakenly believed to exceed the critical pressure of CO_2. In fact, the vapor pressure of CO_2 at 15 °C (temperature of sediments soaked with sea water) is about 50 atmospheres which corresponds to a depth of about 500 m under sea level. This -15 m level may well be of a tectonic nature and correspond, for instance, to an eastern limit of the N-S fracture zone where the volcanic activity has been restricted during historic ages (see above); or, it could correspond to a major fracture of the caldera (see above). It is also probable that steam percolating through soaked sediments condenses as well and that CO_2 and the few non-condensable (dry) components of the fumaroles are readily diluted during their ascent.

The beach and the sea floor of the Baia di Levante are formed by few plates of tuffs and breccias outcropping from a very coarse sand. The tuffs seem to belong to the nearby trachyandesitic "Faraglione". The thickness of the sand layer resting on the tuffs is highly variable (0 to 1 m) near the shore, but increases with depth.

Fig. 4. General view of the fumarolian field of Baia di Levante

When fumaroles are emitted through the tuffs, they are often aligned along an irregular network of diaclases. They seem to be more randomly distributed in the sand. In
both materials they form circular depressions 3 to 40 cm in diameter (Fig. 5a, b).
The strings of bubbles ascend either continuously or intermittently through the water.
There are a dozen exceptionally active fumaroles in which the bubbling keeps the sand
grains and pumice fragments constantly bouncing. The fumaroles are generally grouped within patches of a few square meters, their presence is often made conspicuous
from afar by a yellow layer of colloidal sulfur several cm thick.

One of the most active fumaroles was sampled right on its submarine vent and its
chemical composition (ELSKENS, unpublished) is shown in Table 2; the H_2S content
was found to be 0.04 % dry gas. BERNAUER (1935a) had sampled some other submarine eminations and measured 6 to 52 % H_2S with a 21 % average. Compositional
fluctuations have been observed (ELSKENS, unpublished), the periodicity of which
varies from 1 second to a few hours. The termperature, pH and composition of the
sea water are much altered in a zone about 10 m wide surrounding a give group of
fumaroles; the sea water temperature a few cm from an important group of fumaroles

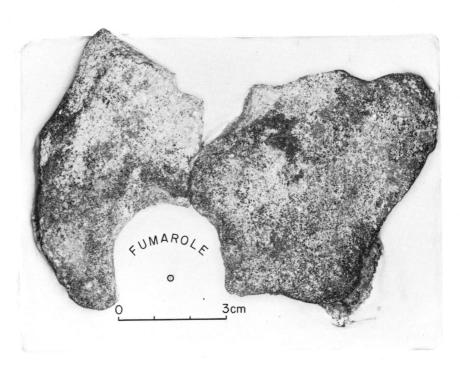

Fig. 5. Erosion funnels dug out by fumaroles in sands (5a) and in tuffs (5b)

Table 2 - Opaque mineral point counting in "heavy fraction" concentrates from sediments of the Baia di Levante (M = samples dredged from the "Mechelen"; E = samples dredged during diving surveys; * sampled directly over a fumarole) (see locations Fig. 3)

1	2	3	4	5	6	7	8
E67/14-1 * (0.5 m)	48	48	-	-	-	52	21
E68/14 * (5-10 m)	4.0	2.5	0.6	-	-	96.9	0.1
E68/20 * (5 m)	8.1	2.6	2.0	-	-	95.4	0.2
M3 (40 m)	1.5	0.3	10.3	0.1	0.6	88.7	ngl.
M20 (65 m)	0.13	0.1	11.4	0.2	0.6	87.7	ngl.
M4 (80 m)	0.5	-	6.5	-	0.3	93.2	-
M6 (105 m)	1.9	0.3	10.6	0.6	10.6	77.9	ngl.
M7 (150 m)	1.0	-	12.0	0.7	0.5	86.8	-
M9 (200 m)	0.3	0.1	16.2	0.9	0.8	82.1	ngl.
M10 (250 m)	n.d.	-	10.8	0.6	2.3	86.3	-
M19 (300 m)	0.2	-	15.5	2.4	2.0	80.1	-
M18 (440 m)	1.9	0.3	14.4	0.2	1.0	84.1	ngl.
M1 (550 m)	3.6	0.1	11.2	0.6	0.6	87.5	ngl.
M13 (600 m)	1.1	-	8.3	0.1	1.3	90.3	-
M14 (700 m)	0.5	0.3	11.6	0.3	0.6	87.2	ngl.

1 = sample number and depth in m
2 = heavy mineral % in total sediments
3 = pyrite + marcasite % in the heavy mineral fraction
4 = Ti-magnetite + ilmenite % in ibid mineral fraction
5 = hematite %
6 = goethite %
7 = transparent mineral %   n.d. = not determined
8 = sulfides % in total sediments   ngl. = <0.01 and negligible

may rise to 65⁰ and its pH drop down to 3.0 - 3.5; these values are not constant as they are continually changed by the surf.

Sedimentology of the Baia di Levante

The fumarolian influence is superimposed on the "normal" sedimentological background of the Baia di Levante; we have first to consider the latter zones deprived of fumarolic activity in order to separate the "normal" features from our observations on the fumarolized sediments.

The major sedimentological features of the sediments off Vulcano Island result from the interaction of the four following acting agents:

1. Four historically active volcanoes characterized by explosive eruptions: "La Fossa" and "Vulcanello" on Vulcano Island and "Monte Pelato" and "Monte La Guardia" on Lipari Island; their explosions projected directly into the sea trachytic and rhyolitic ashes, coarse clastics as well as blocks of older volcanics and xenoliths from their basement (HONNOREZ and KELLER, 1968). It is probably the most important process by which material was introduced and sediments were generated around Vulcano Island. The last eruption of "La Fossa" in 1889-1890 must have been responsible for generating the uppermost sand layer on the sea floor of the Baia di Levante.

2. Wave erosion which pulled down material from the steep cliffs forming 90 % of the Vulcano shore lines. This is the most important process during periods of dormancy like the present time. During this process the finest and lightest debris are washed away to the open sea and the coarser and heavier fraction is accumulated near the shore.

3. Rain and wind erosion along the steep slopes and dells of a very uneven inland relief. The wind action removing dust from weathered and fumarolized volcanics would be quite effective because of the small rainfall on the island. On the contrary, the rain erosion is very secondary; there is only one intermittent stream, the "Rio Grande", which rages after heavy rains between "La Fossa" and "Monte Lentia" and disappears in the isthmus Vulcano-Vulcanello long before it reaches the sea; it is probably the least important process generating detrital material at Vulcano.

4. A prevailing westerly wind (Ministero Difesa Aeronautica, undated) generating a current with the same orientation in "Bocche di Vulcano", a 1 km wide strait between Lipari and Vulcano islands; this current removes solutions and finer sediments from the bay. It is the main reason why the geochemical gradients are so strong around the submarine fumaroles. The geochemical anomalies of the water due to the fumarolic activity are readily diluted and the concentration of heavy metals (more particularly Ti, Cu, Zn, Ag, Pb) is kept below the geochemical threshold over which solutions start precipitating sulfides and other metallic compounds. This is the main difference between the submarine sulfide deposit of Vulcano where only Fe-minerals are formed and those of economic interest which are mined for Zn, Cu, Ag sulfides.

The contribution of material by marine currents from the nearest emerged lands, i.e. Lipari and from Sicily and Calabria (respectively 2.5, 25 and 80 km far from the Baia di Levante) is negligible in the sediments close to Vulcano Island. The sediments in depths below 500 m (i.e. more than 3 km from the Faraglione) are derived not only from Vulcano but other nearby volcanic sources including Lipari.

VALETTE (1969) could divide the sediments around Vulcano Island according to their mineralogy into four petrographical sectors corresponding to the four main volcanic units of the island. In the Baia di Levante trachytic material from "La Fossa" is

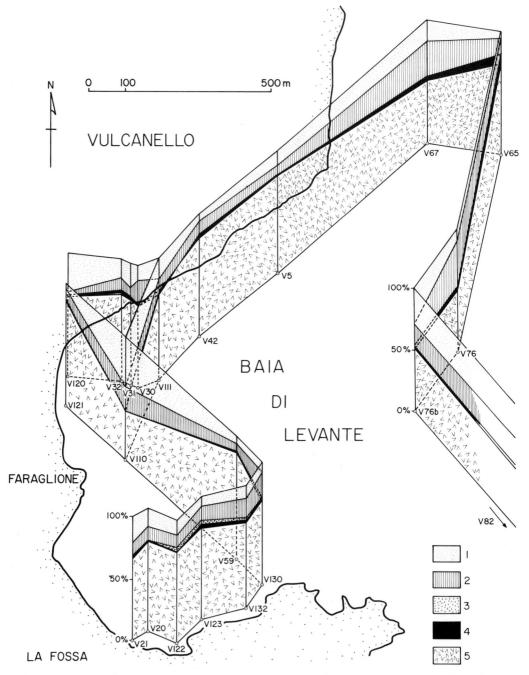

Fig. 6. Volume % frequency variations of the various heavy minerals within the heavy fractions of sediments from the Baia di Levante (after VALETTE, 1969): 1 = opaques, 2 = glass, 3 = micas, 4 = olivine, 5 = augite

mainly found; it grades to the NW into the leucite tephritic zone generated by Vulcanello and to the SE into the basic zone which fringes most of the island shores and
corresponds to the oldest formations.

Out of the fumarolic field, the main mineralogical components of the sediments from
the Bay are the following: augite, plagioclase, sanidine, and volcanic glasses; titano-
magnetite, hematite and olivine are accessories; muscovite, biotite, quartz, zircon,
garnet, apatite and pyrite are casually found. The amount of heavy minerals (augite,
glasses and opaques, and olivine) ranges in most samples from 6 to 14 %. Fig. 6
shows the frequency variations of the various heavy minerals in sediments through
the Bay. Augite is by far the most abundant heavy mineral; its amount ranges from
42 to 88 % of the heavy minerals with a 68 % average. Considering how difficult it
is to distinguish between opaque minerals and volcanic glass in thin sections and
smear slides (except for obsidian and sideromelane which are not common at Vulcano), we can state that the augite % variations are balanced by those of glasses -
opaques. The pyroxene seems roughly to be more abundant close to the shore, and
glasses and opaques more towards the center of the Bay except for one sample
dredged 90 m away from the Faraglione.

As already stated before, the most common minerals of the Bay have been generated
from the trachytic material erupted by the nearby volcanoes "La Fossa" and "Vulcanello". The origin of the accidental components such as micas, garnet, zircon and
apatite cannot be ascribed to marine currents carrying these minerals from Lipari,
Sicily or Calabria. On the other hand, their size is too large to attribute them to airborne material carried by winds. The explanation is simpler: a rhyolitic tuff of south
Lipari called "Unterer Bimshorizont" (KELLER, 1970b) contains zircon; it has been
generated by violent preshistoric eruptions of Monte Guardia volcano (South Lipari).
The same zircon-bearing tuffs have been found on Vulcanello and on the SE part of
Vulcano (KELLER, 1970a). Moreover, granitic xenoliths (K-feldspars, plagioclase,
quartz, apatite and biotite) are common in the rhyolitic tuffs of the Monte Pelato
(HONNOREZ and KELLER, 1968); these tuffs are also known on Vulcanello and Vulcano Island (KELLER, 1970a, b). Carbonate xenoliths bearing garnets, biotite, pyroxene, etc. , have been found at Vulcano (HONNOREZ and KELLER, 1968) and gneis-
sic xenoliths containing garnet, cordierite, sillimanite along with apatite, zircon,
biotite, quartz, etc. , have been described (BERGEAT, 1910; HONNOREZ and KEL-
LER, 1968) in andesites from Monte S. Angelo, at the center of Lipari.

The granulometric fraction larger than 1 mm usually represents 10 % of the Bay
sediments from the littoral and sublittoral zones and sometimes from greater depth
(as M3 from 40 m depth); some samples are so coarse that this fraction varies from
20 - 74 % and gravels and blocks up to several cm are common.

The granulometric curves in Fig. 7 indicate that the origin of the sediments from the
Baia di Levante is complex: uni- and multi-modal sediments are found side by side
and their distribution in the Bay does not follow any kind of pattern close to the shore
line and far away from it.

By measuring medians and quartiles, VALETTE (1969) inferred that down to depths
of -100 to -300 m, mixed medium and coarse-grained sands (mean $\emptyset$ ranging from
80μ to 1 cm) normally run along the N. E and W shores of the island essentially formed there by high tuffaceous cliffs; finer sediment is found at greater depths. The
south shore (old, smooth cineritic relief) is bordered with fine sands (mean $\emptyset$ 120 -
185μ). The distribution of the "lutites" content (sediments composed of fine particles
with $< 40\mu$ mean $\emptyset$ and thus similar to the "silts and clays") in the sediment follows
a rather regular concentric zoned pattern around the island except in the Baia di

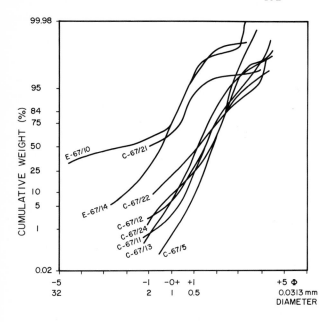

Fig. 7. Granulometric curves (particles Ø in I units vs cumulated weight % frequency) sediments from the Baia di Levante (see location, Fig. 3)

Levante. In the Bay, both fine and coarse fractions are intimately mixed and their distribution is quite erratic.

The rather shallow fine sediments could be generated by winnowing action of the waves on the debris from the tuffaceous cliffs; it carries the finest fraction down the immersed slopes and leaves the coarsest close to the shore. The wind erosion transporting the finest dust from the inland weathered and fumarolized volcanics to the surrounding sea has also to be taken into account. It is worth mentioning (VALETTE, 1969) that the finest fraction is not formed by clay minerals but by an intimate mixture of volcanic glass and feldspars; the clay minerals start appearing below 1000 m depth.

By calculating the S_O (Trask sorting coefficient), the Quartile deviation and Ska (Quartile skewness), VALETTE (1969) inferred that the sediments from the littoral zones surrounding the island shore, excepting the Bay, are generally well sorted (S_O = 0.74 - 0.80) and their coarse fraction is the best sorted. Between 20 and 30 m depths, the sorting ranges from 0.75 to 0.50 and the coarse fraction is still the best sorted; even at about 200 to 250 m depth, the sediments are still fairly sorted (S_O <0.65). But in the Baia di Levante, the sediments are poorly sorted (S_O often <0.25).

In summary, the sediments from the Baia di Levante, contrary to those of the other parts of the island, are poorly sorted, and do not follow any regular pattern of granulometric distribution. Moreover, their mineralogical composition is quite homogeneous and corresponds mainly to that of trachytic pyroclastic sands.

All these characteristics point to a volcanic origin of the sands by explosions of the nearby volcanoes directly throwing the material onto the sea; later the sand grains were not transported very much except those which fell in the littoral zone; there the sand was winnowed by the surf action and enriched with respect to its heaviest components, the Fe-Ti oxides and pyroxenes.

Petrography of the sulfidized tuffs and sands of the Baia di Levante

We have shown (HONNOREZ, 1969) that the Fe-sulfides develop within the immersed tuffs of the Baia di Levante, according to the following processes:

1. By filling in the inter- and intra-granular spaces and fissures; pyrite deposited first on the lapilli and afterwards marcasite cemented them all, or vice versa (Fig. 8a).
2. By replacing the volcanic Fe-Ti oxides inside the lapilli; the initial Ti-magnetite, ilmenite and martitized hematite could no longer be found.
3. As pseudomorphs after silicates (pyroxenes and feldspars) and the glassy ground-mass of the lapilli which had been previously only slightly fumarolized.

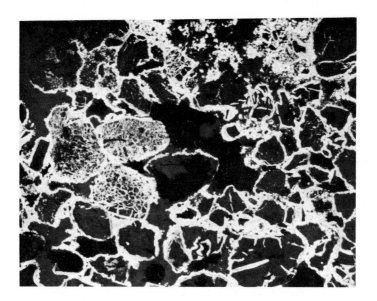

Fig. 8. Microphotographs of polished sections in sulfidized tuffs and sands
8a) Pyrite-marcasite infilling of inter- and intra-granular spaces in tuff. Magn. 12 x

We could not prove that the pattern sequence 1, 2, 3 was the general rule. The same three patterns (see Fig. 8b) have been found in the fumarolized sands but the trans-itional stages of Fe-sulfide pseudomorphs after Fe-Ti oxides were represented (see HONNOREZ-GUERSTEIN and WAUSCHKUHN, in preparation). Among them the "en cocarde" or "onion scales" fabric resulting from the replacement of ilmenite by mar-casite lamellae (Fig. 8c) were quite unknown in the sulfidized tuffs. Sulfidized fos-siliferous sands have been also found (Fig. 8d) where the carbonate shells had been dissolved and their empty inside moulds are observed.

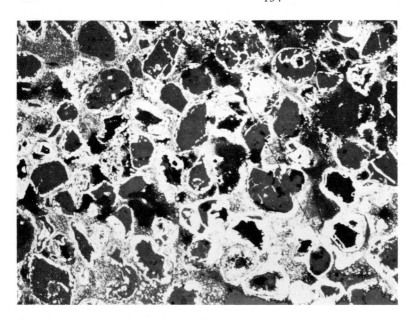

8b) The same in sand. Magn. 6 x

The diaclases along which the fumaroles rise within the tuffs are coated with colloform pyrite crusts up to 5 mm thick (Fig. 8e) which might sometimes alternate with
sulfur, alunite and opal. Five kinds of Fe-sulfides spherules were also described in
the fumarolized tuffs and they were also found in the sulfidized sands (Fig. 8f): a
fibroradial marcasite spherule and various types of framboidal pyrite spherules
with or without concentric "peels" of marcasite or pyrite; the last pattern displayed
all transitional stages between plain homogeneous pyrite spherules and pyrite-marcasite framboids where the peels and the cores successively homogenize. We had
concluded (HONNOREZ, 1969) that the physico-chemical conditions are such in the
submarine field of Vulcano that a fully inorganic process can produce these framboids and that bacteria do not have to be called upon; this hypothesis was confirmed
by the works of BERNER (1969), RICKARD (1969a, b, 1970), ROBERTS (1968), and
ROBERTS et al. (1969).

Marcasite crystals are generally much larger than those of pyrite. Pyrite occurs a
least as two generations: better crystallized light yellow grains (with a diameter
ranging from 1 mm in the case of framboids to 0.2 mm in isolated subhedral crystal
and fine tannish coatings which display the features of the so-called "gelpyrit" or
"colloidal pyrite". The amount of sulfides counted in those tuffs (point counting on
about 1000 points) varies from 9 to 39 % by volume. High porosity is a noteworthy
characteristic of these tuffs and breccias: the present space volume in the mineralized tuffs ranges from 15 to 37 %; taking into account the intergranular infilling by
sulfides, we can assume that the initial porosity of the tuffs was at least 35 %.

Distribution of exhalative sulfides on the sea floor of the Baia di Levante

The superficial sediments from the sea floor of the Baia di Levante have been sampled by hand coring from the littoral zone down to a depth of 52 m during several

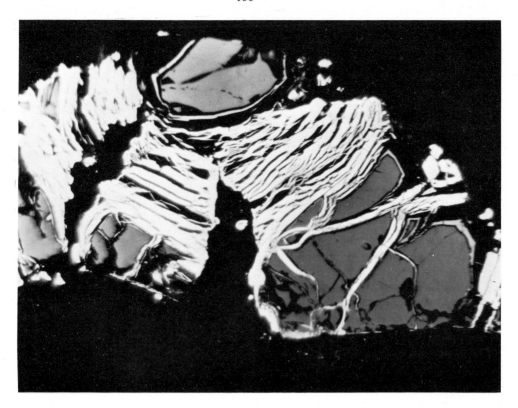

8c) Ilmenite substitution by marcasite lamellæ along (0001) basis of ilmenite; in sand.
Magn. 800 x

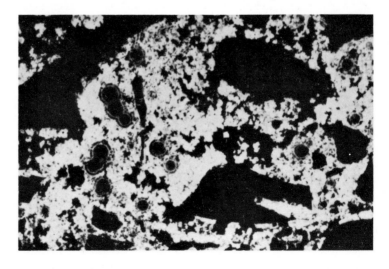

8d) Sulfidized fossiliferous sands. Magn. 64 x

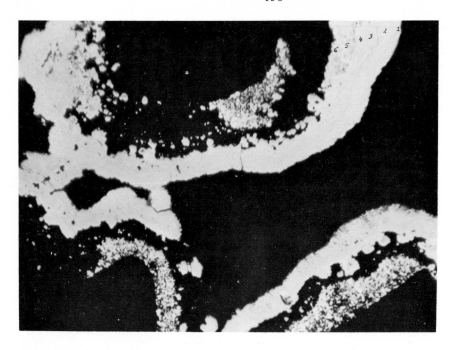

8e) Colloform pyrite crust along fumarolian fissure in tuffs. Magn. 125 x

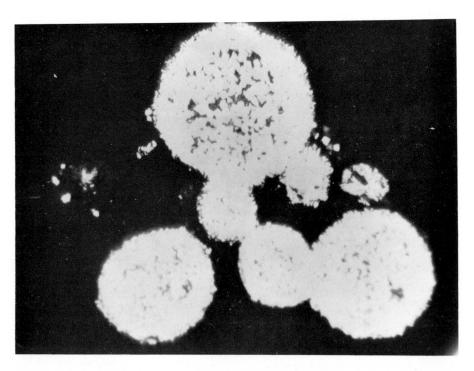

8f) Framboidal pyrite with marcasite coating in sands. Magn. 1800 x

diving surveys, and by dredging between -40 and -700 m during the 1968 cruise of the "Mechelen" of the Belgian Navy. Our various attempts to take longer cores by piston coring from the ship failed because the sands were too coarse and mixed with too many gravels and blocks. Various samples (see locations, Fig. 3) have been processed as follows: the dried sediments were sieved (U.S. standard opening system) and the heavy minerals were concentrated (using diiodo-methane: ds 3.33 at 20°C) from the sand fraction comprised between 1 and 0.062 mm; these concentrates were impregnated while being centrifuged with "araldite" and studied with a reflected light microscope; the various opaque minerals were identified optically and checked by X-ray diffraction and microprobe analyses. Modal analyses are based on point counts of 1000 points per section, 0.3 mm between points and between rows. The few droplets of primary sulfides (mainly chalcopyrite) within the volcanic pyroxenes and feldspars were so small and rare that they have not been counted. The coarser fractions were checked with a binocular microscope and a very few sulfide incrustations and isolated grains were observed; the finer fractions were checked by X-ray diffraction and also found devoid of sulfides.

From Table 2 and the last two observations we can infer that:

1. The superficial sediments from outside the fumarolian field do not contain any notable amount of sulfides but bear amounts of Fe-Ti oxides and hydroxides ranging from 6.8 to 23 % of the heavy mineral fraction.

2. On the contrary, the superficial sediments from the fumarolian field contain up to 21 % Fe sulfides and the Fe-Ti oxides and hydroxides are very minor. The highest amount of Fe hydroxides (10.6 % goethite is 10 times more than the 0.9 % average) was measured in sample M6 dredged from the furrow at about 100 m depth (see p. 4). Since the primary magmatic Fe-Ti oxides % (10.6) of this sample is similar to those of the other samples (11.5 average %) from outside the fumarolian field,

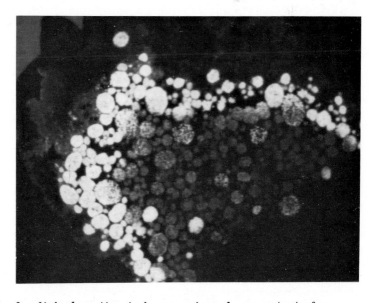

Fig. 9. Microphotograph of polished section in heavy mineral concentrate from dredged sample M6 showing goethite pseudomorphs after framboidal pyrite; white framboids = pyrite; gray framboids = goethite. Magn. 400 x

this goethite does not result only from the hydrolysis of Ti-magnetite. On the other hand, the abundance of goethite displaying textures such as framboidal spherules (Fig. 9) or colloform crusts and the finding of pyrite relicts induces us to ascribe the Fe-hydroxides to the hydrolysis of Fe-sulfides when they are carried out of the reducing domain of the fumaroles where they are stable; we can think that the waves (action down to 15 m depth) and the current in the Bocche di Vulcano carry iron sulfides along with sands out of the shallow fumarolian field; the sand is enriched in its heaviest components (i. e. Fe-sulfides and oxides) by winnowing wave action which partly eliminates the highest (and finest) fraction towards the open sea. The sediment slumps along the steep submarine slope until it is trapped in a topographic depression (the furrow at 100 m depth). Samples from here showed that these abnormal 11 % goethite values would thus have an allochthonous origin and be added to the 12 % Fe-Ti oxides background values of the sediment from the non-fumarolian domain of the Bay.

An alternative explanation would be that these 11 % goethite values are autochthonous resulting from the hydrolysis of Fe sulfides precipitated at 100 m depth by ancient submarine fumaroles which are no longer active. In this case, the 12 % Fe-Ti oxide would be allochthonous and carried there by a slumping mechanism, but the coincidence of the goethite accumulation with a topographic trap induces us to prefer the first explanation.

One can also conclude that the Fe sulfides are unstable out of the fumarolian zone of influence and that the latter is very restricted. We had already observed (HONNORE 1969) that the surface sea water temperature, which is much higher (up to 85 °C) in a shallow submarine fumarole, is normal at some 10 m distance from it; in the same way, the pH of sea water which drops down to 2.5 or even 2 in a submarine fumarole returns to normal value a few meters from the fumarole. Finally, the main result of a water study made in 1968 by the oceanographic team aboard the "Mechelen" (BALLESTER and TONANI, unpublished) indicated that the sea waters are quite normal at a distance less than 10 m from a large group of fumaroles: salinity, chlorinity, nitrites, nitrates, temperature, etc. do not display any unusual values.

Vertical distribution of exhalative sulfides in the sediments of the Baia di Levante

In two one-meter long cores from the Baia di Levante which have been studied in detail (WAUSCHKUHN, 1970), a layer of slightly indurated pyritized sand was found between 20 and 40 cm below the water-sediment interface; these cores were taken directly on fumaroles at 15 m (W-11) and 7 m (W-III) depths (see locations, Fig. 3). Similarly, core penetration was hindered when the barrels were driven 20 cm into the sand during coring on fumaroles at 13 other locations in the Bay bottom between 4 and 10 m depths (WAUSCHKUHN, 1970). No pyrite was found in another one meter long core (W-II) sampled in the sand of the Baia di Ponente (see Fig. 3). Table 3 presents the main results of the detailed study on core W-11 after WAUSCHKUHN (1970). This core was chosen to be studied because it had the highest heavy mineral content (15 %) and it had been taken on one of the deepest fumaroles (15 m water depth). The sand grains were found under the microscope to be encrusted with Fe-sulfides, mainly pyrite, down to 50 cm below the water-sediment interface and framboidal Fe-sulfides were observed down to 100 cm. The sulfides were the most abundant, and their crusts the thickest, particularly between 20 and 40 cm depths.

159

Table 3 - Variations of heavy mineral weight content, S and both sulfidized and oxidized irons vs depth below the sediment-water interface within core W-11 (after WAUSCHKUHN, 1970)

1	2	3	4	5	6	7	8
0-10	15.0	11.7	2.52	+ 9.16	2.9	17.25	9.83
10-20	24.5	12.2	2.12	+10.12	2.4	14.51	10.86
20-30	33.0	15.4	3.18	+12.12	3.7	21.77	13.01
30-40	21.0	14.5	3.33	+11.13	3.8	22.79	11.94
40-50	8.0	9.5	0.92	+ 8.54	1.1	6.30	9.17
50-60	6.3	8.9	0.62	+ 8.28	0.7	4.24	8.89
60-70	7.2	9.0	0.43	+ 8.61	0.5	2.94	9.24
70-80	9.1	9.2	0.76	+ 8.42	0.9	5.20	9.04
80-90	10.0	8.6	0.40	+ 8.22	0.5	2.74	8.82
90-100	10.4	8.9	0.33	+ 8.57	0.4	2.26	9.20
						100.00	100.00

1 = depth in cm below the sediment-water interface
2 = heavy mineral %; 3 = total Fe % in heavy mineral fraction
4 = sulfidized Fe %; 5 = oxidized Fe %; 6 = S %
7 = sulfidized Fe % in the whole sediment with respect to the bulk sulfidized Fe content of the whole core
8 = oxidized Fe % with respect to the bulk oxidized Fe content of the whole core

Sulfidization and supply of Fe

The core W-11 was analyzed for S and for two states of iron: sulfidized Fe which is bound with S within pyrite and/or marcasite, and oxidized Fe which is bound with O within Ti-magnetite, hematite, ilmenite and pyroxene.

One can infer from Table 3 and Fig. 10 that:
1. the meavy mineral content, the S and both states of Fe contents in the heavy mineral fraction and in the whole sediment are higher in the upper 40 cm of the core than in its lower 60 cm;
2. there is simultaneously for all these parameters a maximum at the -20 and -30 cm level; the oxidized Fe content is only slightly above (5 % higher, i.e. 0.63 times more) the average sulfidized Fe content (4 %) of the lower layers.

The increase of iron content in the upper layer and its maximum at 20 - 30 cm has then to be ascribed to:

1) the direct precipitation of Fe-sulfide crusts around the sand grains through fumarolic activity;

2) the sulfidization of iron-rich minerals (mainly Ti-magnetite, ilmenite and hematite);

3) the concentration of the heaviest mineral (Fe and Ti oxides, and Fe-sulfides, preferentially to pyroxene) by the surf action.

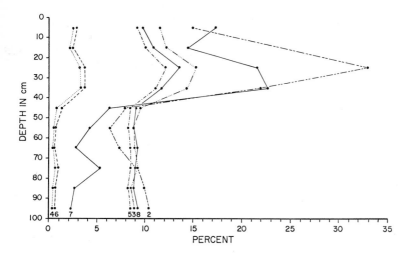

Fig. 10. Variations of heavy mineral weight content, S and both sulfidized and oxidized irons vs depth under sediment-water interface within core W-11 (after WAUSCHKUHN, 1970). Same caption as Table 3

There is good agreement between our observations on the sands and tuffs which are being sulfidized at Vulcano and the experiments of various authors; according to the works of BERNER (1964, 1969, 1970), RICKARD (1969b) and ROBERTS (1968) and collaborators (ROBERTS et al. , 1969), the conditions required for precipitation of Fe-sulfides are met in the fumarolic field of the Baia di Levante. The bacterial activity has thus not to be set forth as a "deus ex machina" to the formation of the Fe-sulfides (see also p. 8). The fumaroles provide the sea water with an unlimited supply of H_2S and heat, and the low pH which is required; an excess of colloidal S resulting from H_2S oxidation (by the O_2 dissolved in sea water) is generally kept in suspension where the sulfidization reactions occur. The Fe is supplied by three sources

1. The original volcanic Ti-magnetite, ilmenite and hematite content of the sands and tuffs; in tuffs, the initial amount of Fe is about 6.5 weight % while in sands it can be strongly increased by the winnowing action of the surf which preferentially accumulates the heavy Fe-Ti oxides.

2. The superheated steam which generally represents the major component of fumaroles is at depth a very active dissolving agent; it is able to leach many elements out of the wall rocks of the fumarolic fissures during its ascent and to carry them toward the surface; a kind of "chromatographic spreading pattern" results from the solubility variation of the various components with the progressive temperature decrease, and only the most soluble components reach the surface. It is reasonable to think that the major part of the components of fumarolian deposits, and among them the metallic elements such as Fe, are due to precipitation of recycled components from the country rocks, the volcanic pile and the crustal rocks of the basement. In the fumarolic activity, the magmatic part might be mainly to supply the thermal energy necessary to leach out and carry the major components. In addition, it contributes sulfur and few other minor components such as As, Hg, Bi, Te, etc. Formation of pyritized tuffs at Vulcano required a large supply of Fe since the initial Fe content of the unaltered pyroclastics does not exceed 6.5 weight % (i. e. 8.5 total FeO weight %) and we have collected hand specimens with up to 16.5 Fe weight % (i. e. 22 total FeO weight %). This large Fe supply in sulfidized tuffs cannot be ascribed to "per des-

censum" migrations of iron leached from a superimposed thick pile of terrain which would have been since eroded, for we know that they have not been overlain by any other deposits. Nevertheless, a "per descensum" enrichment is quite possible around the Faraglione because the two hills are an actual stock of easily soluble sulfates which is continuously reworked through marine and atmospheric erosion and corrosion.

3. A precipitation of Fe dissolved in sea water. It is very difficult to estimate how important the sea water contribution is in bringing in Fe during the sulfidization process. An Fe analysis (by a quick colorimetric method) on an unfiltered sample of sea water taken right above a submarine fumarole displayed a 0.970 mg/l Fe content; after filtration (paper filter with 50 - 100 porosity) no more iron could be detected which infers an Fe content lower than the 0.050 mg/l detection limit of the method; the Fe content of the sea water in the fumarolic field of Vulcano is almost 20 times higher than that of normal sea water (i. e. approx. 0.010 mg/l). At the surface of the sea floor, sea water is constantly renewed by tides and currents, and thus some Fe supply is continually brought to the fumarolic H_2S input; deeper in sediments and tuffs, the water circulation is much slower and this process is probably not efficient in supplying Fe.

There is only one major discrepancy between our observations from Vulcano and the experiments of the above-mentioned authors: they usually got various Fe-monosulfides (mackinawite or greigite) as transitional products in their reactions between ferrous ions and H_2S. In the sulfidized rocks of Vulcano, no such Fe-monosulfides could be observed under the microscope and their presence could also not be checked by X-ray diffraction. Observations of black turning to rusty films coating some samples of "pyritized" sand (which were kept soaked with sea water in unsealed containers) 5 weeks after they had been collected led us to suspect that some kind of unstable phase might have existed; but the many possibilities prevented us from drawing a general rule.

Extension of the exhalative sulfide deposit of Vulcano

The area covered by the fumarolic field of the Baia di Levante is roughly 45000 m^2. We know that in the submarine portion of the deposit a 20 cm thick sand layer indurated by sulfidization was found during most of the hand corings, and that the 20 cm superficial sand layer also contains sulfides. There is then on the bottom of the Baia di Levante a 40 cm thick layer of sulfidized sand, the lower half of it being a higher grade ore with an overlying 20 cm layer of disseminated sulfide ore. Without any exploration by drilling several tens of meters deep, it is difficult to estimate accurately the extension at depth of the sulfidized rocks; we only know, thanks to DESSAU's corings (1934), that the thickness of sulfidized sand is greater than 6 m in the subaerial part of the isthmus Vulcano-Vulcanello; we also know that in these low relief areas the water table regulates the uppermost limit of sulfide deposition which has to follow the seasonal motion of the meteoric water in the ground; above the water table, H_2S and the sulfides are readily oxidized to sulfates and sulfur by the air seeping through the porous rocks; this oxidation is accompanied by production of H_2SO_4 which leaches the rocks above the water table. The sulfide content of the country rocks there (DESSAU, 1934; BERNAUER, 1935a) is much smaller than in the submarine portion of the same fumarolic field: we find a low grade ore with a disseminated texture. In higher areas of relief, atmospheric oxygen penetrates a thicker layer of rock before being stopped by the water table; O_2 might not even reach the ground water in locations where the relief is high enough such as around the crater.

In such places the upper limit of pyrite deposition does not coincide with that of the water table; we have rocks dotted with few sulfides (BERNAUER, 1940) and the ore grade might be the lowest.

Sulfide ores with the highest grade are found there where the fumaroles percolate through rocks which had initially high porosity such as volcanic breccias, both explosive and talus, or coarse and unsorted tuffs and sands. These rocks correspond to the most active periods of acidic volcanoes which are characterized by frequent eruptions; the material ejected by the successive cycles of explosions falls to sea and great amounts of blocks, scoriae and ash are deposited on the submerged slopes of the volcano. Fumaroles percolate more easily through these kinds of formations and react much more with their components because of their large specific surface. Layers of fine-grained, well sorted tuffites or alluvium corresponding to long period of dormancy of the volcano have low porosity and are not expected to contain much sulfide. Clay or carbonate layers resulting from deposition of deeply weathered volcanics would be almost completely free of fumarolian sulfides.

Exhalative pyrite accumulates then in volcanic sediments and forms a mineralized layer in the uppermost 40 cm of the sea floor in sublittoral and littoral zones; a successive paroxismal eruption will bury the old sea floor under a new thick layer of pyroclastics and the sulfide deposition level will automatically migrate to be located again a few decimenters below the new sea floor. This shifting of the mineralized horizon would occur after each important eruption like the last one in 1889-1890; after a longer series of weaker eruptions, the shift would not be as sharp and the mineralization would appear more diffuse. These alternations of decimeter-thick mineralized levels with thicker barren layers, as well as the abrupt changes in sulfide content or the gradual zones of diffuse mineralization are all characteristic of volcanic sedimentary sulfide deposits.

The extension of the exhalative sulfide deposit of Vulcano and its space variations is schematically represented in Fig. 11.

Acknowledgements

We are thankful to Profs. G. C. Amstutz, G. Müller and P. Ramdohr for advising the three of us (J. H. , B. H-G. , and A. W.) at the Mineralogisch-Petrographisches Institut der Universität Heidelberg. We thank Prof. C. Doboul-Razavet (Centre de Recherches Sédimentologiques Marines de Perpignan) and Ing. J. Goni (Bureau de Recherches Géologiques et Minières de Orléans) for helping and guiding one of us (J. V.) during her work in their respective institutions.

We thank Prof. A. Capart for the 10 days' cruise of the "Mechelen" in 1968 organized off Vulcano by the Institut Royal de Sciences Naturelles de Belgique (Bruxelles).

We thank the Comm. Ceux commanding the "Mechelen", the officers and the crew whose dedicated work and patience made successful the dredgings and bathymetric surveys at Vulcano.

We remember the population of Vulcano who helped us during our field work on and off the island.

This research was supported by the Alexander von Humboldt-Stiftung, the Deutscher Akademischer Austauschdienst and the Deutsche Forschungsgemeinschaft (for, respectively, J. H. , B. H-G. , and A. W.) of the German Federal Republic and by the assistance of the French Ministère de l' Education Nationale.

163

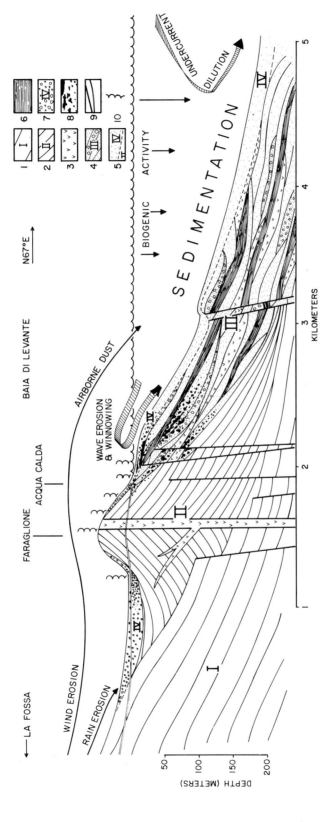

Fig. 11. Schematic representation of the extension and space variations of the exhalative sulfide deposit of the Baia di Levante, at Vulcano.
1 = "La Fossa" volcano formations; 2 = "Vulcanello"volcano formations;
3 = massive lava flows; 4 = recent tuffs and breccias; 5 = volcanic sands;
6 = tuffites and fine grained sediments; 7 = alluviums and talus; 8 = sulfides;
9 = fault (heavy line) and water table (double thin line); 10 = fumarole

Bibliography

ARNORSSON, S. , HAWKES, H. E. , TOOMS, J. S. : Present day formation of pyrite in hot springs in Iceland. Applied Earth Science, Trans. Inst. Mining and Metallurgy, Sec. B, 76, (England), p. 115-117 (1967).

BALLESTER and TONANI: Personal communication (1967).

BARTH, T. F. W. : Volcanic geology, hot springs and geysers of Iceland. Carnegie Inst. Wash. , Publ. No. 587 (1950).

BERGEAT, A. : Die äolischen Inseln geologisch beschrieben. Abh. kgl. bayer. Akad. Wiss. II cl. , 20, I. Abt. , (München), p. 1-274 (1899).

— Der Cordieritandesit von Lipari, seine andalusitführenden Einschlüsse und die genetischen Beziehungen zwischen dem Andalusit, Sillimanit, Biotit, Cordierit, Orthoklas und Spinell in den letzteren. N. Jb. Min. , Beil. , 30, (Stuttgart) p. 575-627 (1910).

BERNAUER, F. : Rezente Erzbildung auf der Insel Vulcano. Fortschr. Min. Krist. Petr. , 17, 28 p. (1933).

— Rezente Erzbildung auf der Insel Vulcano. Teil I, N. Jb. Miner. , Beilage 69, p. 60-91 (1935a).

— Rasche Verkiesung organischer Reste an vulkanischen Schwefelquellen. Sonder Abdruck aus dem Zentralblatt f. Min. etc. , Abt. A, no. 11, p. 343-344 (1935b).

— Primäre Teufenunterschiede, Verwitterungs- und Anreicherungsvorgänge am Krater von Vulcano. Fortschr. Min. Krist. Petr. , 20, 31 p. (1936).

— Rezente Erzbildung auf der Insel Vulcano. Teil II, N. Jb. Miner. , Beilage 75, p. 54-71 (1940).

BERNER, R. A. : Iron sulfides formed from aqueous solution at low temperatures and atmospheric pressure. J. Geology, 72, p. 293-306 (1964).

— The synthesis of framboidal pyrite. Econ. Geol. , 64, p. 383-384 (1969).

— Sedimentary pyrite formation. Am. J. Sci. , 268, p. 1-23 (1970).

DEGENS, E. T. , ROSS, D. A. : Hot Brines and Recent Heavy Metal Deposits in the Red Sea. Springer-Verlag New York Inc. , 600 p. (1969).

DE FIORE, O. : I fenomeni avvenuti a Vulcano dal 1890 al 1913. Part I and II. Z. Vulk. , I, p. 57-73, and II, p. 16-66 (1914-1915).

— I fenomeni eruttivi avvenuti a Vulcano (Isole Eolie) nel 1916. Boll. Soc. Sism. It. , XXVI, p. 246-262 (1920).

— Di un solfuro de ferro delle fumarole sottomarine di Vulcano formatosi nel 1916. Rend. Cont. Acad. Lincei, Ser. 5 XXX, p. 142-146 (1921).

— Vulcano (Isole Eolie). Riv. Vulcan. Friedländer Inst. , suppl. III, 393 p. (1922)

— Brevi note sull'attivitá di Vulcano (Isole Eolie). Bull. Volc. II, 7 p. (1924).

DESSAU, G. : Nuovi studi su Vulcano. La Ricerca Scientifica, I, p. 620-633 (1934). (Published also in Mem. Ist. Geol. Appl. , 4, Napoli, p. 31-48 (1951))

ELSKENS: Personal communication (1967).

HJELMQVIST, S. : Resa till Lipariska Öarna. Geol. Foren. Stockholm Forh. , 73, p. 473-491 (1951).

HONNOREZ, J. : La formation actuelle d'un gisement sous-marin de sulfures fumerolliens à Vulcano (mer tyrrhénienne). Partie I: Les minéraux sulfurés des
tufs immergés à faible profondeur. Mineral. Deposita, 4, p. 114-131 (1969).

HONNOREZ, J. , KELLER, J. : Xenolithe in Vulkanischen Gesteinen der Äolischen
Inseln (Sizilien). Geologische Rundschau, 57, p. 719-736 (1968).

HONNOREZ-GUERSTEIN, B. , WAUSCHKUHN, A. : Sulfidization of iron oxides.
(in preparation).

IMBO, G. : Sulle osservazioni termiche di fumarole nell'isola di Vulcano. Ann. R.
Osservatorio Vesuviano, Ser. 4, 3, p. 153-161 (1935).

JAMES, A. L. : Preliminary reconnaissance of Punta Banda, Baja California, Mexico.
(Unpublished) (1970).

KELLER, J. : Die historischen Eruptionen von Vulcano und Lipari (Deutungen alter
Berichte aufgrund neuer geologischer Befunde). Z. Deutsch. Geol. Ges. , 121,
p. 150-155 (1970a).

—   Datierung der Obsidiane und Bimstuffe von Lipari. N. Jb. Geol. Paläont. Mh,
p. 90-101 (1970b).

MARTINI, M. , TONANI, F. : Geochimica dei gas vulcanici per fini di previsione
dell'attività vulcanica esplosiva. Rapp. no. 1, rilevamento idrogeochimico
di Vulcano. Rapp. no. 1A, rilevamento idrogeocheimico di Vulcano. Rapp.
no. 5, relazione riassuntiva generale sulle mission effettuate nell'isola di
Vulcano nel 1969. Unpublished, Palermo (1970).

MATTEUCI, R. : Sull'attività dei vulcani Etna, Vesuvio, Vulcano, Stromboli e San-
torino nell'autunno 1898. Soc. Sism. It. , 5, p. 132-144 (1898).

MORI, A. : Un'escursione al cratere di Vulcano. Publ. Ist. Geografia Fisica e Vul-
canologia R. Univ. Catania, No. 12, p. 11, Firenze (1919).

RICKARD, D. T. : The microbiological formation of iron sulphides. Stockholm Contr.
Geol, 20, p. 49-66 (1969a).

—   The chemistry of iron sulphide formation at low temperatures. Stockholm
Contr. Geol. , 20, p. 67-95 (1969b).

—   The origin of framboids. Lithos, 18, p. 269-293, Amsterdam (1970).

ROBERTS, W. M. B. : The formation of pyrite from hydrated iron oxide in aqueous
solution at 20 °C. Mineral. Deposita, 3, p. 364-367 (1968).

ROBERTS, W. M. B. , WALKER, A. L. , BUCHANAN, A. S. : The Chemistry of Pyrite
in Aqueous Solution and its Relation of the Depositional Environment. Mineral.
Deposita, 4, p. 18-29 (1969).

SCHULZ, O. : Unterwasserbeobachtungen im sublitoralen Solfatarenfeld von Vulcano
(Äolische Inseln, Italien). Mineral. Deposita, 5, p. 315-319 (1970).

STANTON, R. L. , BAAS-BECKING, L. G. M. : The formation and accumulation of
sedimentary sulphides in seaboard volcanic environments. K. Akademie Weten-
schappen, Proc. , Ser. B, 65, p. 236-243, Amsterdam (1962).

TROVATINI, G. M. : Dissertazione chimico-fisica sull'analisi della acqua minerale
dell'isola di Vulcano nel porto di Levante detto volgarmente l'Acqua del Bagno.
(1786).

VALETTE, J. : Etude sédimentologique et géochimique des dépôts littoraux entourant
l'île Vulcano (Sicile). Thesis presented to "La Faculté des Sciences de Paris
pour l obtention du doctorat de 3e cycle". Unpublished, Paris, 175 p. (1969).

WAUSCHKUHN, A.: Untersuchungen über die Bildung einer submarinen Sulfidlager-
    stätte bei Vulcano, Italien. Thesis presented to Mineralogisch-Petrographisch
    Institut der Universität Heidelberg, eingereicht als Diplomarbeit. Unpublished
    Heidelberg, 114 p. (1970).

ZAMBONINI, F., DE FIORE, O., CAROBBI, G.: Su un Solfobismutito di piombo di
    Vulcano (Isole Eolie). Rendiconti Acc. Sci. Fisica Mat. Napoli, 31, p. 24-29
    (1925).

Addresses of the authors:

J. HONNOREZ   Rosenstiel School of Marine and
B. HONNOREZ-GUERSTEIN   Atmospheric Science,
                             Miami, Florida 33149 / U.S.A.

J. VALETTE   Centre de Recherches de Sédimentologie Marine
                             66 Perpignan / France

A. WAUSCHKUHN   Mineralogisch-Petrographisches Institut
                             der Universität Heidelberg
                             69 Heidelberg / Germany

Iron-bearing Oolites and the Present Conditions of Iron Sedimentation in Lake Chad (Africa)

J. Lemoalle and B. Dupont

Abstract

The oolites are found at a depth of 1 to 3 meters of water, in a zone of about 2700 sq. km, off the Chari delta. The grain size is around 0.250 mm. Montmorillonite nuclei are surrounded by goethite and silica, the iron content attains a maximum value of 49 % Fe_2O_3. Pollen analyses show a lacustrine formation. The present chemical conditions lead to a scheme of iron behaviour in which colloidal or adsorbed iron from the solid load of the incoming rivers (94ooo tons reactive iron in 1970-71) is separated from kaolinite, which becomes unstable in the lake, and forms a coprecipitate with silica.

Résumé

Les oolithes se trouvent sous 1 à 3 mètres d'eau, sur une surface de 2700 km^2 environ, au large du delta du Chari. Leur taille est de 0,250 mm environ. Les noyaux de montmorillonite sont entourés de goethite et de silice, la teneur en fer atteignant 49 % en Fe_2O_3. Les analyses polliniques indiquent une formation lacustre. Les conditions chimiques actuelles dans le lac permettent de présenter un schéma du comportement du fer dans lequel le fer colloidal ou adsorbé apporté par le fleuve (94000 tonnes de fer réactif en 1970-71) se sépare de la kaolinite, qui sort alors de son domaine de stabilité, et forme un coprécipité avec la silice.

- : -

The Chad basin provides an ideal example of a closed continental basin which allows us to follow, in time and space, the different steps of erosions, transportation and sedimentation of the dissolved and particulate elements. Some observations and results dealing with the transportation and sedimentation of iron in the lower part of the basin are presented here.

Observations

Lake Chad lies between parallels 12°20 and 14°20 North, and is divided between four states: Niger, Nigeria, Cameroun and Chad. Its largest dimension from the Chari delta is 230 km; its area 24000 sq.km, and its average depth 3.5 m. Islands and "bank-islands", shallows on which reeds and papyrus have grown, cover its north-eastern part; they are the tops of some dunes of an ancient and partly immersed erg (fig. 1).

The lake is a closed basin fed by the Chari River. Its water amounts to 83 % on the average of the total influx; rains and other tributaries contribute 13 and 5 % respectively. The Chari and its main affluent, the Logone, originate in a humid tropical climate on the northern side of the Central ridge which, in the Central African Republic, divides the waters between the Chad and the Congo basin. From west to east one can distinguish several zones in the upper part of the basin according to the nature

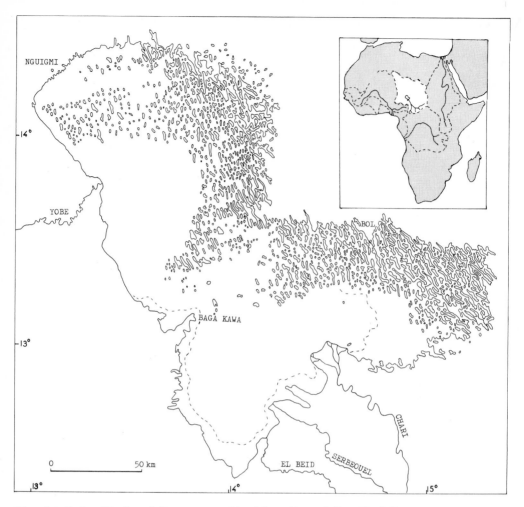

Fig. 1. Lake Chad and the geographical location of the Chad Basin

of the substratum; the river system is dense on granites and migmatites, moderately dense on gneiss, mica-schist and quartzite, and rather open on sandstones.

Waters reaching the lake are fresh (25 °C conductivity: 60 micromhos/cm on the average); the most important solids brought by the river are quartz and clay, mainly kaolinite, with montmorillonite being absent from the analyzed samples (GAC, private communication). The fine fraction of the lake sediments, on the other hand, is composed of montmorillonite; kaolinite and illite appear only as traces.

The lake sediments (DUPONT, 1970) can be divided into four main types, often related to the emerged landscape (fig. 2):

- Mud, rich in organic matter (13 % dry weight) is present in the whole band-islands zone.

- Clay appears under several aspects according to its degree of compaction and covers the bottom of the open-waters near the bank islands zone. Clay and mud cover the bottom of the whole archipelago.

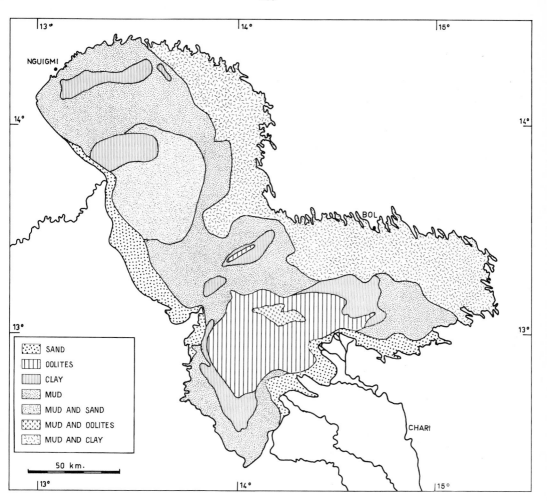

Fig. 2. The sediments of Lake Chad

- Sands are fine and well sorted and originate from two different stocks. The first
  one derives from the erg lying north-east of the lake and is almost entirely quartz-
  ous. The second one is micaceous and was brought by the rivers. It constitutes
  the sediments along the shore near the deltas of the Chari and Yobé rivers. The
  iron content of the sand is about 0.5 % (Fe_2O_3), but locally, in the zone of oolites,
  the grains are covered with a film of oxide and the amount of iron can reach 27 %.

- The iron-bearing oolites with a size similar to that of the sands are found on the
  bottom of a large zone of open-waters off the Chari delta.

The oolites

First described by GUICHARD (1957) as "pseudo-sable", the oolites of Lake Chad
are small, round shaped, polished or cracked grains of brown color when dry (fig.
3a, 3b). When wet, they look like coffee grounds. They form a layer of variable thick-

ness, up to 40 cm, including one or two local clay intercalations. They always lay on clay and are sometimes covered with mud. The oolites are found at a depth of 1 to 3 meters of water in a zone of about 2700 sq. km, westwards off the Chari delta (fig. No oolites have been found towards the east, off the fossil estuaries of the Chari, the latest of which has been flowing till about 150 A. C. (PIAS, 1970). This geographical location seems to assign them a recent origin related to the present course of the Chari River. This origin may be allochthonous, brought as detrital oolites by the Chari, or autochthonous by precipitation of the iron brought by the river.

The nature and origin of these oolites has been investigated by some analyses and observations.

The grain size distribution has a single mode at 0.250 mm, medians ranging from 0.205 to 0.283 mm. TRASK's sorting index, close to 1.2, indicates a well sorted material.

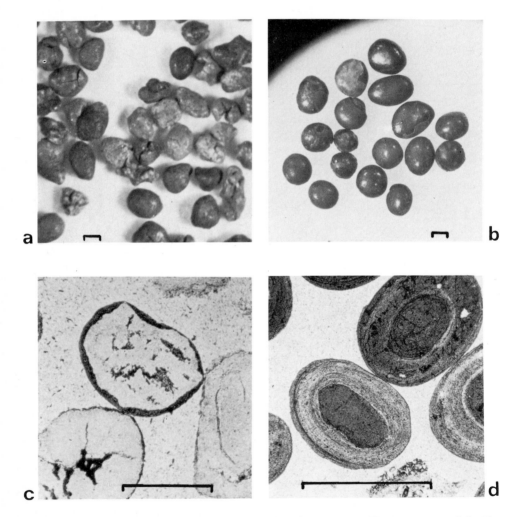

Fig. 3. Oolites. Top, a and b; the line is 0.2 mm long. Bottom, c and d, thin sections; the line is 0.5 mm long

Thin layer observations show several grain types according to a more or less matured state of the oolites. In the first state, light yellow-brown nuclei of clay, and sometimes quartz, are covered by a thin, i. e. a few microns thick dark cortex. Some nuclei had small internal cracks into which oxides entered; others display bare cracks (fig. 3c). The most matured grains have a clay nucleus covered by a thick reddish cortex (10 to 15 microns) exhibiting a clear oolitic structure (fig. 3d).

X-ray diffraction analysis has shown that the main crystallized minerals are montmorillonite and goethite. Kaolinite, quartz and calcite peaks also appear. The presence of calcite can be accounted for by the large quantity of molluscs (Corbicula africana) living on the oolite deposits.

By chemical analysis other features of the oolites were brought out: Concentrated hydrochloric acid released the iron into solution, but did not alter the shape of the oolites. After rinsing, the cortex (which may have split) appeared as a white, porous and siliceous framework revealing the association between iron and silica (table I).

A comparison between the load of solids of the river, lacustrine clays and oolites showed that the titanium to alumina ratio remained fairly constant while the ratios iron to alumina and silica to alumina increased considerably, the former passing from 0.5 to 30, and the latter from 2.5 to 36, in the river and the oolites respectively. Manganese is also concentrated in the oolites, but more irregularly than iron.

Consequently, a very large increase in silica, iron and manganese in regard to alumina is seen in the oolites and, to a lesser extent, also in some samples of granular "clay" taken from the border along the zone of oolites.

A pollen analysis was made [1] on two samples of washed oolites. These were treated according to the standard method followed by acetolysis (FAEGRI and IVERSEN, 1964).

When referring to the present vegetation (MALEY, 1970) and to other pollen analyses of recent and subrecent sediments of the Lake (MALEY, unpublished), it appears that the pollen spectra of the oolites are very similar to those of the sediments in the same region of the Lake. They are characterized by the prevalence of the local vegetation (Graminaceae) surrounding the water and of Lake Chad vegetation (Cyperaceae, Typha, Aeschinomene elaphroxylon) and also by the very low proportion of the allochthonous pollen grains carried from the sudanian zone by the river (table II).

From these observations some conclusions can be drawn

- in favour of the autochthony of the oolites: pollen spectra pointing to a lacustrine formation, absence of such oolites in the Chari alluvial deposits;

- in favour of the recent nature of the deposit: the oolites are located only off the present Chari delta and almost no other sediment is found on top of the oolite layer.

Present conditions of iron sedimentation

These observations led us to the analysis of some elements of the present transportation and sedimentation of iron in the lower Chari and in Lake Chad, i. e. the nature and quantity of the incoming iron, and chemical evolution of the waters and the particles in the Lake.

---

1) Analysed by J. MALEY, O. R. S. T. O. M. , Fort Lamy

Table I - Chemical analyses of some sediments of Lake Chad

No.	Nature	Ignition loss %	Quartz and non solubles %	Total phosphorus ‰	SiO_2 %	Al_2O_3 %	Fe_2O_3 %	TiO_2 %	MnO_2 %	CaO %	MgO %	K_2O %	Na_2O %
010	Solid load, Logone	11.2	22.3	2.52	31.2	21.4	9.25	1.56	0.045	0.30	0.67	0.60	0.26
013	Solid load, Logone	11.2	26.1	2.26	30.1	20.1	8.50	1.38	0.045	0.28	0.65	0.61	0.22
011	Solid load, Chari	13.4	8.9	1.76	40.0	26.3	7.20	1.08	0.035	0.25	0.46	0.60	0.22
012	Solid load, Chari	12.9	18.1	1.80	34.3	23.5	7.40	1.08	0.040	0.34	0.46	0.64	0.32
969	Clay, granular	12.4	11.0	1.80	40.0	14.4	17.00	0.55	0.050	1.16	1.04	0.56	0.41
1432 A	Clay, blue	11.9	12.5	0.89	43.3	19.5	7.50	1.00	0.085	0.65	0.95	0.80	0.23
1432 F	Clay, blue	16.3	16.3	8.95	40.3	21.3	8.00	0.93	0.045	0.65	0.94	0.75	0.23
1432 K	Clay, blue	14.3	17.3	0.87	40.1	17.6	6.75	0.64	0.035	0.60	0.84	0.68	0.25
1467	Clay, granular	15.7	17.2	1.41	34.3	11.0	17.50	0.85	0.075	1.17	0.98	0.51	0.18
1469	Clay, granular	18.5	6.15	0.78	36.8	5.3	28.50	0.29	0.125	1.50	0.98	0.39	0.19
1454	Clay, blue	8.1	54.9	1.25	18.2	9.7	6.50	1.04	0.075	0.54	0.49	0.42	0.18
1094	Oolites (large)	8.70	0.40	4.36	35.0	1.75	49.5	0.09	0.288	1.18	1.13	0.18	0.16
1455	Oolites	7.30	13.90	2.18	33.0	2.15	39.1	0.10	0.520	1.27	1.05	0.14	0.18
1457	Oolites	7.50	2.35	1.68	43.8	2.70	39.1	0.13	0.400	1.46	1.25	0.19	0.21
1459	Oolites	8.55	2.20	0.95	43.1	4.00	37.9	0.20	0.225	1.47	1.25	0.22	0.20
1461	Oolites	7.40	2.25	0.63	45.1	3.15	37.9	0.15	0.145	1.47	1.27	0.16	0.17
1463	Oolites	7.35	4.05	0.86	45.2	3.00	36.0	0.15	0.155	1.44	1.31	0.16	0.13
1465	Oolites	8.55	6.90	0.76	41.8	3.70	34.5	0.23	0.110	1.32	1.16	0.40	0.19
1479	Oolites	7.50	1.45	0.78	43.5	2.00	40.5	0.08	0.225	1.69	1.39	0.17	0.13
1483	Oolites	7.70	4.80	1.35	38.8	1.85	41.0	0.08	0.235	1.54	1.31	0.20	0.20

Table II - Pollen analysis of some oolites

	Sample 931	Sample 1094
Lacustrine vegetation		
Cyperaceae	59	31
Typha	23	2
Aeschynomene elaphroxylon (Papil.)	2	3
Total	84	36
Per cent	57 %	47.3 %
Local sahel vegetation		
Graminaceae	40	31
Phyllanthus rotundifolius (Euph.)	-	1
Compositae	-	1
Combretum sp.	2	-
Mitracarpus scaber (Rub.)	1	-
cf. Celosia trigyna (Amaran.)	1	-
Total	44	33
Per cent	29.3 %	43.4 %
Allochthonous sudanian vegetation		
Uapaca sp.	1	-
Alchornea cordifolia (Euph.)	1	-
Oleacea type	2	-
cf. Adina sp. (Rub.)	1	-
Ulmaceae	-	1
Total	5	1
Per cent	3.3 %	1 %
Undetermined	16	6
Total observed	149	76

It has been established by filtration experiments (LEMOALLE, 1969) that the concentration of dissolved iron is very low, i.e. lower than $50\mu g/l$, in the Chari and in the lake waters as well. The pH and oxidation conditions and the very low quantities of dissolved organic matter do not allow the presence of large concentrations of soluble or complexed iron. Soluble iron is therefore negligible compared to the iron present as solid sludge particles, and is neglected in the value of total iron. The charge of the sludge particles containing iron has been determined by electrophoresis. It is either negative or zero.

Dissolved iron being neglected, we have looked for that part of particulate iron which was able to react rather easily. This reactive iron (extracted in boiling 4 % HCl during 10 minutes, STRICKLAND and PARSONS, 1965) closely corresponds here to free iron as determined by DEB (1950), that is, particulate or adsorbed oxides and hydroxides (FRIPIAT and GASTUCHE, 1952) and represents about half of the total iron determined after alkaline fusion or triacid attack.

Weekly measurements on the Chari and Logone Rivers before their confluence allowed to measure that reactive iron amounts to 3.6 % of the solid load.

Maximum reactive iron concentration occurs at the beginning of the flood (fig. 4) and reaches very high values (30 mg/l) in the Logone. During 1970-71, when an average amount of water was measured, 94000 metric tons of reactive iron reached the Lake

Subjected to evaporation, the Lake water increases in salinity progressively from the delta and reaches values which are five and ten times higher in the east and in the

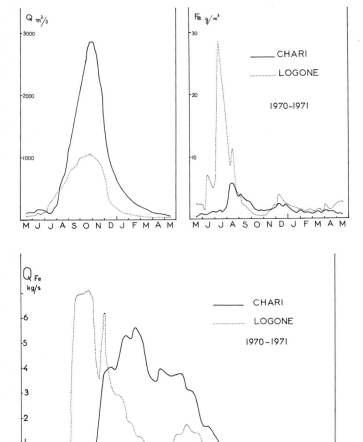

Fig. 4. Hydrograms (m³/s), reactive iron concentrations (g/m³) and iron flow (kg/s) in the Chari and Logone Rivers at Fort Lamy. Their sum is the inflowing quantity of iron received by the Lake

north, respectively, with some modifications in the ionic composition (CARMOUZE, 1970).

The mean composition of two stations are given here, i. e. the Chari Delta and Bol station located in the archipelago of the eastern zone where the ionic concentration is about twice as high.

	C μmhos/cm	pH	HCO_3^- me/1	Na^+ me/1	K^+ me/1	Ca^{++} me/1	Mg^{++} me/1	SiO_2 mg/1
Delta	60	6.9	0.70	0.16	0.06	0.27	0.21	24
Bol	120	7.8	1.39	0.37	0.10	0.53	0.40	34

The importance of biological factors, mainly photosynthesis in the pH regulation, must not be overlooked, the pH decreasing quickly when photosynthesis ceases (LE-MOALLE, 1969).

Reactive iron is more or less bound to clay particles. It is therefore important to know how the clay particles behave in the Lake. Equilibria between kaolinite and montmorillonite, which are the two main clay mineral species, are presented in two diagrams (fig. 5) corresponding with the following reactions:

kaolinite + SiO_2 + $Na^+ \rightleftharpoons$ montmorillonite Na + H^+

kaolinite + SiO_2 + $Ca^{2+} \rightleftharpoons$ montmorillonite Ca + $2H^+$

The equilibrium constants are, at a temperature of 25 $^{\circ}$C (ROBIE and WALDBAUM, 1968; TARDY, 1969):

$$pK = 4 \log \left[H_4SiO_4\right] - \log \frac{\left[Na^+\right]}{\left[H^+\right]} = -9.31$$

$$pK = 8 \log \left[H_4SiO_4\right] - \log \frac{\left[Ca^{2+}\right]}{\left[H^+\right]^2} = -15.70$$

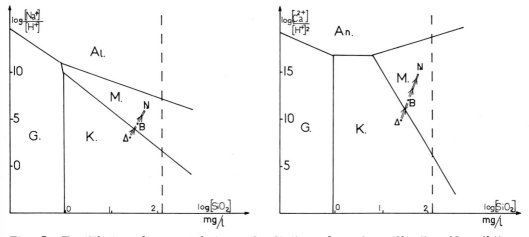

Fig. 5. Equilibrium diagrams between kaolinite and montmorillonite. Al = albite, G = gibbsite, K = kaolinite, M = montmorillonite, An = anorthite. Arrow indicates the water concentration starting from the estuary (Δ) to Bol (B) and to the most concentrated waters in the northern part of the Lake

The Chari water reaching the Lake is represented by a dot inside the kaolinite stability field. From the delta the lake water concentrates by evaporation. The representative point in the diagram shifts and crosses the kaolinite-montmorillonite equilibrium line rather rapidly, even before the concentration is double.

Although not dealing with the kinetics of the reactions, these thermodynamic relation allow us to say that there may be transition in the Lake from kaolinite to montmorillonite. This transformation or instability may be of some importance in the separation of adsorbed oxides and hydroxides from clays.

A scheme of iron behaviour

The silica of rivers precipitates by adsorption and reaction with inorganic particles when pH and salinity increase through mixing with seawater in estuaries (BIEN et al. 1958; LISS and SPENCER, 1970). When flowing in Lake Chad, the pH and electrolyte concentration of the waters increase. Clay particles become unstable and iron, whic settles rapidly in such conditions (COONLEY et al., 1971) is distributed between several types of sedimentary material.

The oxidized structure of the oolite cortex suggests a coprecipitation of silica and colloidal iron. The affinity of these two substances for each other is high (MORTIMER, 1941, 1942; FLEHMIG, 1970) and the molar composition of such a precipitate is a function of the medium pH and of the isoelectric points of its components (PARK 1967). When A and B are the isoelectric points of the iron oxides and of silica which react, the molar composition of the precipitate is a function of the pH and will be ne the following expression:

$$\% \ SiO_2 = \frac{A - pH}{A - B}$$

Uncertainty on the pH values of the zero point of the charge of silica and iron does not allow an exact calculation of the precipitate composition. For a pH 7.0 to 7.5 the theoretical ratio of silica to iron is close to 1. In the cortex of the oolites, this ratio is 0.86 on an average, with extreme values of 0.72 and 1.15.

Iron and silica reacting together, there must be, elsewhere in the Lake, some sediments in which the alumina content is higher than in the solid load of the Chari; mor analyses are being performed.

Conclusion

The first results of this work on the oolites of Lake Chad led to the conclusion that the deposit is autochthonous and recent. A study of the present conditions of iron sec imentation in the Lake made it possible to define its behaviour. In the proposed sche iron which is in particulate form undergoes an evolution in which it is dissociated fr clay particles during the water concentration, and coprecipitates with silica. The oo litic structure might then form on small clay grains and sometimes quartz which is found on the Lake bottom in shallow, well aerated warm waters (18 to 30 $^{\circ}$C) stirred all the year long by the wind.

With the available data, the oolitic iron, on an area of 2700 sq. km and 5 cm mean thickness, is evaluated to be 30.10^6 metric tons, expressed in pure iron. This is equal to the load of free reactive iron of the Chari during 320 years, given a constan yearly load of sediment.

A comparative study of numerous deposits of iron-bearing oolites in the post-Eocene formations of Africa has led FAURE (1966) to define the physico-chemical conditions of oolite formation as follows: agitation during concentration, with low depth and warm temperature. Our observations corroborate these conclusions and those of BU-BENICEK (1961) who writes, "iron deposition in its oxidized form, oolites growing in dynamic conditions before settling". The mechanism which is given here can also apply to the iron-bearing oolites found by LENEUF (1962) in Ivory Coast and, more generally, should be helpful when assessing the chemical conditions of the formation of iron and silica oolite beds.

- : -

The help of Mrs. Maley in the translation of this paper is gratefully acknowledged.

Bibliography

BIEN, C.S. , CONTOIS, D. E. , THOMAS, W. H. : The removal of silica from fresh-water entering the sea. Geochim. Cosmochim. Acta, 14, 1, p. 35-54 (1958).

BUBENICEK, L. : Recherches sur la constitution et la répartition du minerai de fer dans l'Aalénien de Lorraine. Sc. de la Terre 8, no. 1-2, p. 5-204 (1961).

CARMOUZE, J. P. : La salure globale et les salures spécifiques des eaux du lac Tchad. Cha. ORSTOM, sér. Hydrobiol. 3, 2, p. 3-14 (1970).

COONLEY, L. S. , BAKER, E. B. , HOLLAND, H. D. : Iron in the Mullica River and in Great Bay, New Jersey. Chem. Geol. 7, p. 51-63 (1971).

DEB, B. C. : The estimation of free iron oxides in soils and clays, and their removal. J. Soil Sci. 1, 2, p. 212-220 (1950).

DUPONT, B. : Distribution et nature des fonds du lac Tchad. (nouvelles données). Cah. ORSTOM, sér. Géol. , 2, 1, p. 9-42 (1970).

FAEGRI, K. , IVERSEN, J. : Textbook of pollen analysis, 2d ed. , Munksgaard, Copenhague, 237 p. (1964).

FAURE, H. : Reconnaissance géologique des formations sédimentaires post-paléo-zoïques du Niger Oriental. Mém. BRGM, no. 47, 632 p. (1966).

FLEHMIG, W. : Zum Vorkommen von SiO_2 in Nadeleisenerzooiden. Contr. Mineral. and Petrol. 28, p. 19-20 (1970).

FRIPIAT, J. J. , GASTUCHE, M. C. : Etude physico-chimique des surfaces des argiles. Les combinaisons de la kaolinite avec les oxydes de fer trivalents. Publ. INEAC, Sew-Scient. , no. 54, 60 p. (1952).

GUICHARD, E. : Sédimentation du lac Tchad. O. R. S. T. O. M. , Fort-Lamy, 46 p. multigr. 1 carte H. T. (1957).

LEMOALLE, J. : Premières données sur la production primaire dans la région de Bol (avril-octobre 1968) (Lac Tchad). Cah. ORSTOM, sér. Hydrobiol. 3, 1, p. 107-120 (1969).

— Premières données sur la répartition du fer soluble dans le lac Tchad. O. R. S. T. O. M. , Fort-Lamy, ronéo, 10 p. (1969).

LENEUF, N. : Les pseudo-oolithes ferrugineuses des plages de Côte d'Ivoire. C. R. Som. Soc. Géol. France 5, p. 145-146 (1962).

LISS, P. S. , SPENCER, C. P. : Abiological processes in the removal of silicate from sea water. Geochim. Cosmochim. Acta, 34, p. 1073-1088 (1970).

MALEY, J. : Contributions à l'étude du Bassin Tchadien. Atlas de pollens du Tchad. Bull. Jard. Bot. Nat. Belgique, 40, p. 29-48 (1970).

MORTIMER, C. H. : The exchange of dissolved substances between mud and water in lakes. J. Ecology, 29, p. 280-329 (1941).

— The exchange of dissolved substances between mud and water in lakes. J. Ecology, 30, p. 147-201 (1942).

PARKS, G. A. : Aqueous surface chemistry of oxides and complex oxide minerals. in: Equilibrium Concepts in natural water systems, W. STUMM, ed.. Am. Chem. Soc. Washington, p. 121-160 (1967).

PIAS, J. : Les formations sédimentaires tertiaires et quaternaires de la Cuvette Tchadienne et les sols qui en dérivent. Mém. ORSTOM, Paris, no. 43, 407 p. (1970).

ROBIE, R. A. , WALDBAUM, D. R. : Thermodynamic properties of minerals and related substances at 298. 15 ^{O}K (25.0 ^{O}C) and one atmosphere (1013 bars) pressure and at higher temperatures. U. S. Geol. Surv. Bull. 1259, 256 p. (1968).

STRICKLAND, J. D. H. , PARSONS, T. R. : A manual of sea water analysis. Fish. Res. Bd Canada, Bull. 125, 203 p. (1965).

TARDY, Y. : Géochimie des altérations. Etude des arènes et des eaux de quelques massifs cristallins d'Europe et d'Afrique. Thèse Fac. Sc. Univ. Strasbourg, 274 p. ronéo. (1969).

Address of the authors:

O. R. S. T. O. M.
B. P. 65
Fort - Lamy / Tchad

Physical Sedimentation in Precambrian Cherty Iron Formations of the Lake Superior Type

Joseph T. Mengel

Abstract

Sand size iron-rich granules in cross bedded and ripple marked strata containing other clasts establish the particulate nature of the original materials of six Precambrian cherty iron formations of the North American Lake Superior region. Granules are of medium sand size, moderately well sorted, with a mesokurtic size distribution skewed toward an excess of fine material. Packing densities are similar to arenites, but packing proximities are lower. Granule specific gravities average about that of opal. Large grain size and low density may favor selective transport and explain the usual absence of detrital quartz and aluminous materials.

Iron formation layering is lenticular in character with bed thickness a function of the coarseness of the original materials. Type of iron mineralogy or degree of metamorphism do not influence bed thickness, nor does presence or absence of volcanic materials.

Introduction

The Precambrian cherty iron formations are accumulations of particulate matter whose physical characteristics can be studied quantitatively to establish the physical conditions of sedimentation and form an objective basis for comparative studies of the many formations throughout the world. In this paper are presented measurements to characterize certain common and widely distributed textures and structures of six Precambrian cherty iron formations of the North American Lake Superior region. Measurements were made on two prinicpal physical types of iron formation: (1) Even bedded, largely non-granule-bearing varieties, with alternating layers of chert-rich and iron-rich material, and (2) Irregularly thicker bedded, granule-bearing varieties with sand-size iron rich granules set in a chert matrix. JAMES (1954, 1955) has described the range of iron formation mineralogies which are present in these formations. It is reasonably certain that neither the iron mineralogy nor the rank of metamorphism seriously effects the features measured.

Regional Setting

The structure, extent, and meximum thickness of the formations studied are shown in Figure 1 and their map relationships in Figure 2. Classical descriptions of the formations and districts are given by VAN HISE and LEITH (1911) and recent views of regional chronology and relationships are given by GOLDICH et al.(1961).

With the single exception of the Soudan, the iron formations are considered to be approximately correlative and of Middle Precambrian age. They belong to the "Hu-

180

ronian" (or "Animikean") subdivision of the Lake Superior Proterozoic succession. The Soudan formation is associated with an older, more volcanic rich succession classically considered to be of Archean age. It is perhaps upwards of a billion years older than the 1.7 - 1.9 billion year age of the Middle Precambrian rocks. All were

District	Structure	Extent	Iron Formation	Thickness
Menominee	Faulted syncline	25 miles long	Vulcan	650'
Felch Mountain	Narrow faulted syncline	10 miles long	Vulcan	50'
Marquette	Complex synclinorium	15 miles long	Negaunee	1000'
Gogebic	Northwest dipping (60 degrees) faulted monocline	70 miles long	Ironwood	400'
Gunflint	Southeast dipping (15 degrees) monocline	110 miles long	Gunflint	400'
Vermillion	Complexly faulted synclinal bodies in greenstones	Lenses up to 10 miles long	Soudan	100'
Mesabi	Southeast dipping (15 degrees) monocline	100 miles long	Biwabik	700'

Fig. 1. Iron-bearing districts studied

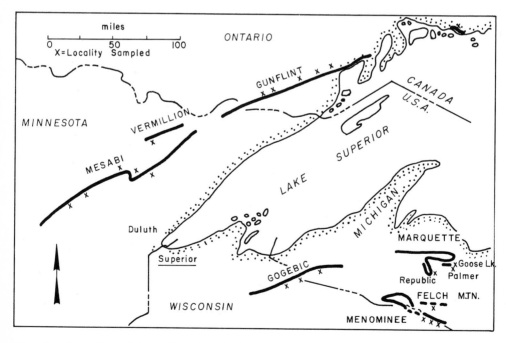

Fig. 2. Location of iron-bearing districts studied

181

affected to some extent by the Penokeean orogeny, which followed the deposition of the younger iron formations, and by intrusion of the vast Duluth Gabbro body and related rocks about 1.1 billion years ago. Metamorphic zones on the south shore of Lake Superior have been described by JAMES (1955) and in the Mesabi district by GUNDERSON and SCHWARTZ (1962) and FRENCH (1968).

Iron Formation Composition

The composition range of the iron formations is limited, consisting of about one fourth to one third ferrous and/or ferric iron by weight, together with minor amounts of Al_2O_3, CaO, and MgO, and major amounts of SiO_2 in the form of chert. They contrast with most younger ironstones which, while they contain about the same percentage of iron, also contain much larger amounts of the elements listed, and lack essential chert (cf. LEPP and GOLDICH, 1964).

The chert content of the Precambrian iron-bearing rocks is an "original" constituent of the formations because:

1) its distribution is not controlled by metamorphism, fracturing, folding, depth of burial, nearness to intrusive rocks or proximity to the weathering surface.

2) chert clasts occur in sandstones interbedded with the iron formation (Figure 3); and chert beds occur in these same sandstones (Figure 4, Isabella mine, Marquette district).

3) diverse lithologies occur in intraformational conglomerates and as lenses or channels within the iron formation (TYLER and TWENHOFEL, 1952).

4) the chert contains delicate microfossils (BARGHOORN and TYLER, 1965), and stromatolite algal structures.
The iron content of the formations is "original" for many of the same kinds of reasons, as are also the features measured in this study.

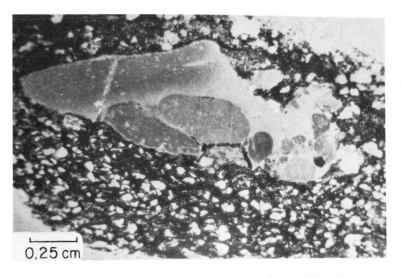

0.25 cm

Fig. 3. Chert clasts in sandstones interbedded with iron formation, Isabella Mine, Palmer area, Marquette District

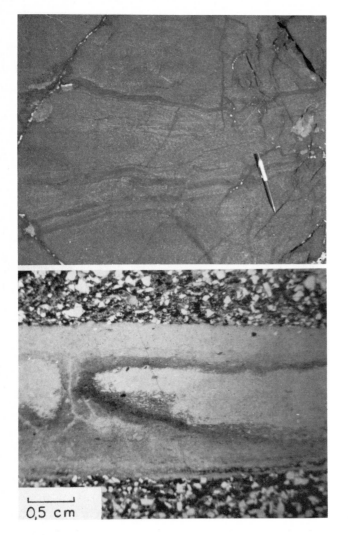

Fig. 4. Chert bed interlayered with sandstone within iron formation, Isabella Mine, Palmer area, Marquette District

The Particulate Nature of Iron Formation

Iron-rich granules in a chert matrix occur in thick, irregularly bedded strata which form a characteristic lithosome of some members of the Gunflint, Biwabik, Ironwood and Vulcan formations. The association of granules with chert and carbonate pebbles (Figure 5), fragments of algal structures, oolites and detrital quartz proves that they may be regarded as particulate detritus, and that the chert matrix which separates them was particulate at the time of accumulation. The same conclusion can be reached from the frequent occurrence of cross-bedding in granule-bearing layers. Particularly fine examples of cross-bedding in the Gunflint formation are illustrated by GOODWIN (1956) and are to be seen in the Walker and other mines at the western end of the Mesabi district. A small scale example is shown in Figure 6, taken of a specimen from the Auburn mine in the central part of the Mesabi district. Cross beds also

Fig. 5. Iron-bearing carbonate pebble in granule-bearing iron formation, Embarrass Lake, Mesabi District

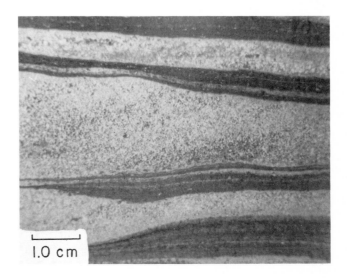

Fig. 6. Small scale cross-bedding in granule-bearing iron formation, Auburn Mine, Mesabi District

occur in non granule-bearing iron formation, as at the Athens mine in the Marquette district. Ripple marks, indicative of the finely particulate character of the original materials of the iron formation, are also known. Figure 7 illustrates ripple marks in a sample from the Quinnesec mine in the Menominee district. These ripples are seemingly also rain drop (?) imprinted.

Iron Formation Layering

Typical non granule-bearing iron formation layering is shown in Figure 8, from the Kloman mine near Republic in the Marquette district. Notice that the chert beds (light) are actually lenses and that thicker beds can be formed by stacking these lenses. The Kloman mine is an unusual exposure because the iron formation, being almost as brit-

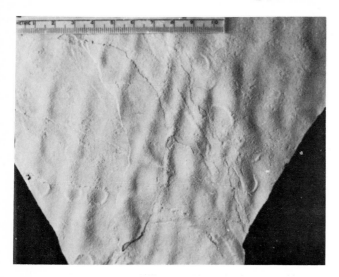

Fig. 7. Ripple marks with possible rain-drop imprints (?), Quinnesec Mine, Menominee District

Fig. 8. Typical non granule-bearing iron formation layering, Kloman Mine near Republic, Marquette District

tle as glass, is pervasively jointed in most localities, making the tracing of individua layers difficult and leading to an incorrect belief in the lateral continuity of bedding.

Related to this belief is the suggestion made by several authors (SAKAMOTO, 1950, for example), that the alternation of iron-rich and chert-rich layers in non granule-bearing iron formation is a phenomenon similar to varving. Figure 9 is a plot of a sequence of 196 successive chert and iron beds at the Kloman mine illustrating that there is virtually no correlation between superjacent layers, such as might be expected if they were seasonally deposited. Also bearing on this problem are the typical thicknesses of both the iron-rich and the chert-rich layers. If these deposits are seasonal, accumulation at the high rate of about an inch per year is indicated. These data, taken in conjunction with the overall facies relationships demonstrated by workers such as WOLFF (1917), HOTCHKISS (1919), ALDRICH (1929), WHITE (1954), and GOODWIN (1956), suggest that energy distribution patterns, rather than seasonal patterns are reflected in the deposits. Obviously a seasonal rhythm to the introduction of material to the basins in which the iron formations accumulated is not at all ruled out by this evidence - but it is doubtful whether the non granule.bearing facies records such a pulse in easily recognizable form.

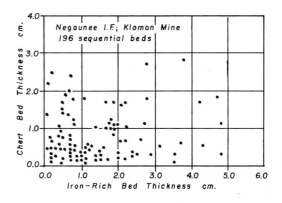

Fig. 9. Chert/iron-rich bed sequential relationships, Kloman Mine near Republic, Marquette District

Five thousand non granule-bearing chert bed thickness measurements taken throughout the region bring to light other interesting information:

1) The bed thickness of the non granular chert is the same throughout the region. A thickness of one centimeter is typical.

2) The bed thicknesses are the same for "older" and "younger" iron formations regardless of the fact that the Soudan formation occurs in association with volcanic materials, while the younger rocks are part of a shelf-type association. (Figure 10)

3) Non granular chert bed thicknesses are log-normal. (Figure 11)

4) Bed thickness is not dependent on associated iron-rich layer mineralogy. (Figure 12)

5) Bed thickness is not influenced by metamorphic rank. (Figure 13)

This data justifies measurement of layers as a valid record of primary phenomena and indicates a similarity to the physical environments of accumulation for this physical facies throughout the region.

Figure 14, a plot of data from the Munro mine in the Menominee district illustrates that both granular and non granular iron formation layering is lenticular in character. Also evident is the fact that both types of layering tend to approach a maximum thickness value which is greater for the granule-bearing type than for the non granular. Figure 15, representing measurements from all localities sampled in the Menominee district, records the same phenomenon. This data graphically demonstrates the fact that, in general, the iron formation layer thickness has a functional relationship to the average grain size of the original material, as is the case with detrital material

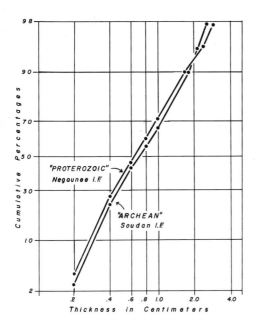

Fig. 10. Chert bed thickness comparison between older (Archean) and younger (Proterozoic) iron formations

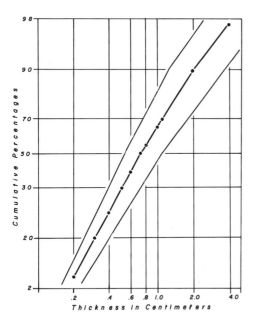

Fig. 11. Mean bed thickness distribution, 5000 non-granular chert layers from the six iron formations studied

187

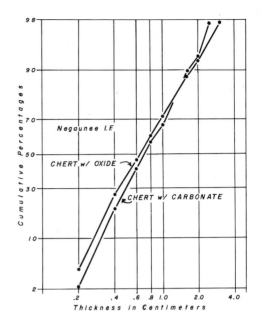

Fig. 12. Chert bed thickness comparison between beds associated with iron oxide mineralogy and those associated with iron-bearing carbonate mineralogy

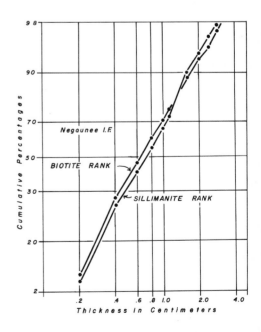

Fig. 13. Chert bed thickness comparison between beds in iron formation in biotite rank and those in iron formation in sillimanite rank

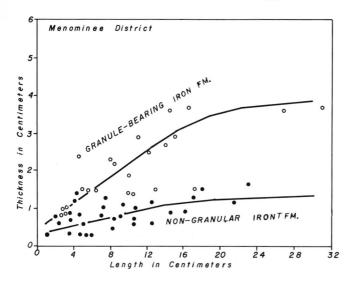

Fig. 14. Lenticular character of bedding of chert beds in non-granular iron formation and beds in granule-bearing iron formation

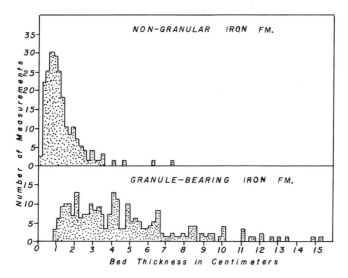

Fig. 15. Bed thickness comparison between chert beds in non-granular iron formation and beds in granule-bearing iron formation, Menominee District

Granules

As noted previously, iron-rich granules in a chert matrix (Figure 16) form a characteristic lithosome of some members of the Gunflint, Biwabik, Ironwood, and Vulcan formations. They are also known in portions of the Negaunee formation. Granule have a mean grain size at the coarse end of the medium sand size range, moderately good sorting, and a nearly symmetrical, mesokurtic grain size distribution skewed toward an excess of fine material. This data is based on measurements of a selection of 108 thin sections chosen as representative after examinations of the collections made for the U.S. Geological Survey monographic studies in the Lake Superior Region (1885-1910), the private collection of Professor S.A. Tyler, and about 600 sections made expressly for this study. A total of about 3000 slides were examined, as well as some material which had been sawed into slabs but not made up into slides. Such a large and diverse body of samples would seemingly represent most of the more common phenomena to be seen in the formations under study, but the extent and thickness of the formations, coupled with the metamorphic and other changes they have suffered in some areas stand as a warning that the information presented is but a first step toward characterization.

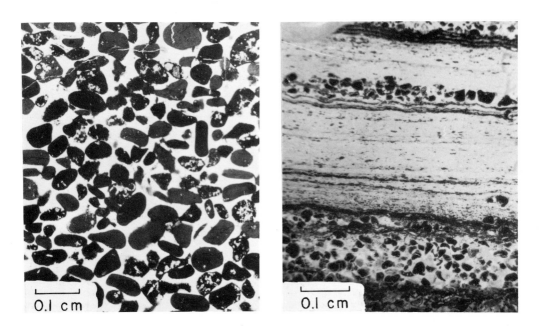

Fig. 16. Iron-rich granules in (a) chert matrix and (b) cherty iron-bearing carbonate matrix

The apparent long diameter of 250 randomly chosen grains in each slide selected for detailed examination was measured using a micrometer ocular. Results were cumulated and plotted to get the following parameters based on the work of FOLK and WARD (1957):

Folk mean:
$$\frac{P_{16} + P_{50} + P_{84}}{3}$$

Inclusive graphic standard deviation:
$$\frac{P_{84} - P_{16}}{4} + \frac{P_{95} - P_{5}}{6.6}$$

Inclusive graphic skewness:
$$\frac{P_{16} + P_{84} - 2 P_{50}}{2 (P_{84} - P_{16})} \qquad \frac{P_{5} + P_{95} - 2 P_{50}}{2 (P_{95} - P_{5})}$$

The typical granule size distribution is shown in Figure 17, which is very nearly a straight line plot between $P_5 = 0.2$ and $P_{95} = 1.9$ on log probability paper (Keuffel and Esser 259-23). Limited material from the Belcher Islands and Labrador suggest that this curve is apt to be widely applicable in characterizing grain size distribution Mean granule size ranges from 1/4 to 1 mm, i.e. from fine to very coarse sand. A plot of P_{84} against P_{16} gives a simple visual appreciation of the data (Figure 18).

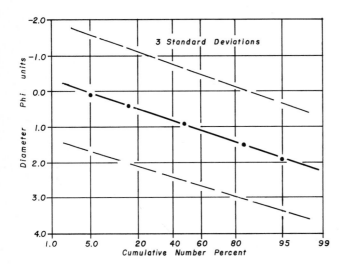

Fig. 17. Typical size distribution iron-bearing granules

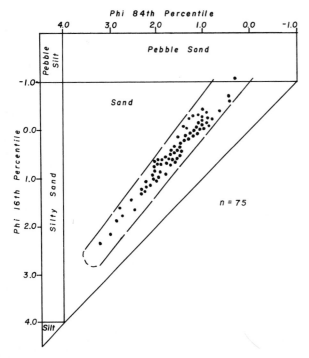

Fig. 18. Grain size distribution iron-bearing granules

Coarser grain size samples are known but no samples in the collection established the existence of finer granules. The finer materials may well be represented in the very fine, even-bedded "slaty" facies which is a prominent component of the Biwabik and other formations.

Plots of mean size versus skewness (Figure 19) shows a skewness range from -0.3 to +0.3, i.e. coarse skewed to fine skewed, but most samples are within -0.2 and +0.2 indicating distributions which do not depart strongly from symmetricality. The tendency to negative skewness in fine sizes is similar to that reported for fine clastics (INMAN, 1949). A tendency to accentuate positive skewness has also been noted from diagenetically altered sandstones (FRIEDMAN, 1962).

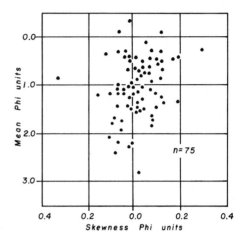

Fig. 19. Granule mean size versus granule distribution skewness

Plots of mean size versus standard deviation (Figure 20) indicate a mean size range between about 1/4 and 1 mm, i.e. between fine and coarse sand. The standard deviations are 0.5 ± .3 with the majority between .5 and .8. The plot suggests a tectonically stable environment with a low rate of deposition in which considerable reworking took place. This conclusion is an independent confirmation of interpretations reached from stratigraphic studies by WHITE (1954) and JAMES (1954).

Moderately well sorted fine grained detrital quartz occasionally found in association with the granules in the same chert matrix permits estimation of the specific gravity

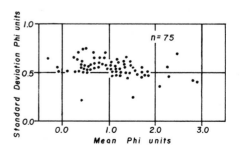

Fig. 20. Granule standard deviation versus mean size

of the granules at the time of deposition. If it is assumed that the quartz and the granules had about the same shape and were deposited from the same current, it is possible to estimate the specific gravity of the granules using the work of McINTYRE (1959). The functional relationship is:

$$\emptyset A - \emptyset B = \frac{1}{1.5} \ (Log_2 \ D_A - Log_2 \ D_B)$$

where: $\emptyset$ = phi mean grain size
         D = density of the mineral minus the density of water
      1.5 = constant for sand size mineral.

Results indicate a specific gravity for the granules which is about 2.0, i.e. similar to that for opal. Such a low specific gravity, coupled with a rather coarse grain size may account for the almost complete lack of quartz in the granule facies, even when it is interbedded with sandstones, and for the segregation of the granules from aluminous material, resulting in the typically low aluminum content of much cherty iron formation.

Packing densities (KAHN, 1956 a and b) for granule-bearing iron formations are similar to, and show the same decrease with grain size as do those of other arenites. Packing proximities, i.e. the number of grain to grain contacts among the granules range from about 5 - 45 % and are thus lower than for arenites generally. This may reflect at least two things about the granules:

1) a tendency to mutual repulsion of grains of colloidal character having adsorbed similar cations,

2) a low rate of granule production or deposition.

Perhaps this latter alternative is realistic in view of a characterization of large, light granules being moved more easily than the finer, denser particulate matter which is now chert.

References

ALDRICH, H.R.: The geology of the Gogebic iron range of Wisconsin. Wisconsin Geol. and Nat. History Survey, Econ. ser. 24, Bull. 71 (1929).

BARGHOORN, E.S., TYLER, S.A.: Microorganisms from the Gunflint chert. Science v. 147, no. 3658, p. 363-375 (1965).

FOLK, R.C., WARD, W.C.: Brazos River bar: A study of the significance of grain size parameters. Jour. Sedimentary Petrology, v. 27, p. 3-26 (1957).

FRENCH, B.M.: Progressive contact metamorphism of the Biwabik iron formation, Mesabi Range, Minnesota. Minnesota Geol. Survey Bull. 45 (1968).

FRIEDMAN, G.M.: On sorting, sorting coefficients, and the lognormality of the grain-size distribution of sandstones. Jour. Geology, v. 70, p. 737-753 (1962).

GOLDICH, S.S., NIER, A.O., BAADSGAARD, H., HOFFMANN, J.H., KRUEGER H.W.: The Precambrian geology and geochronology of Minnesota. Minnesota Geol. Survey Bull. 41 (1961).

GOODWIN, A.M.: Facies relations in the Gunflint iron formation. Econ. Geology, v. 51, p. 565-596 (1956).

GUNDERSEN, J.N. , SCHWARTZ, G. M. : The geology of the metamorphosed Biwa-bik iron formation, Eastern Mesabi district, Minnesota. Minnesota Geol. Survey Bull. 43 (1962).

HOTCHKISS, W. O. : Geology of the Gogebic range and its relation to recent mining developments. Eng. Mining Jour. , v. 108, p. 443-452, 501-507, 537-541, 577-582 (1919).

INMAN, D. L. : Sorting of sediments in the light of fluid mechanics. Jour. Sediment-ary Petrology, v. 19, p. 51-70 (1949).

JAMES, H. L. : Sedimentary facies of the Lake Superior iron formations. Econ. Geology, v. 49, p. 235-292 (1954).

— Zones of regional metamorphism in the Precambrian of northern Michigan. Geol. Soc. America Bull. , v. 66, p. 1455-1488 (1955).

KAHN, J. S. : The analysis and distribution of the properties of packing in sandstones. Jour. Geology, v. 64, p. 385-395 (1956a).

— The analysis and distribution of packing properties in sand-size sediments. Jour. Geology, v. 64, p. 578-606 (1956b).

LEPP, H. , GOLDICH, S. S. : Origin of Precambrian iron formations. Econ. Geology, v. 59, p. 1025-1060 (1964).

McINTYRE, D. D. : The hydraulic equivalence and size distributions of some mineral grains from a beach. Jour. Geology, v. 67, p. 278-301 (1959).

SAKAMOTO, T. : The origin of the Precambrian banded iron ores. Amer. Jour. Science, v. 248, p. 449-474 (1950).

TYLER, S. A. , TWENHOFEL, W. H. : Sedimentation and stratigraphy of the Huron-ian of Upper Michigan. Amer. Jour. Science, v. 250, p. 1-27, 118-151 (1952).

VAN HISE, C. R. , LEITH, C. K. : Geology of the Lake Superior region. U. S. Geol. Survey Monograph 52 (1911).

WHITE, D. A. : The stratigraphy and structure of the Mesabi range, Minnesota. Minnesota Geol. Surv. Bull. 38 (1954).

WOLFF, J. F. : Recent geologic developments on the Mesabi iron range, Minnesota. Amer. Inst. Mining and Metallurgical Eng. Trans. , v. 56, p. 142-169 (1917).

Address of the author:

Department of Geology
Wisconsin State University
Superior, Wis. / U. S. A.

Reef Environment and Stratiform Ore Deposits (Essay of a Synthesis of the Relationship between Them)

G. Monseur and J. Pel

Abstract

Comparative analysis of Givetian reef complexes from Belgium and Aptian reef complexes from the Santander province in Spain displays an identical sedimentary rhythm. The study of the stratiform Reocin lead-zinc deposit, located in Aptian dolostones from this Santander province, has disclosed genetic relations between mineralization and location in the reef complex. These relations are dependent on the sedimentary rhythm.

This observation affords a basis for comparing several stratiform deposits of varying mineralogy and different ages, also associated with reef complexes. Sedimentological analysis leads to an examination of other deposits related with peri-reef facies and eventually to a conclusion concerning the paleogeographical distribution of stratiform mineralization within reef complexes.

Résumé

L'analyse comparée de la sédimentation en milieu récifal du Givétien belge et de l'Aptien de la province de Santander (Espagne) a mis en évidence un même rythme sédimentaire et l'étude des minéralisations plombo-zincifères du gisement stratiforme de Reocin, localisé dans les dolomies aptiennes de cette province, a permis de faire apparaître des relations génétiques étroites - régies par le rythme sédimentaire - entre la minéralisation et certains horizons récifaux.

Sur cette base, la comparaison a été étendue à plusieurs gisements stratiformes, de nature minéralogique et d'époques différentes, également associés à des récifs. L'analyse sédimentologique conduit à examiner d'autres gisements en relation avec des faciès périrécifaux et à proposer en conclusion une distribution paléogéographique des minéralisations stratoïdes dans l'environnement récifal.

- : -

The comparative analysis of thick sedimentary series of the cyclic type belonging to the Middle Devonian and to the Aptian permits us to display the influence of the sedimentary rhythm on the location of stratiform mineralizations in certain facies of reef environments.

Other stratiform mineralizations associated with reef complex facies can be included in the rhythmic scheme explained hereafter, whether they have been summarily described or thoroughly studied by authors like MAUCHER and SCHNEIDER (1964, 1967), SAMSON (1965), BAZIN and LEBLANC (1968), PALTRIDGE (1968), CASSEDANNE (1969). Thus, it will be possible to propose a first synthesis of repartition of the mineralizations in reef environments.

Many publications have dealt with the cyclic sedimentation in reef complexes without mineralization nor dolomitization of the Givetian stage of the Dinant Synclinorium in Belgium (figure 1) (PEL, 1961, 1965, 1967a and 1967b); others have dealt with the Aptian stage of the Province of Santander (Northern Spain), rich in galena - sphalerite and in dolomites (MONSEUR, 1966, 1967, 1970), and with the principal conclusions on a comparison of the analyses (MONSEUR and PEL, 1970, 1971). Let us begin by remembering their principal features.

Fig. 1. Map of location of Givetian (I) and Aptian (II) complexes

Givetian and Aptian Series

The Givetian sedimentary series, set up on the same model, comprise three major units whose lithological and faunal characteristics are indicated in table I. The basic unit I is constituted by limestones of a large proportion of terrigenous material deposited in a marine environment, as shown by its fauna.

The intermediate unit II is characterized by an intensive coralline activity whose result is the formation of bioherms or biostromes. There we have to distinguish between two sub-units based on the reef building organisms: the lower one contains solitary or lamellar corals and the upper one massive corals.

The top unit III contains five lithofacies always arranged in the same manner and typical of a back-reef environment. The limestones and the primary thinly bedded dolomites, crowning at the same time the unit and the sedimentary series, have been formed in a supratidal flat as shown by the mud cracks.

The constitution of the numerous series forming the Givetian is not always identical because tectonic movements were contemporaneous with the sedimentation and because of the differences between the rates of sedimentation and of subsidence that sometimes take away from the series its basis or top, or produce recurrences of units or parts of a unit. However, the sequence of the units and of the facies in thes

Table I

Units	GIVETIAN REEF — Lithology	GIVETIAN REEF — Fauna	APTIAN REEF — Lithology	APTIAN REEF — Fauna
III	Thinly bedded dolomite and limestone Cryptocrystalline limestone Oolitic limestone Pseudo-oolitic limestone Bioclastic limestone	Foraminifers Ostracods Calcispheres Algae Gastropods *Stringocephalus burtini*	Cryptocrystalline limestone Oolitic limestone Pseudo-oolitic limestone (more or less completely dolomitzed at Reocin)	Foraminifers Nerineidae Orbitolinids
II	II b	Massive stromatoporoids Compound rugose corals *Alveolites* (numerous species) Tabulate and solitary rugose corals (often in fragments)		Orbitolinids
II	Reefal limestone II a	Tabulate corals Simple rugose corals Disphyllidae Lamellar stromatoporoids Brachiopods Crinoids	Reefal limestone (completely dolomitized at Reocin)	Rudistids Madreporian corals and rudistids
I	Alternating bioclastic limestone and shale Silty limestone	Bryozoans Tentaculitids Trilobites Small brachiopods and gastropods Echinoderms Conodonts	Dark colored bioclastic limestone (black dolomite at Reocin) Marly limestone	Foraminifers Orbitolinids *Exogyra latissima*

units remains always the same in all series and will be the clue of the interpretation of the sedimentary rhythm in mineralized formations.

In the Aptian reef complex (Province of Santander) (figure 1), the fore-reef facies are situated on the eastern side of the town, whereas the rudistid-madreporian coral - orbitolinid reefs and the associated back - reef facies are broadly developed in the west.

In spite of the intensive dolomitization the mineralized Aptian complex presents itself as a sedimentary series overlain by three recurrences of the upper units II and III. It contains the same units and the same lithological facies (table I) set up according to the same sequence as those of the Devonian series. Moreover, the two series contain zoological groups identical in the back-reef and similar, by convergence, in the reef unit. So it became interesting to specify the location of mineralizations in the so synthesized sedimentary complex.

Galena-sphalerite stratiform mineralizations of Reocin - which bear no direct relationship to a magmatic source - are confined to three horizons of the dolomitized unit II and in particular to the facies contiguous to the madreporian coral-rudistid reefs (figure 2) that were sufficiently rich in organic matter to permit the precipitation of metals from solutions "impregnating" the loose sediments.

This relationship appears so close that very often the mineralization presents the same recurrences as the madreporian coral-rudistid levels themselves; the orbitolinid horizons being practically deprived of it (figure 2).

The compared study of the two Devonian and Aptian series underscores the similarity of environment having facilitated, during the Upper Aptian, the deposition of the mineralizations confined to certain dolomitized reef levels that have experienced the variations owing to the depositional conditions and the phenomena of subsidence.

The Sedimentary Series of Stratiform Ore Deposits

The sedimentological method that we have just described too briefly has been applied to about thirty formations containing stratiform mineralizations of different ages and mineralogical nature, whether in association with reef complexes or located in back-reef facies.

In the beds enclosing more or less widely the ore deposits, we have defined the different units based upon the above mentioned lithological and faunal criteria; and as far as possible we have differentiated unit III in its different facies (a = bioclastic; b = oolitic; c = cryptocrystalline; d = supratidal).

By comparison with virtual series defined in the preceeding chapter, table 2 shows the succession of the units in one or more series being more or less complete, often differentiated by recurrences, and sets off a same succession and a same constitution of the units as in the virtual series of the Givetian and the Aptian stages.

Among the analysed series four types must be distinguished:
1. Complete series and of standard type corresponding to reef complexes widely spread out under the shelter of which the back-reef facies are developing. This type is illustrated by the ore deposits of Missouri, Pine Point, Canning Basin,

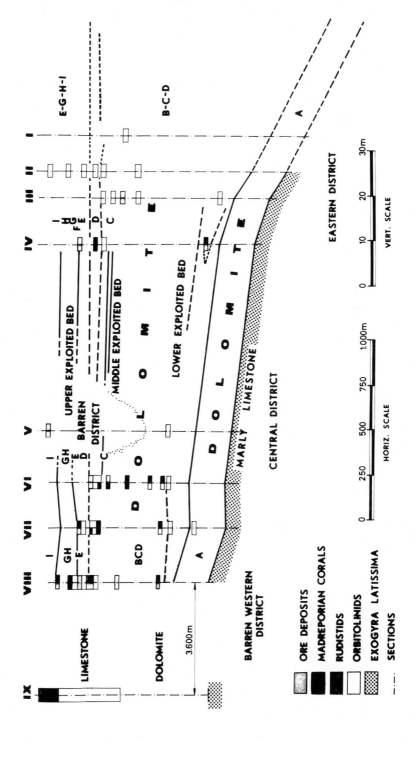

Fig. 2. Cross section of Reocin deposit

Table 2

ORE DEPOSITS	UNITS					
	I	II	III			
			a	b	c	d
REOCIN (Spain)	▬	▬	▬	▬	▬	
MESLOULA (Algeria)	▬	▬				
OUENZA–BOU KHADRA (Algeria)	▬	▬				
BOU JABER* (Tunis)	▬	▬				
SIDI AMOR*, SLATA,* AZERED*, TROZZA* (Tunis)	▬	▬				
SIDI ET TAÏA*, DJEBEL RESSAS* (Tunis)	▬					
DEGLEN, DOMINIQUE LUCIANI (Algeria)	▬	▬				
EL ABED (Algeria)	▬					
TOUISSIT (Morocco)		▬			▬	
BOU DAHAR (Morocco)			▬			
GRAND ZAGHOUAN* (Tunis)	▬	▬				
TAGOUNT (Morocco)			▬			
BOU ARFA (Morocco)				▬		
BOU ARHOUS (Morocco)			▬			
TIHARATINE et YOUDI (Morocco)		▬	▬	▬	▬	
EASTERN ALPS (Italy, Austria ,Yougo-Slavia)	▬	▬				
CANNING BASIN (Australia)	▬	▬	▬	▬	▬	▬
PINE POINT (Canada)	▬	▬				
MISSOURI (United States)		▬	▬	▬	▬	▬
CAMPO PISANO (Sardinia)	▬	▬				
TAMJOUT (Morocco)		▬	▬	▬	▬	▬
MUFULIRA (Zambia)		▬				
LUKULA and LUBI (Zaïre)	▬	▬				
BAMBUI Series (Brazil)		▬		▬	▬	▬

* Non stratified deposits located in reef complexes.

Eastern Alps, Ouenza-Bou Khadra, Mesloula, and Reocin. Certain of them have excellent recurrences of the reef and/or back-reef facies.

2. Series only containing the units I and II and especially those of Grand Zaghouan, Sidi and Taïa, Djebel Ressas, Sidi Amor, Slata, Azered, Trozza, Bou Jaber and partly Lukula and Lubi. The marine terrigenous material (fore-reef and basin) alternating with coralling biolithites attests a relatively deep environment and a regular subsidence not very favorable to the development of back-reef facies.

3. Series constituted by the units II and III differing from the preceeding ones by shallow water conditions where the reef facies pass laterally to widely developed back-reef facies. The series of Bambui, Tamjout, Tiharatine-Youdi and El Abed illustrate this type.
To this category belong the series containing transgressive reef complexes on a basement; the basis of those series is constituted by back-reef facies in the reverse order of the standard series, the middle part by reef facies and the top by back-reef facies in standard order.

4. Series being too incomplete to give an exact interpretation of the sedimentation: Bou Dahar, Tagount, Bou Arfa, Bou Ahrous. Only the units are characteristic.

The list of ore deposits is not at all exhaustive and could be completed by a great number of stratiform deposits of disseminated mineralization of little economic value; technical literature does not mention them. Moreover, this list can be completed by "crack-zone" deposits that would merit a detailed study of the mineralizations, of the associated facies and of the tectonic structures.

Paleogeographic Distribution of the Mineralizations in Reef Environment

The studied deposits are synthesized in table $3^{x)}$ indicating the age and the nature of the associated rock (dolomite or limestone), the possible presence of intense silicification phenomena, the main reef building organisms of unit II and,on the basis of the sedimentological analysis, the location of the ore deposits in the reef and the back-reef facies.

The examination of this table requires the following remarks:

1. The age of the stratified mineralizations comprises the period from the Precambrian to the Cretaceous. We have no knowledge of younger important deposits although the Tertiary of Tunis has interesting deposits, but we have not yet sufficient information about them.

2. The associated rock is:
   - dolomitic in the Precambrian, Infracambrian and Cambrian,
   - dolomitic as well as calcareous in the Devonian, Triassic, Liassic and the Lower Jurassic,
   - especially calcareous in the more recent stages.

3. The reefs enclose mineralizations rich in Cu, Zn, Pb and Fe, and poor in Ba, F and Mn. Copper can frequently be found in the Precambrian and Infracambrian reefs, whereas Zn (mainly) and Pb are prevailing in the more recent reefs.

---

x) We owe to Mr. du Dresnay to add to this table the Djebel Daït ore deposit, where mineralizations, rich in galena, are located at the top of the reef dolomites of the Middle Liassic.

Table 3

AGE	ORE DEPOSITS	Dol	Lst	Silic	REEF BUILDING ORGANISMS	II	III a	III b	III c	III d
LOWER CRETACEOUS (Aptian)	BELOCIN (Spain)	+			Rudistids Madreporian Corals Orbitolinids	Zn Fe Pb ←	►			
	MESLOULA (Algeria)	+	+	+	Rudistids (Orbitolinids)	Pb(Zn,Cu)				
	BOU KHADRA (Algeria)		+		Rudistids (Algae)	Fe				
	OUENZA (Algeria)		+		Rudistids Corals (Algae,Orbitolinids)				Fe	
	DJERISSA* (Tunis)		+		Reef building organisms Orbitolinids	Fe				
	BOU JABER* (Tunis)		+		Rudistids (Orbitolinids) Corals	Pb Zn(Ba,Fl)				
	SIDI AMOR* (Tunis)		+		Orbitolinids Corals	Pb(Zn,Cu,Ba)				
	SLATA* (Tunis)		+		Orbitolinids Corals	Fe (Pb,Cu)				
	AZERED* (Tunis)		+		?	Zn Pb(Ba)				
	TROZZA* (Tunis)		+	+	?	PbZn (Ba,Fe)				
UPPER JURASSIC	SIDI ET TAÏA* (Tunis)		+		Rudistids Corals Algae	Pb Zn(Ba,Fl)				
	DJEBEL RESSAS* (Tunis)		+		Rudistids	Pb Zn				
LOWER JURASSIC (Upper Liassic)	DEGLEN (Algeria)	+			?	Zn Pb ←	► ?			
	DOMINIQUE LUCIANI (Algeria)	+		+	?	Zn Pb ←	► ?			
	EL ABED (Algeria)	+			?	Zn Pb (Fe) ► ?			Pb	
(Middle Liassic)	TOUISSIT (Morocco)	+			Algae Stromatactis ?	Zn Pb (Fe) ► ?				
	BOU DAHAR (Morocco)		+	+	Opisoma	Pb ←	► Pb			
	GRAND ZAGHOUAN* (Tunis)		+		Rudistids Corals Algae	ZnPb(Ba,Fl)				
	TAGOUNT (Morocco)		+	+	?	← Pb(Fe,Zn) →				
	BOU ARFA (Hamarouet Lev)(Morocco)	+	+		?			Mn (Fe)		Mn
(Lower Liassic)	TAGOUNT (Morocco)		+		?	Pb ←	► Pb			
	BOU ARHOUS (Morocco)	+		+	Gervillia Branching corals	Pb ←	► Pb			
	BOU DAHAR (Morocco)	+			Opisoma			Pb		
	TIHARATINE and YOUDI (Morocco)		+		Red Reef			Mn Fe		
MIDDLE TRIASSIC	EASTERN ALPS (Austria, Ital., Yougos)	+	+		Algae (Dasycladaceae) Corals	Pb Zn (Fe)			PbZn	
MIDDLE DEVONIAN	CANNING BASIN (Australia)		+		Stromatoporoïds	Pb Zn				
	PINE POINT (Canada)	+			Corals (Thamnopora Amphipora)	Zn Pb				
UPPER CAMBRIAN	S.E. MISSOURI (United States)	+			Collenia Algae(Girvanella)	Pb(Zn,Cu,Fe) →				
MIDDLE CAMBRIAN	CAMPO PISANO (Sardinia)	+				Zn Fe(Pb) ← - - - - →				Zn Fe(Pb)
UPPER INFRA CAMBRIAN	TAMJOUT (Morocco)	+			Collenia	← Cu(Pb,Zn,Fe) →				
UPPER PRE-CAMBRIAN	BAMBUI Series (Brazil)	+	+	+	Collenia Algae (Solenoporaceae)	Pb (Zn,Fl,Ba,Cu)	Fl (Pb,Zn,Cu)	Pb (Zn,Fl,Ba,Cu)	Pb	
	MULULEIRA (Zambia)	+			Collenia	Cu				
	LUKULA and LUBI (Zaire)	+			Stromatolites (Collenia, Conophyton, Lomamia)	Pb(Cu, MnZn)				

* Non stratified deposits located in reef complexes.

4. The crust zone of the reef and the bioclastic sediments, formed close to reefs, are mainly mineralized in Pb.

5. The back-reef facies contain Fe, Mn, Pb, and in a lesser proportion Zn.

Consequently, we can propose a plan of paleogeographical distribution of the Cambrian and post-Cambrian stratified ore deposits in reef environment where we can

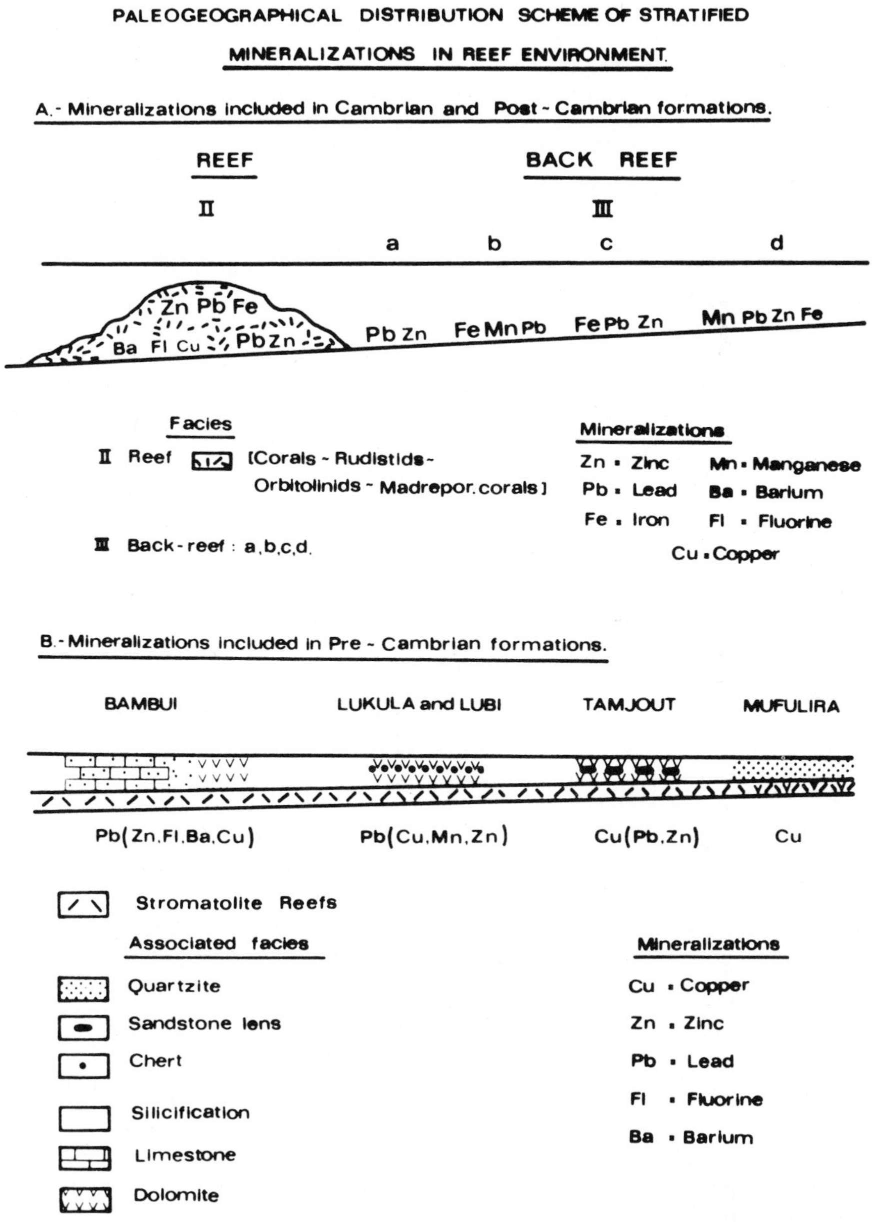

Fig. 3. Paleogeographical distribution scheme of stratified mineralizations in reef environment

observe an excellent differentiation of the mineralizations Mn - Fe - Pb - Zn beginning from the border of the lagoon to the reef.

We immediately conceive the interest of this plan from the point of view of the strategic prospection of the minerals.

Based upon four cases of mineralizations included in Precambrian formations, the analysis of the associated facies permits to come to the conclusion that the Collenia reefs built under shallow water are:

1. copper and dolomite bearing deposits in the littoral zones (Mufulira, Cu),
2. progressively becoming richer in Pb to the offing (Tamjout, Cu - Pb; Lukula and Lubi, Pb - Cu), and
3. containing only Pb in a more marine environment, whereas the dolomitization and silicification phenomena diminish.

The sedimentological analysis of about thirty deposits leads to the conclusion - of highest importance for ore research - that the facies of reef complexes have their own ore-bearing deposits. In the reef itself exists also a differentiation in the mineralizations: we underlined it not only in the Collenia biostromes, but also in the comparison between the central and the crustal zone of more recent reefs.

On the other hand, the mineralizations are sensitive to the movements of the contemporaneous events affecting the bottom in such a manner that small movements of subsidence are responsible of recurrence phenomena of the reef banks and at the same time of the associated mineralizations. Indeed, these are many arguments in favour of the syngenetic origin of the mineralizations.

From the point of view of the prospection the sedimentological and metallogenic methods applied to stratified ore deposits associate harmoniously in order to define a valuable strategy: The detailed analysis of the facies leading to a paleogeographical reconstitution of the basin guides the research of the favorable isopic zones (fig. 3), whereas the evolution of the facies in the sedimentary series resulting from marine transgression or regression movements earmarks the total valuation of the ore deposits.

Bibliography

AGARD, J. , du DRESNAY, R. : La région minéralisée du Jbel Bou-Dahar, près de Beni-Tajjite (Haut Atlas oriental) : étude géologique et métallogénique. In: Colloque sur des gisements stratiformes de plomb, de zinc et de manganèse du Maroc (2 mai - 14 mai 1962). Notes Mém. Serv. Géol. Maroc, Rabat, 181, p. 135-166 (1965).

AMSTUTZ, G. C. : Sedimentology and ore genesis. Devel. in Sedimentology, Elsevier Amsterdam, 2, 184 p. (1964).

BAZIN, D. , LEBLANC, M. : Récifs et minéralisations plombo-zincifères. Exemples empruntés au Haut-Atlas Oriental (Maroc). Chron. Mines Rech. min. , Paris, 372, p. 115-120 (1968).

BLONDEL, F. , MARVIER, L. : Symposium sur les gisements de fer du Monde. Congrès géologique international, XIX session, Alger, I, 638 p. (1952).

205

BUROLLET, P. F. : Contribution à l'étude stratigraphique de la Tunisie Centrale.
Ann. Mines Géol. , Tunis, 18, 352 p. (1956).

CAMPBELL, N. : Tectonics, reefs and stratiform lead-zinc deposits of the Pine
Point Area, Canada. Symposium intern. sur les gisements stratiformes de
Pb, Zn, Ba, Fl, New York. Econ. Geol. , Lancaster, Monograph 3, p. 71-89
(1967).

CASSEDANNE, J. : Biostrome à Collenia dans le calcaire de Bambui. An. Acad.
brasil. Ci, Rio de Janeiro, 36, p. 49-58 (1964).

—   Révision des gisements de plomb et de zinc du nord-est et du centre du Brésil.
An. Acad. brasil. Ci, Rio de Janeiro, 36, p. 151-158 (1964).

—   Découverte d'algue dans le calcaire de Bambui (Etat de Minas Gerais, Brésil).
An. Acad. brasil. Ci, Rio de Janeiro, 37, p. 79-81 (1965).

—   Niveau à galets mous dans le calcaire de Bambui. Importance paléogéographique.
An. Acad. brasil. Ci, Rio de Janeiro, 38, p. 281-288 (1966).

—   Métallogénie du plomb et du zinc dans l'état de Bahia. An. Acad. brasil Ci,
Rio de Janeiro, 38, p. 465-474 (1966).

—   Les minéralisations plombo-zincifères du Groupe Bambui. An. Acad. brasil.
Ci, Rio de Janeiro, 41, p. 549-563 (1969).

CASTANY, G. : Etude géologique de l'Atlas tunisien oriental. Ann. Mines Géol. ,
Tunis, 8, 632 p. (1951).

CHAZAN, W. : Les gisements stratiformes plombo-zincifères de l'Infracambrien de
l'Anti-Atlas occidental. Notes Mém. Serv. Géol. Maroc, 8, no. 120, p. 98-
126 (1954).

DUBOURDIEU, G. : Les mines de fer de l'Ouenza et du Bou-Khadra. In: F. BLONDEL
et L. MARVIER (op. cit.), p. 70-72 (1952).

—   Etude géologique de la région de l'Ouenza (confins Algéro-Tunisiens). Publ.
Serv. Carte géol. Algérie, Alger, n. Sér. Bull. 10, 664 p. (1956).

—   Esquisse géologique de Jebel Mesloula (Algérie orientale). Publ. Serv. Carte
géol. Algérie, Alger, n. Sér. Bull. 21, 162 p. (1959).

du DRESNAY, R. : La stratigraphie du J. Bou Arfa (Haut-Atlas marocain oriental).
C. R. Acad. Sci. , Paris, 256, p. 461-464 (1963).

FLEISCHER, R. , ROUTHIER, P. : Quelques grands thèmes de la géologie du Brésil.
Miscellanées géologiques et métallogéniques sur le Planalto. Sci. Terre, Nancy,
XV, p. 47-102 (1970).

GOTTIS, C. : Les gisements de fer en Tunisie. In: F. BLONDEL et L. MARVIER
(op. cit.), p. 211-220 (1952).

JACKSON, S. A. , FOLINSBEE, R. E. : The Pine Point lead-zinc deposits, N. W. T. ,
Canada. Introduction and Paleoecology of the Presqu'île Reef. Econ. Geol. ,
Lancaster, 64, p. 711-717 (1969).

LOMBARD, J. , NICOLINI, P. : Gisements stratiformes de cuivre en Afrique, (Symposium). Lithologie, Sédimentologie, Ass. Serv. Géol. Afr. , Paris, 212 p. ,
(1962).

MALAN, S. P. : Stromatolites and other algal structures at Mufulira, Northern Rho-
desia. Econ. Geol. , Lancaster, 59, p. 397-415 (1964).

MAUCHER, A. , SCHNEIDER, H. J. : The Alpine lead-zinc ores. Symposium intern. sur les gisements stratiformes de Pb, Zn, Ba, Fl, New York. Econ. Geol. , Lancaster, Monograph 3, p. 71-89 (1967).

MONSEUR, G. : Contribution à l'étude sédimentologique et génétique du gisement plombo-zincifère de Reocin (Espagne). Acad. Roy. Sc. Outre-Mer, Bruxelles Cl. Sc. Techn. , N. S. , XVI - 5, 87 p. (1966).

— Synthèse des connaissances actuelles sur le gisement stratiforme de Reocin (Province de Santander, Espagne). Symposium intern. sur les gisements stratiformes de Pb, Zn, Ba, Fl, New York. Econ. Geol. , Lancaster, Monograph p. 278-293 (1967).

— Medios arrecifales y mineralizaciones estratiformes. Universidad de Madrid (in print) (1970).

MONSEUR, G. , PEL, J. : Sédimentologie et métallogénie. Revue Universelle des Mines, Liège, no. 3, p. 131-140 (1970).

— Reef facies, dolomitization and stratified mineralization. Mineral. Deposita (Berl.), 7, p. 89-99 (1972).

NICOLINI, P. : Symposium sur le plomb et le zinc en Afrique. Assoc. Serv. Géol. Afr. Ann. Mines Géol. , no. 23. Serv. Géol. Tunisie, Tunis, 477 p. (1968).

— Gîtologie des concentrations minérales stratiformes. Gauthier-Villars, Paris, 792 p. (1970).

PALTRIDGE, I. M. : An Algal biostrome fringe and associated mineralization at Mufulira, Zambia. Econ. Geol. , Lancaster, 63, p. 207-216 (1968).

PEL, J. : Observations sur le Givetien de la région de Hotton-Hampteau. Bull. Cl. Sc. Acad. r. Belg. , Bruxelles, 5e Sér. , XLVII, p. 640-650 (1961).

— Etude du Givetien à sédimentation rythmique de la région de Hotton-Hampteau (Bord oriental du Synclinorium de Dinant). Ann. Soc. Géol. de Belgique, Liège 88, no. 8, p. 471-521 (1965).

— Interprétation nouvelle du Givetien de Givet (Mont d'Haurs). C. R. Acad. Sci. , Paris, 264, Sér. D, p. 1961-1964 (1967a).

— Etude sédimentologique du Givetien au Nord-Est de Givet. International Symposium on the Devonian System. Alberta Soc. Petrol. Geol. , Calgary, p. 441-45 (1967b).

PLAYFORD, Ph. E. , LOWRY, D. C. : Devonian Reef Complexes of the Canning Basin Western Australia. Geol. Surv. W. Austral. , Perth, Bull. 118, 150 p. (1966).

POUIT, G. : Les gîtes de manganèse marocains encaissés dans les formations carbonatées : éléments pour une synthèse. Chron. Mines, Paris, p. 331-343, p. 371-380 (1964).

RAT, P. : Les pays crétacés basco-cantabriques (Espagne). Publ. Univ. Dijon, Presses Univ. France, XVIII, 525 p. (1959).

RAUCQ, P. : Contribution à la connaissance du Système de la Mbujimayi (Congo Belg Ann. Mus. Roy. Congo Belge, in 8°, Sc. géol. , Tervuren, XXVIII, XII, 427 p. (1957).

ROUTHIER, P. : Rapport de synthèse. In: G. C. AMSTUTZ (op. cit.), p. 167-176 (1964).

SAINFELD, P. : Les gîtes plombo-zincifères de Tunisie. Ann. Mines Géol. , Tunis, 9, 285 p. (1952).

SAMSON, Ph. : Le gisement plombo-zincifère de Touissit : monographie et interprétation géologique. Colloque sur des gisements stratiformes de plomb, zinc et manganèse du Maroc (2 mai - 14 mai 1962). Notes Mém. Service Géol. Maroc, Rabat, <u>181</u>, p. 69-91 (1965).

SCHNEIDER, H. J. : Facies differentiation and controlling factors for the depositional lead-zinc concentration in the Ladinian geosyncline of the Eastern Alps. In: G. C. AMSTUTZ (op. cit.), p. 29-45 (1964).

SYNDER, F. G. , EMERY, J. A. : Geology in development and mining Southeast Missouri lead belt. Mining Engn. , New York, 8, no. 12, p. 1216-1224 (1956).

Address of the authors:

Laboratoires de Géologie Générale et
Appliquée de l' Université de Liège
4000 Liège / Belgium

Ore Deposition in Karst Formations with Examples from Sardinia

G. Padalino, S. Pretti, D. Tamburrini, S. Tocco, I. Uras, M. Violo, and P. Zuffardi

Abstract

Economic concentrations of bauxite, barite, fluorite and ores of iron, lead, and zinc were investigated in karst formations in Sardinia; they occur in addition to the known strata-bound and vein-type deposits of this major mining region. Three main periods of karst development and karst ore accumulation are recognized: the earliest of Cambro-Ordovician age, a second one of post-Hercynian age; the last of Alpine and post-Alpine age. Reworking of earlier karst ore was observed in different localities and took place during different periods.

A Hercynian granite intrusion has locally converted some karst fillings into skarn. Details on the ore sedimentation in the karst of different periods and areas are presented on all for scales (regional, mine, hand-specimen, and mineral or microscopic).

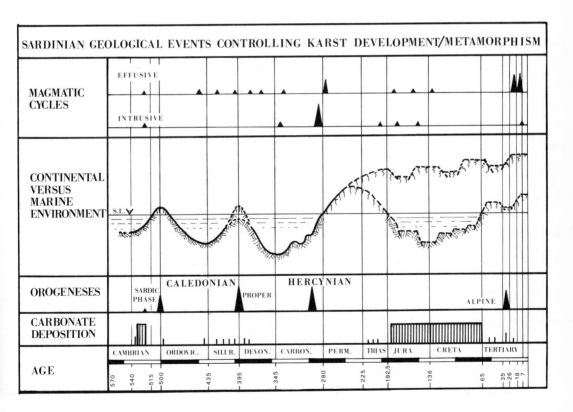

Fig. 1

Geologic Outline

Some geologic events that controlled karst development in Sardinia are schematically
outlined in the diagrams of Fig. 1. The types of karsts and related ores are listed in
Fig. 2. The four points which follow refer to the tables and diagrams of these two
figures.

1) Only two large carbonate formations are known in Sardinia: the so-called "Metallifero" of Middle Cambrian age, and the carbonates of the Jurassic-Cretaceous Systems. They are quite similar from the stand-point of macro-composition and consist both of limestones and dolomites, with some marls in the Jurassic-Cretaceous
   Systems. They have very different minor metal contents: the background in lead -
   zinc of the Jurassic-Cretaceous strata is 5 to 10 ppm, whereas in the "Metallifero"
   it is almost 50 ppm and some horizons contain as much as 0.5 to 1 % dispersed
   zinc (basal dolomite) and others (top limestones) have, in spots, 0.5 to 0.8 % dispersed Pb and even higher values. Some Cambrian limestones conceal variable
   amounts of Ba, sometimes grading up to more than 1 %, even with small lenticular interbedded workable concentrations of barite.
   As a consequence of these differences between "Metallifero" and the Jurassic-
   Cretaceous Systems, the possibilities for autochthonous karstic accumulations of
   ores are also very different; as a matter of fact, we know of only bauxite accumulations in Jurassic-Cretaceous strata (with the exception of two small smithsonite-
   cerussite-limonite ore bodies), whereas a number of large ore deposits of Pb, Zn,
   $BaSO_4$ and some CaF_2 concentrations are exploited in Cambrian rocks.

SYNOPSIS OF ORES RELATED TO KARSTS IN SARDINIA						
TYPE OF KARST / TYPE OF ORE	RELATED TO THE CAMBRO-OR- DOVICIAN UNCONFORMITY			RELATED TO THE HERCYNIAN PENEPLANE		RELATED TO THE ALPINE UPLIFT
	FOSSIL	REJUVENATED	THERMOMETA- MORPHIC	FOSSIL	CONTINUOUSLY REJUVENATED	
BAUXITE						▭ OLMEDO
LIMONITE + Zn & Pb OXIDATES					▭ IGLESIAS	
HAEMATITE	▭ FLUMINIMAGGIORE					
GALENA+CERUS- SITE +BARITE	▭ ARENAS IGLESIAS-SULCIS	▭ ARENAS ORIDDA	▭ ARENAS ORIDDA	▭ NORTH SULCIS	▭ SILVER RICH IGLESIAS-SULCIS	▭ IGLESIAS
BARITE	▭ SULCIS	▭ SULCIS		▭ SA BAGATTU	▭ BAREGA	▭ BAREGA EAST
Fe,Zn,Pb MIXED SULFIDES	▭ CONCAS DE SINUI					▭ IGLESIAS
FLUORITE	▭ NARCAO					▭ ORIDDA
TOTAL TONNAGES	▭ <100 000 tons		▭ < 1 000 000 tons		▭ millions of tons	

Fig. 2

2) Sardinia, or a part of it, was subjected to <u>four main periods of continental uplift</u>, namely:
   a) during the Cambro-Ordovician orogeny;
   b) between the Silurian and Devonian periods;
   c) after the end of the Hercynian orogeny, with particular intensity during Permo-Triassic times;
   d) after the Alpine uplift.

   Karst development took place during three of these periods of uplift: (a), (c) and (d) (see above). No karst was formed during (b), because the only great carbonate complex existing at this time (the Metallifero) was protected by a thick, impervious Ordovician-Silurian cover. No time break between (c) and (d) exists in many parts of the Island, so that karst development and ore accumulations could have occurred during a time span of 280 million years.

3) The Cambro-Ordovician orogeny produced large open folds and little uplift. In contrast, the subsequent Caledonian uplift and the Hercynian orogeny were responsible for intense folding and faulting with marked uplift. The Alpine orogeny was essentially disruptive and gave rise to the present horst-and-graben structure of Sardinia, the uplift being pronounced in many places. As a consequence, the depths of Cambro-Ordovician fossil paleo-karsts is very shallow (a few meters); in contrast, the depths of Alpine and of post-Hercynian karsts are frequently great (up to several hundred meters), especially in those areas of continuous horst development.

4) <u>Two main magmatic cycles</u> were active in Sardinia: the Hercynian intrusive acidic cycle and the Alpine cycle, which was effusive. As a consequence well developed thermometamorphic effects can only be observed in some Cambro-Ordovician paleo-karsts; post-Hercynian and post-Alpine karsts are practically unaffected by such alteration.

Examples

In the text which follows the examples are listed according to their ore composition, with particular attention given to those ores (i. e. galena, sphalerite, pyrite, marcasite, barite, fluorite) which are perhaps somewhat unusual or new for karsts.

We think that further investigations may demonstrate that more deposits, the genesis of which till now has been described in different ways, should probably be interpreted as karst accumulations, although, at the moment, insufficient factual observations are available to fully prove the new interpretation.

A) <u>Bauxite</u>: Bauxite deposits occur in the north-western part of Sardinia at Olmedo (close to Alghero) in fossil paleokarsts developed in the Cretaceous series. According to COCCO and PECORINI (1959), bauxite is derived by weathering of marls of the same series.

B) <u>Limonite-smithsonite-hemimorphite-cerussite</u>: Huge deposits of these ores have accumulated as "iron caps" on top of strata-bound and vein-type lead-zinc-pyrite bodies occurring in the Metallifero of the well known Iglesias area. They were described by ZUFFARDI (1964); in the present report, we wish to point out two characteristics of these deposits. They are 1) their extreme size (several hundred meters are not uncommon), and, 2) their transition in depth to re-generated sulfides of the types mentioned below under items D and F (MÜNCH, 1960; ZUFFARDI, 1968 and 1970).

C) Hematite: The type locality is the mountain range south of Fluminimaggiore, where a number of small pockets of very pure hematite with interbedded phyllites occur along the unconformable boundary between Cambrian limestones and Ordovician basal conglomerates.

The shape of the footwall, the presence of short veinlets and/or of diffused spots of hematite within the footwall and the geometric relationships between footwall, bedded hematite and Ordovician basal conglomerate (see fig. 3) are evidence for karst deposition.

Limestones are very pure, especially as regards their iron contents: this condition speaks against an autochthonous deposition. On the other hand, the continuation of hematite deposition into the basal conglomerates, and the gradual transition from high grade Fe concentration to low grade Fe-bearing sediments is evidence for allochthonous deposition.

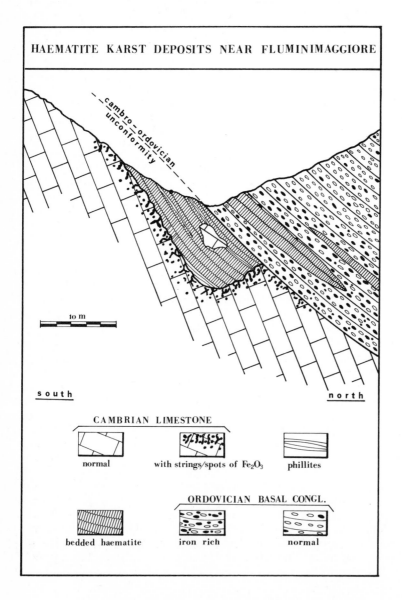

Fig. 3

In hand-specimens and polished or thin sections the ore shows a banded texture, sometimes carrying recrystallized oolites. We classify these deposits as fossil paleokarsts of Cambro-Ordovician age with allochthonous fillings, regionally recrystallized.

D) <u>Barite-galena-cerussite</u>: There are two groups of these deposits,

1) related to the Cambro-Ordovician unconformity;
2) related to the post-Hercynian peneplane.

Arenas is the best locality for an investigation of all three types of deposits. This area was described in detail by BENZ (1964, 1965) and is schematized in fig. 4. In summary the following features are noted:

1-a) The <u>fossil paleokarsts</u> are a number of generally small pockets lying in the same stratigraphic position as the hematite deposits described above. The filling, however, is quite different, being made (see fig. 4) of various types of "terra rossa", overlain by a 2-4 meter thick bed of (continental) silica. Some galena is present at the base of each pocket; some cerussite occurs in the "terra rossa"; barite forms a rather continuous bed directly over the silica crust. Some lenticular interbedded deposits of silica, very often carrying sulfides and barite, occur in the first ten to thirty meters of the basal Ordovician, thus proving that, at least partially, the filling is allochthonous.
In addition to Arenas, hundreds of kilometers along the Cambro-Ordovician unconformity in Sardinia show this same situation. Lead grade is generally

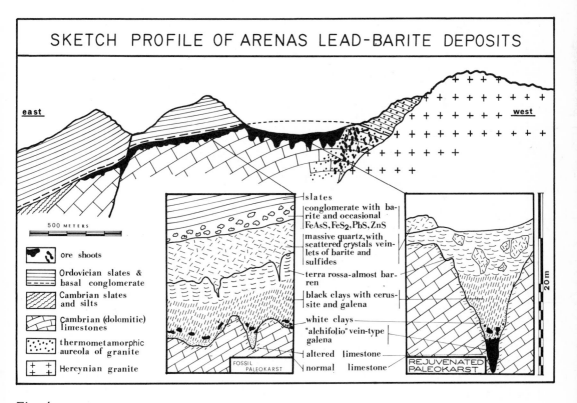

SKETCH PROFILE OF ARENAS LEAD-BARITE DEPOSITS

east

west

500 METERS

- ore shoots
- Ordovician slates & basal conglomerate
- Cambrian slates and silts
- Cambrian (dolomitic) limestones
- thermometamorphic aureola of granite
- Hercynian granite

slates
conglomerate with barite and occasional FeAsS, FeS₂, PbS, ZnS
massive quartz, with scattered crystals veinlets of barite and sulfides
terra rossa-almost barren
black clays with cerussite and galena
white clays
"alchifolio" vein-type galena
altered limestone
normal limestone

FOSSIL PALEOKARST

REJUVENATED PALEOKARST

20 m

Fig. 4

low (less than 1.5 %); at places it reaches 4 or 5 % or even more. As we previously mentioned, intense folding took place following deposition over the Cambro-Ordovician unconformity; as a consequence, all these deposits are now nearly vertically tilted, thus simulating vein-type deposits; they have been misinterpreted for a long time in the past (see fig. 5).

1-b) <u>Rejuvenation</u> of the previous deposits took place, in more or less recent times, where the Ordovician was eroded and meteoric waters trickling through the fissures of the siliceous crust could rework the above-mentioned deposits.

A deeper karst (some 20 meters size) is thus produced, which yields accumulations of cerussite and galena in black clay below an almost barren collapse breccia cemented by "terra rossa".

Deposition of small amounts of "alchifolio"-type galena is not infrequent at the bottom.

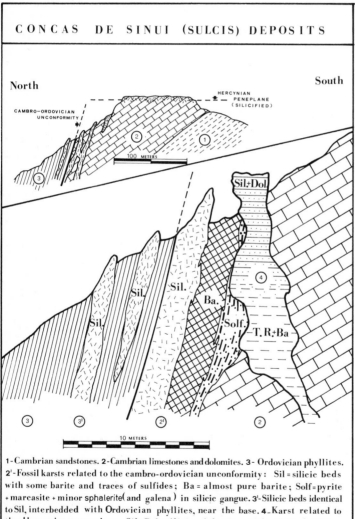

CONCAS DE SINUI (SULCIS) DEPOSITS

1-Cambrian sandstones. 2-Cambrian limestones and dolomites. 3- Ordovician phyllites. 2'-Fossil karsts related to the cambro-ordovician unconformity: Sil = silicic beds with some barite and traces of sulfides; Ba = almost pure barite; Solf=pyrite + marcasite + minor sphalerite(and galena) in silicic gangue. 3'-Silicic beds identical to Sil, interbedded with Ordovician phyllites, near the base. 4-Karst related to the Hercynian peneplane.-Sil+Dol= silicic dolomite with some barite. -T.R.+Ba=" terra rossa" with detrital barite.

Fig. 5

The importance of the described rejuvenation as a controlling factor for concentration and consequently for cheaper exploitation is sometimes so essential, that the D-1-a deposits are often uneconomic, whereas D-1-b deposits derived from them are generally of commercial interest.

1-c) A granite intrusion and related <u>thermometamorphism</u> affected a part of the D-1-a deposits, giving rise to a skarn mainly consisting of garnets with veinlets of galena cementing newly formed silicates.
Increased lead grade, up to 10 % for the whole garnetiferous ore is fairly common.
Structures, textures and composition, both at the scale of hand-specimens and in polished and thin sections, are quite similar to those which are described in older literature as having formed by pyrometasomatic and pneumatolytic-hydrothermal processes (URAS, 1957; DI COLBERTALDO, 1958).

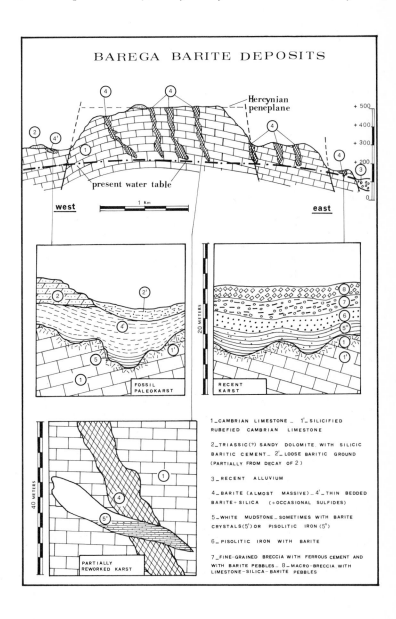

Fig. 6

2) Two types of this group are known. They include the following:

2-a) Horizontally bedded rhythmic deposits, made up of millimetric bands of galena, barite, black silica.
Their depth is generally small (a few meters). Sometimes the original Triassic (?) overburden is still preserved (Sa Bagattu) (fig. 6, 7a, 7b, 8a, 8b).
They should be classified as fossil paleokarsts of post-Hercynian (Triassic ?)
age (TAMBURRINI, 1968; TAMBURRINI and ZUFFARDI, 1969).

2-b) Pipe-like bodies with crustified cockade fillings and vertical zoning (pure
barite at the top, barite-black silica-aragonite-galena-silver minerals at
lower levels). The lengths range up to several hundred meters. Typical are
the so-called "silver rich deposits" of S. Giovanni - S. Giorgio (south of
Iglesias) (BRUSCA and DESSAU, 1968).

2-c) Pipe-like bodies with stock-work texture. The filling is made of a barite-
galena-calcite-dolomite-aragonite suite of minerals. The depths are variable, ranging from some tens of meters to up to more than 130 meters; a
strong enrichment in galena was observed in the deepest zones of some
"pipes" (Giuenni mine).

It should also be remembered that the process forming the post-Hercynian karstic
reconcentrations was one of the main (if not the main) controlling factor in forming the Monteponi high grade galena ore shoots, as has been described in previous
papers (SALVADORI and ZUFFARDI, 1964; ZUFFARDI, 1968, 1969, 1970).

E) Barite: Barite filled karsts are quite similar to barite-galena-cerussite karsts,
into which they change gradually. The type area is Barega as shown in fig. 6.
The main differences are:
a) almost no barite (at least in commercial amounts) is present in metamorphosed
karsts;
b) massive fillings, as local ore-shoots in all types, are not infrequent;
c) vein-type fillings of siliceous barite, occurring along faults (Barega) and fractures, are not infrequent; they occur also in non-carbonate country rocks.

The presence of thick crusts of barite-bearing silica, both above the Cambro-
Ordovician unconformity and above the Hercynian peneplane, suggests that at
least a part of barite is allochthonous.

F) Pyrite-marcasite-sphalerite-galena: Two cases of karstic accumulations with a
combination of sulfides are known:
1) in the bottom part of karst cavities related to the Cambro-Ordovician unconformity. Black silica is the gangue mineral (Concas De Sinui: fig. 5).

2) some recent crevices in the deepest level of the Iglesias area are partially
filled with horizontally stratified earthy deposits of these minerals (Zn grade
is up to 12 %), mixed with fresh black clays and/or reddish cherts.

Both F-1 and F-2 have limited tonnages.

Fig. 7a and 7b. Hand-specimen of rhythmic barite (whitish) - silica (black) - galena
(not visible in photo 7a; white brilliant in photo 7b). Photographed with oblique light
(7a) and with perpendicular light (7b)
The sample is in its top-bottom position; actual vertical dimension: about 7 cm.
A diagenetic fracture is visible close to the right side

Fig. 7 a

Fig. 7 b

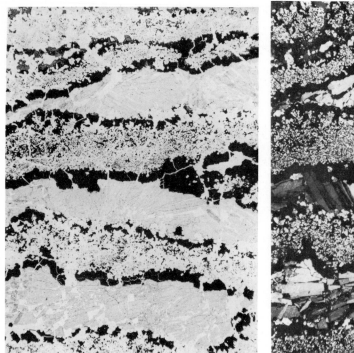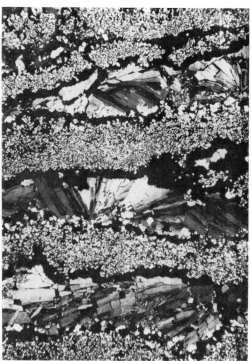

Fig. 8a and 8b. Thin section of the rhythmite of Fig. 7a, b. The sample is in its top-bottom position; actual vertical dimension: about 2 cm. Photo 8a is taken with parallel Nicols; photo 8b with crossed Nicols.
Barite (light greyish) is well crystallized, foliated, very pure. Galena is black, both in seams and in spots in the silica, that appears whitish and it fine grained

G) <u>Fluorite</u>: Two cases of fluorite accumulation in karsts are reasonably recognizable. The first one occurs at Narcao and is related to the Cambro-Ordovician unconformity. It is made up of a stock-work of barite, fluorite and calcite veinlets cutting across highly silicified Cambrian dolomites, immediately below the base of the Ordovician. It may be considered as local (fluorite-bearing) variant of the E type; the source of the fluorite remains uncertain.

The second example occurring at Oridda, Mines of Perda Niedda and Nebidedda, is more puzzling. Many generally small karsts (2000 to 50000 tons) in this area are filled with high purity fluorite with minor quantities of calcite and barite. The sizes are, at the most, 20 - 40 meters from the present earths surface, which is quite close to the Cambro-Ordovician unconformity.

The bottoms of such karsts are always "terra rossa"; they are very often covered with a lateritic-hematitic overburden. Other interesting factual observations are:

1) Scattered fluorite crystals and stringers are present in particular (highly ferruginous, sometimes manganiferous-chloritic-amphibolitic-garnetiferous-sulfide bearing) portions of the metamorphic aureola of a neighbouring granitic intrusion. This suggests that fluorite was generated not later than Hercynian metamorphism, but it gives no evidence whether fluorine was introduced <u>durin</u> metamorphism <u>or</u> preexisted <u>before</u> metamorphism.

2) The above-mentioned portions of metamorphic aureolas are deeply weathered and partially changed into a thick crust of high grade hematite (at the top surface: limonite).
A one to two meters thick bed of fluorite and silica occurs at the base of hematite, directly over the unweathered rock. This suggests high mobilization of fluorite during weathering under particular climatic conditions, more severe than the ones existing in recent times (formation of hematite instead of limonite).

3) Fluorite karst deposits are quite close to the above described complexes, thus suggesting a (genetic) link between them.

In conclusion we propose the hypothesis that a part of the fluorite could escape from the weathered zone in solutions of supergene waters, and regenerate fluorite in karsts, as soon as these solutions reached carbonatic rocks.

The origin of the primary fluorite (Hercynian? pre-Hercynian?) is, at the moment, an unsolved problem.

Bibliography

BENZ, J. P. : Le gisement plombo-zincifère d'Arenas (Sardaigne). Thèse de doctorat, Fac. Sciences de l'Université Nancy, 123 p. (1964).

— Nouvelles observations sur le gisement d'Arenas. Symposium Problemi geominerari Sardi, p. 331-342 (1965).

BRUSCA, C. , DESSAU, G. : I giacimenti piombo-zinciferi di S. Giovanni (Iglesias) nel quadro della geologia del Cambrico Sardo. L'Industria Mineraria, XIX, p. 470-489, 533-552, 597-609 (1968).

COCCO, G. , PECORINI, G. : Osservazioni sulle bauxiti della Nurra (Sardegna Nord-occidentale). Atti Acc. Naz. Lincei, serie VIII, vol. V, Sez. II, Fascicolo 7, anno CCCLVI, p. 175-214 (1959).

DI COLBERTALDO, D. : Il giacimento piombo-zincifero di Arenas nello Iglesiente. Rend. Soc. Min. Ital. , XIV, p. 172-203 (1958).

MÜNCH, W. : Ricerche geo-giacimentologiche in Sardegna. Giornate di studio sulle ricerche geo-giacimentologiche, I, 2, 8 p. Rome: AMMI (1960).

SALVADORI, I. , ZUFFARDI, P. : Supergene sulfides and sulfates in the supergene zones of sulfide ore deposits. Development in sedimentology - vol. 2: sedimentology and ore genesis. Elsevier (Amsterdam-London-New York), p. 91-99 (1964).

TAMBURRINI, D. : I giacimenti baritici sardi: caratteri geogiacimentologici e minerari. Symposium sulle bariti della Sardegna, p. 17-30 (1968).

TAMBURRINI, D. , ZUFFARDI, P. : Field evidences of supergene remobilization of Barium (and possibly of Barite) in Sardinia. Remobilization of Ores and Minerals: a Symposium, p. 305-314 (1969).

URAS, I. : Il giacimento piombifero di Tiny-Arenas. Studio geominerario. Res. Ass. Min. Sarda, LXI, 7, p. 3-32 (1957).

ZUFFARDI, P. : Considerazioni sulla composizione e sulla distribuzione dei minera calaminari sardi. Symposium sull'arricchimento dei minerali calaminari, p. 1-20 (1964).

—   Transformism in the Genesis of Ore Deposits: Examples from Sardinia Lead-Zinc Deposits. XXIII Int. Geol. Congr. , 7, p. 137-149 (1968).

—   Remobilization in Sardinian Lead-Zinc Deposits. Remobilization of Ores and Minerals: a Symposium, p. 283-292 (1969).

—   La Métallogenèse du Plomb, du Zinc et du Barium en Sardaigne: un Exemple de Permanence, de Polygénétisme et de Transformisme. Annales Société Géo Belgique, 92, p. 321-344 (1970).

Received: August 1971, March 1972.

Address of the authors:

Università di Cagliari
Istituto di Giacimenti Minerari,
Geofisica e Scienze Geologiche
09100 Cagliari / Italy

On the Anisotropy of Ore-bearing Series in Stratiform Deposits

V. M. Popov

Abstract

The concept of anisotropy of ore-bearing rocks is prominent in formulating the hypothesis of ore deposits and as an important factor in the formation of sulphide deposits
of the stratiform type. The supporters of this hypothesis believe that the essence of
the above phenomena is reflected in a selective association of mineralization to well
stratified complexes of sedimentary rocks having various physico-mechanical and
chemical properties such as hardness, fragility, plasticity, porosity, permeability,
jointing, granulometric composition, degree of chemical activity, etc.. It is also
considered that in uniformly thick rock series without anisotropic elements no ore
concentrations can be formed (SATPAEV, 1958; TYCHINSKY, 1961; and others).

The theory of anisotropy serves as a basis in this case for a concept on "favourable"
and "screening" horizons in stratiform ore deposits. In addition, the role of periodicity (rhythmic textures) is discussed.

Attracting the anisotropy of the host rocks and favourable strata is groundless in proving the hydrothermal-metasomatic hypothesis.

- : -

A critical analysis of formations conditions of numerous stratiform deposits showed
that anisotropy of ore-bearing series and favourable horizons failed to explain the
genesis of these deposits utilizing a hydrothermal-metasomatic concept. The theory
of anisotropy, regarded as one of the most universal regularities of space location
of endogenic metallogenic formations proved to be invalid after being analyzed and
tested on concrete objects. The same was observed on "favourable" and "screening"
horizons, their real participation having been previously studied (POPOV, 1962). A
stratigraphical-lithological control over mineralization in the deposits concerned,
regularly resulting from their sedimentary nature, is misinterpreted by the supporters of the endogenic concept and has been substituted by ideas of anisotropy of ore-
bearing series.

At present, a general regularity has been established for stratiform deposits expressed by a selective association of mineralization to heterogeneous types of sections of
sedimentary and volcanogenic-sedimentary formations characterized by a non-uni-
form structure, and the actual absence of mineralization in sections of homogeneous
strata.

The genetic nature of heterogeneous sections lies, first of all, in the inconsistency
of sedimentation conditions that is characterized by instability, rapid and frequent
change of facies and deposit types, fluctuation of physico-chemical environment of
sedimentation (pH, Eh), and unstable paleotectonic regime of the sedimentation area.
The latter is responsible for rhythmic fluctuations of the basin floor and related dis-
placements of a coast line together with a zone of ore formation which causes a displacement of facies in time, their rejuvenation, etc. Such restless sedimentation

conditions usually take place in the near-shore belt of sedimentary basins and are reflected in a rhythmic structure of ore-bearing sedimentary rocks, their heterogeneity and in the shallow-water character of sediments. Oscillating motions of the sedimentation areas and wash down cause a periodical change of the degree of salinity of basin waters and climatic conditions. Such an environment of sedimentation is characteristic of all deposits of the "copper sandstones" type associated with variegated beds. These specific conditions of sedimentation were the most favourable for the formation in sediments of syngenetic ore concentrations the distribution of which in the given strata obeyed the laws of sedimentation.

Thus, for instance, in the Dzhezkazgan copper deposit the mineralization is associated to that part of the Upper Paleozoic series that is a regular alternation of members of red beds (argillites, aleurolites, sandstones) with those composed of grey sandstones and aleurolites; the latter are mineralization carriers. This productive part of the general section of sedimentary formations of the region is distinguished under the name of "Dzhezkazgan" suite occurring between thick ore-free strata with a more uniform colour and composition.

Such a structure of an ore-bearing suite reflects the cyclic recurrence of sediment formation related closely to mineralization. The latter, being associated to certain elements of cycles, forms nine ore-bearing horizons within which the ore mineralization is controlled by a smaller rhythm and certain facies of sediments (DRUZHININ, 1964; POPOV, 1960). Thus, physico-mechanical properties of rocks of the Dzhezkazgan suite for primary accumulation of ore elements in sediments were of no importance. Migration of ore matter within the ore horizons took place later in relation to tectonic movements and general metamorphism of both the ore-bearing rocks and the ores themselves, accompanied by the formation of metamorphogenic veins of the "Alpine type" and enrichment of their selvages (POPOV, 1963).

The predominant association of mineralization with the sections of sedimentary formations that are characterized by frequent alternation of rocks having various lithological composition has been reported in many ore areas with lead-zinc and polymetallic mineralization of a stratiform type. The supporters of the hydrothermal-metasomatic concept regarded this peculiarity as manifestation of "anisotropy" of ore-bearing series and applied it to polymetallic deposits of the East Transbaikal area, Mountain and Ore Altai, and many others. However, this traditional concept used for many decades has not been confirmed by actual data of subsequent studies. Thus a detailed lithologic-geochemical study of polymetallic deposits of the Priargunie (East Transbaikalia) carried out by ALEXEEV (1970) enabled him to distinguish among Precambrian and Paleozoic formations two types of ore-bearing rocks.

The first type is represented by alternations of limestones and graphitic, siliceous, sericite-chlorite schists in the section of the Byrkino suite. To this series are associated the following deposits: Savinskoie (ALEXEEV, 1970), Pochekuevo, Chupino, etc. Four cycles can be distinguished here, and each cycle consists of two main rhythms composed of some elementary rhythms. The second type is presented by alternation of dolomites, limestones, coaly shales and variegated rocks located in the lowermost parts of the Kadainsk and Altachinsk suites. This ore-bearing series is also characterized by a cyclic structure; two cycles and smaller rhythms distinguished in it formed by a frequent alternation of carbonaceous and terrigenous rocks.

Distribution of ore mineralization in both types of ore-bearing rocks is strictly subjected to their cyclic-rhythmic structure that is closely related to multi-stage mineralization in a section of strata, a multi-stage character of structure of certain ore fields and deposits, as well as banded and laminated ore textures. The first type of ore-bearing rocks is characterized by a higher content of zinc, lead and arsenic in siliceous slates arranged at the basement of rhythms of the lower rhythmic series. A gradual increase of mineralization intensity upwards is correlated with an increase of siliceous slates in the same direction. There is a definite relation between higher concentrations of fluorine and limestones, as well as between vanadium and graphitic slates.

In the second type of ore-bearing deposits developed within the Kadain group of deposits, three main ore horizons are clearly established corresponding to the sections characterized by most frequent alternation of carbonate and terrigenous rocks. Higher concentrations of lead, zinc, arsenic and antimony are localized in the zones of transition from regressive to transgressive elements of cycles and rhythms.

Thus, the location of polymetallic mineralization in deposits of Priargunie is related to the process of sedimentation of productive strata. Syngenetic mineralization is concentrated in three parts of the sections of ore-bearing rocks that are characterized by the most frequent change of facies, fine alternation of various lithological types of sediments and an abrupt fluctuation in physico-chemical conditions of sedimentation (Eh, pH, etc.).

The polymetallic deposits of the Altai Mountains are selectively associated to volcanic-sedimentary Devonian deposits, due to physico-mechanical and layer-by-layer anisotropy of ore-bearing series (TYCHINSKY, 1961). A stratigraphic-lithological factor of mineralization control is interpreted by TYCHINSKY on subscribing to the "anisotropy" theory; but as a matter of fact, it is obliged to specific conditions of sedimentation of productive volcanogenic-sedimentary series. The formation of stratiform deposits of polymetallic ores was proceeding synchronously with these series. A synchronous volcanic activity was a source of metals.

The polymetallic deposits of the Ore Altai, and the Tyshinsk deposit in particular, were also characterized by association of mineralization to the sections of frequent alternation of acid volcanogenic and sedimentary rocks. The similar heterogeneous sections are regarded by the supporter of the hydrothermal-metasomatic concept as a favourable environment for ore deposition. Lenses and interbeds of acid volcanogenic rocks, as well as sericitic-quartzite, sericitic slates and microquartzites were subjected to ore metasomatosis and formed at the expense of limy-clay shales as a result of "pre-ore hydrothermal processes" (MAN'KOV, 1964, 1969).

It is difficult to imagine that the formation of continuous sulphide ore could have taken place as a result of metasomatic replacement of the above rocks by ore-bearing solutions; because these rocks have rather small chemical activity to metasomatic processes. It is still more difficult to assume that the initial limy-clay shales would have been transformed into quartz-sericite and sericitic shales and microquartzites, and further that metasomatic processes of ore formation could have taken place on these rocks. In the given case, the products of general regional metamorphism to which the rocks of the whole Devonian section were subjected, regardless of mineralization, were considered as "periore" thermal changes of ore-bearing rocks related to a hydrothermal process of an assumed "periore" stage. We think that, in fact, no hydrothermal changes of ore-bearing rocks took place. Syngenetic ore deposits together with enclosing rocks (acid effusive, carbonaceous-clay shales) participated in all the processes of metamorphism and tectogenesis and shared the fate of productive series in the course of their long natural historical development from the Devonian up to the present day. Stratiform and lense-like ore deposits, coinciding with enclosing rocks, repeat surprisingly exactly all folded and fault dislocations including the zones of schist-forming processes. The post-ore character of these tectonic disturbances is beyond any doubt.

Banded and laminated textures peculiar to the ores appeared not through "metasomatic replacement of hydrothermally altered volcanogenic-sedimentary rocks", but
are the initial sedimentary forms of sulphide formation. Only under the conditions
of basin sedimentation and as a result of a fine chemical differentiation of the matter
thin-banded and laminated, often rhythmic ore textures, could have occurred, that
are a natural and regular expression of a sedimentary way of their formation.

The problem of genesis of sulphide ores of the Ore Altai on the example of the Zmei
nogorsk region was studied by DERBIKOV (1962) who gives the evidence of their effusive sedimentary genesis. Denying the idea of "anisotropy", he pays special attention to diversity of facies of productive series, frequent and rapid changes of thicknesses of certain members, and paleogeographical conditions of sedimentation peculiar to the areas of island arcs where an intense volcanic activity took place. This
explains why (but not relative to "anisotropy") the commercial polymetallic mineralization is concentrated in the deposits of the Eifelian.

The examples cited of the stratiform deposits of base metals testify that the selectiv
association of mineralization with the sections of laminated complexes characterizec
by frequent alternation of various rocks and a rapid change of facies is accounted fo
not by physico-mechanical properties of productive series and their "anisotropy", bu
by specific conditions of sediment formation of ore-bearing formations. The ore mat
ter, regardless of its source, supplied in one form or another into the basin, is an
integral component of a sediment; it shares the fate of the sediment, being subjected
to all laws of sedimentation. Thus, the whole process of ore formation and development of deposits will prove to be under control of sedimentation of enclosing rocks.

Of great importance in this case is the factor of time; duration and a slow rate of
sedimentation processes make possible realization of fine differentiation of the matter in general, and metallogenic elements in particular.

Physico-mechanical properties of ore-bearing rocks played a certain role in the
post-sedimentary period of formation of stratiform deposits in the course of the pro
cesses of metamorphism, tectonogenesis, and sometimes subsequent magmatism
accompanied by redistribution and redeposition of the ore matter of stratiform deposits with formation of vein-shaped ore bodies at the expense of primary ore concentrations occurring during the post-sedimentary stage. An important role in redistribution and redeposition of the ore matter of stratiform deposits belonged to
the latest exogenic processes related to circulation of subsurface waters and karst
formation.

Bibliography

ALEXEEV, D. N. : Lithologo-geochemical peculiarities of Precambrian and Paleozoic of Priargunie and polymetallic mineralization. In: State and Tasks of
Soviet Lithology. Moscow, Izdat. "Nauka" (1970).

DERBIKOV, I. V. : On tectonic and paleogeographic conditions in Ore Altai. Trudy
SNIIGIMS, vyp. (issue) 25, Gosgeologtekhizdat (1962).

DRUZHININ, I. P. : Regularities of cyclic structure of variegated deposits of the
Dzhezkazgan suite. Izv. Vuz. Ser. Geologiya i Razvedka, No. 7 (1964).

KORMILITZIN, V. S. : Periore hydrothermally altered rocks of lead-zinc deposits of the East Transbaikalian area. In: Problems of geology and genesis of lead-zinc deposits of Transbaikalian area". Trudy IGEM, Akad. Nauk SSSR, vyp. 83 (1963).

MAN'KOV, B. V. : Morphology of bodies and distribution of mineralization in the Tishinsk deposit (Ore Altai). Izv. Akad. Kazakhsk. SSR, Ser. Geol. , No. 1, (1964).

— Structural peculiarities of the Tishinsk polymetallic deposit in Ore Altai. Izv. VUZ. Geologiya i Razvedka, No. 9 (1969).

POPOV, V. M. : On rhythmicity in sedimentation of copper red suites. Izv. Akad. Nauk Kirg. SSR, Ser. Estestv. and Tekhnich. Nauk, v. 2, vyp. 1 (1960).

— On favourable and screening horizons in stratiform deposits of non-ferrous metals. In: Regularities of Location of Mineral Resources, v. 5, Izdat. Akad. Nauk SSSR (1962).

— On nature of stratiform mineralized fissures enriched with selvage and banded textures in ores of Dzhezkazgan. In: Geochemistry, Mineralogy and Petrography of Sedimentary Formations. M. Izdat. Akad. Nauk SSSR (1963).

SATPAEV, K. I. : Complex metallogenic prognostic maps of Central Kazakhstan. In: Materials of the Scient. Session on Metallogeny and Prognostic Maps (Reports). Alma-Ata, Izdat. Akad. Nauk Kazakhsk. SSR (1958).

TYCHINSKY, A. A. : On participation of stratigrapho-lithological control in formation of polymetallic deposits. Geologiya i Geophyzika, No. 4 (1961).

Address of the author:

Geological Institute,
Academy of Sciences Kirghiz. SSR
Frunze / USSR

Recent Iron Sediment Formation at the Kameni Islands, Santorini (Greece)

H. Puchelt

Abstract

Iron-rich sediments are now forming in several bays of the Kameni Islands within the Santorini caldera by precipitation from warm submarine springs which represent a phase of late volcanic activity. From one bay of Palaea Kameni several cores from the soft, up to 3 m thick sediment have been taken and the analyses of four of them are reported in this paper. The composition of the gel-like silica-rich substance is not homogeneous. In the uppermost part, ferric hydroxide prevails. Down to approximately half a meter ferrous carbonate occurs. Underneath this layer, iron occurs as ferrous hydroxide, basic sulfate, or water-rich ferrous silicate gel. Only in the uppermost parts of the layers ferrous sulfide forms locally by the activity of sulfate reducing bacteria. Possibilities of the origin of the material precipitated in this bay, mode of transport, and stability relationships relative to Eh and pH values are discussed.

Introduction

Submarine formation of iron-rich sediments has been reported from the Santorini caldera by BEHREND (1936). Investigations and descriptions of the iron sediments of Nea Kameni were published by LIPPERT (1953), HARDER (1960), and BUTUZOVA (1966, 1969). BONATTI et al. (1971) reported on their investigations on iron-rich sediments of Nea Kameni. The more spectacular iron sediments from a bay of Palaea Kameni have not been studied so far.

The geological and geographical situation is as follows: The Santorini Islands, consisting of the remnants of a large-size volcano which collapsed after a tremendous pumice eruption in the 15th century B. C., belong to the island-arc which extends

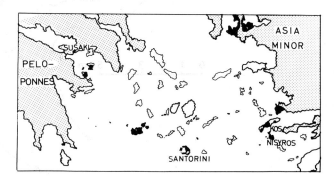

Fig. 1. Map of the Aegean Sea (volcanic areas marked black)

228

from the Isthmos of Corinth over the islands of Methana, Aegina, Milos, Santorini, Kos, and Nisyros to Asia Minor (Fig. 1). The caldera bottom submerged below sea-level and only 197 B. C. a new volcano appeared within the caldera. Several small islands were formed during the past centuries which were united to the two islands Nea and Palaea Kameni during the volcanic activity in 1925-28 (RECK, 1936). Fig. 3 shows the shape of these two islands and the locations of the submarine thermal activities. Solfataric activity with exhalations of elemental sulfur, sulfurdioxide, carbondioxide, and water vapour occurs on Nea Kameni (PUCHELT et al., 1971). The position of Nea and Palaea Kameni islands within the Santorini caldera is shown in Fig. 2

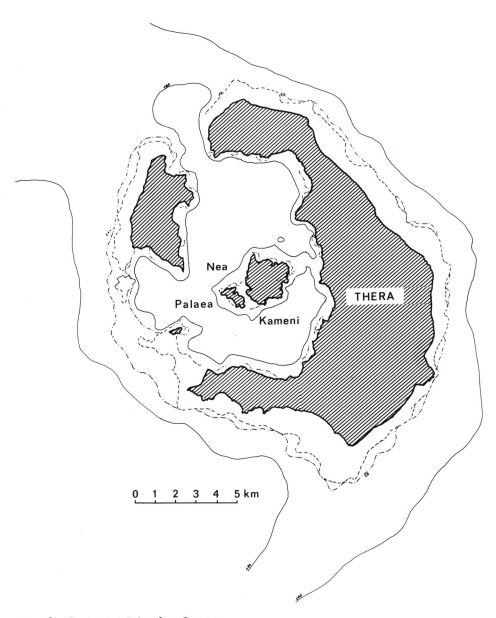

Fig. 2. Santorini Islands, Greece

Both the solfataric activity on Nea Kameni and the emergence of submarine late-vol-
canic solutions in the Kameni bays are connected with distinct thermal activity as
shown by PUCHELT et al. (1972a), by infra-red remote sensing.

Although several places on Nea Kameni are known where iron sediments are forming
in these days, only the bay of Palaea Kameni where iron sediments have formed the
largest thickness is dealt with in this paper

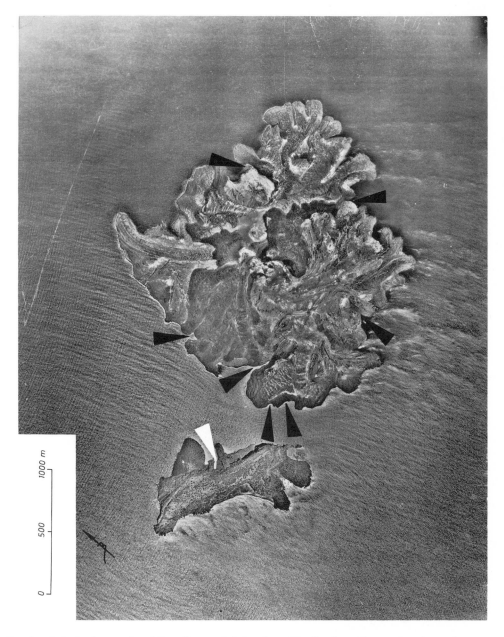

Fig. 3. Aerial photograph of the Kameni Islands with the locations of submarine
thermal activity

230

Description of the Bay

The bay marked on Fig. 3 has a length of 60 m. Its width is approximately 20 m, and the depth of water down to the sediment-water interface is approximately half a meter. A general view of the bay is given on Fig. 4.

Fig. 4. Iron-ore bay on Palaea Kameni, Santorini, Greece (photographer: H. W. Hubberten)

Below the free water, a sediment of up to 3 m thickness of soft consistency was precipitated. Cross sections through this bay have been measured at several places. A schematic drawing of depth profiles is given in Fig. 5. From this figure, it can be seen that the bay is protected from wave action by a barrier of boulders of volcanic rocks at its outlet.

At the west end of the bay, gases emerge from the water at hundreds of locations (Fig. 6). According to investigations with DRAEGER test tubes, they consist of more than 75 % CO_2; neither SO_2, H_2S, HCl, nor AsH_3 were found. An estimation of the amount of emerging gases in this area gives about 5 to 10 m^3/h. The water temperature in this narrow bay, which is surrounded by dark rocks, is up to 35 °C. The distribution of temperature of the surface water within the bay is indicated in Fig. 5.

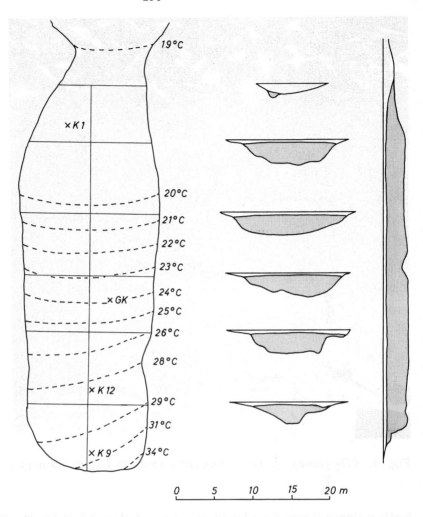

Fig. 5. Profiles and sampling points of the iron-ore bay, Palaea Kameni

The sediment itself is of rusty brown colour at the sediment-water interface. Locally, patches of black sediments are exposed where ironsulfides have been formed as a consequence of the activity of sulfate reducing bacteria. In greater depth, the sediments are of olive and dark grey colour. Locally, layers of almost white colour are inserted (Fig. 7). A selection of cores from the bay representing the different sediment layers is shown in Fig. 8. Within the sediment variable amounts of organic substance, mostly plant remains, are found. Close to the outlet of the bay the sediments are oxidized in various layers due to a better aeration of the iron compounds which are stirred up more actively at this place by wave action.

Sampling

Core samples have been taken at 16 points all over the bay. It has been tried to get as long cores as possible from all stations, but only at point GK (Fig. 5) and two

Fig. 6. CO_2 emerging from submarine vents in Palaea Kameni iron-ore bay

further places it was possible to get cores of almost 2 m length. From the other points cores of shorter length could only be obtained.

Cores K 1 and 9, taken in transparent, 3.5 cm diameter tubing, are reproduced in Fig. 7 together with a scale. The cores have been pushed out of the tubing and pH and Eh measurements were taken immediately from the water-rich sediment. After this, each section of the core was divided into two equal parts. One part was transferred to a special centrifuge to separate the pore solution, the other part was weighed and dried for one night at 105 °C. Thus the water content of the gel was determined. In a few instances particles of harder rock were included in the drying residue. They were crushed together with the softer material in an agate mortar. It was generally found that pumice pieces were the softer, the deeper they occurred in a core. Locally they formed almost white layers of gel substance as demonstrated at the depth of about 45 cm below sediment-water interface of core K 9.

Analyses

The materials dried at 105 °C were subjected to further chemical treatment. All the results are calculated to the material dried at 105 °C which was not previously wash-

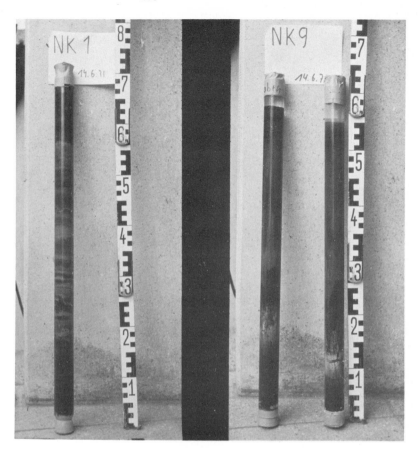

Fig. 7. Cores NK 1 and NK 9 from Palaea Kameni

ed free from the salts of the pore solution. A modified SHAPIRO and BRANNOCK-method was applied for the analyses. SiO_2 was determined gravimetrically. Al_2O_3 was precipitated together with TiO_2 iron-free with benzoate in the presence of thio-glycolic acid. Organic carbon was determined as difference: total CO_2 - as obtained by combustion in a LECO instrument - minus acid-decomposition-CO_2 times conversion factor. Na, K, Mg, Pb, and Cu were obtained using atomic absorption spectrometry. Rb and Sr were determined by X-ray-fluorescence. The sulfate and chloride content of the samples were determined after sufficient leaching of the material by distilled water. Chloride was titrated with silver nitrate using chromate as indicator, sulfate was precipitated as the insoluble barium compound.

In the separated pore solutions which were absolutely clear immediately after separation ferric hydroxide precipitated after a few minutes due to oxidation of ferrous iron. In the pore solutions iron and chloride were determined.

The results of all analyses are given in Tables 1 to 5 and in Fig. 7.

The following facts can be stated:

1. All sediments of this bay consist of a paste with 58 to 80 % water below 105 ^{O}C. Since this relatively immobile pore solution contains the constituents of sea water,

Fig. 8. Selection of cores from iron-ore bay, Palaea Kameni

the elements Na, Cl, K, Mg, SO_4 are included in the drying residue (to 105 °C) of the sediment in the proportion of their content in the sea of this area (PUCHEL et al. , 1971).
Chloride will at the present time almost exclusively be brought into the sediment by the sea water. Since the ratios of the other sea water constituents differ from the normal sea water ratio, other sources must be sought for these components (Na, K, Mg, Ca, SO_4).

2. All material consists of a gel-like substance, sometimes with organic remainder (rootlets etc.).

3. More oxidized brownish parts are always located close to the sediment-water interface.

4. A narrow banding caused by more or less oxidized layers occurs close to the mouth of the bay (core K 1) where wave actions often stir the sediments up and oxidize the ferrous compounds.

5. If H_2S or iron sulfide occurs, it is always found in the uppermost layers of the cores and restricted to very narrow areas. From the pore solutions of all sulfide containing sections living sulfate reducing bacteria were isolated.

6. A general increase of carbonate is observed in all cores towards their top. Since Ca or Mg are not present in the necessary amounts to bind this component, the

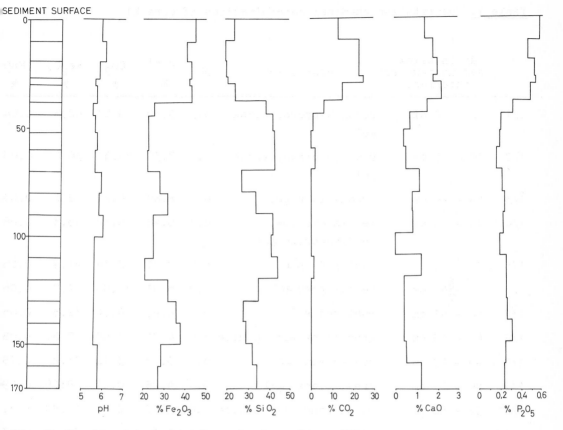

SEDIMENT SURFACE

Fig. 9. Graphic plot of main element values of core GK

only possibility to fix carbonate is the formation of $FeCO_3$. This compound has been observed in a few delicate crystals under the microscope. Most of it is in the gel state so that no X-ray-diffraction patterns are observable.

7. In all deep layers of the bay SO_4 is considerably enriched relative to chloride. The ratio being 0.14 in sea water increases to 0.72 in the deep sediment gels.

8. Pore solutions from core GK contain chloride in the concentration of the surrounding sea water (PUCHELT et al., 1971). Iron occurs in the ferrous state in much higher concentrations than in sea water and also calcium and manganese are concentrated relative to sea water.

Discussion

Although no genuine solfataric solutions have been sampled directly from any Kameni bay, certain qualities can be attributed to them. From the infrared photographs (PUCHELT et al., 1972a) it can be learnt that from all locations where iron sediments are precipitated a fan of warm water spreads over the sea. This water is at all times flowing out of the respective bays proving at least a constant flow of heat. From the colour of the different layers within the cores and from their analyses it can be seen

Table 1. Quantitative chemical determinations of core K1

	depth below sediment-water interface	description	pH	H_2O^+ %	CO_2 %	Fe_2O_3 %	MnO %
1/1	0 - 10 cm	reddish-brown, black gel	6.3	9.30	9.21	42.3	0.04
1/2	10 - 19 cm	reddish-brown, green gel	6.0	7.39	14.30	36.0	0.05
1/3	19 - 24 cm	green-black gel	6.1	8.80	4.29	18.5	0.03
1/4	24 - 30 cm	reddish-yellow, reddish-brown gel	6.1	8.48	6.32	35.3	0.08
1/5	30 - 36 cm	rust-red gel	5.6	9.90	2.34	41.9	0.06
1/6	36 - 40 cm	yellow-grey gel	6.0	6.01	14.09	29.7	0.06
1/7	40 - 41 cm	rust-red gel	5.6	n.b.	4.74	32.3	0.05
1/8	41 - 45 cm	grey-black with pumice	6.0	4.71	1.09	7.82	0.08
1/9	45 - 49 cm	grey-green gel	6.1	5.58	13.12	33.2	0.18
1/10	49 - 52 cm	grey, black gel	6.1	4.28	0.37	8.15	0.09
1/11	52 - 58 cm	black sand with pumice	n.d.	1.17	0.64	14.80	n.d.
1/12	58 - 64 cm	black gel	6.1	5.82	0.31	31.85	0.03

63 cm   total length of core

that low values of the redox potential prevail in deeper parts of the cores. The redox potential assumes values which do not allow formation of hydrogensulfide from sulfate but covers the field of existence for ferrous iron compounds. Only the ironsulfide patches close to the sediment surface have more negative Eh. The measured pH might be somewhat too high due to loss of carbondioxide in the samples (cf. GARREL and CHRIST, 1965). In the lowest layers of all cores carbonate is very rare. It occurs only in the uppermost 50 cm of each core, where it is bound to ferrous iron. This mode of occurrence can be explained as follows:

In larger depth, the penetrating carbondioxide has a higher pressure and consequently a higher solubility and activity in the sea water. Here, the concentration of HCO_3 is high enough to dissolve all the precipitated ferrous carbonate. This is transported closer to the surface. Here, CO_2 pressure is low enough (down to $2.3 \cdot 10^{-2}$ atm.) to allow precipitation of siderite gel. In low layers of the cores, sulfate is considerably enriched. The increase of this compound is much higher than can be deduced from a pure pore solution. This is a hint for the fact that the solutions have been of sulfuric-acid-character. Solfataric activity on Nea Kameni brings SO_2-gases which disproportionate the elemental sulfur and sulfuric acid (PUCHELT et al., 1971). By this mechanism obviously large amounts of sulfuric acid are produced. The same

CaO %	MgO %	TiO$_2$ %	Al$_2$O$_3$ %	P$_2$O$_5$ %	SiO$_2$ %	Na$_2$O %	K$_2$O %	Cl$^-$ %	SO$_3$ %	org.C %	minus oxygen for Cl
1.82	1.57	0.19	n.d.	0.35	22.40	6.46	0.42	7.94	1.07	1.37	1.79
2.21	1.55	0.15	n.d.	0.35	25.66	5.92	0.39	7.56	0.64	0.75	1.71
1.23	1.71	0.16	n.d.	0.28	42.91	9.04	0.66	10.81	0.83	1.37	2.44
1.68	1.47	0.28	n.d.	0.52	n.d.	5.92	0.69	6.23	0.60	1.30	1.41
1.01	1.25	0.26	n.d.	0.70	28.06	6.00	0.63	6.71	0.88	0.85	1.52
1.40	2.17	0.19	n.d.	0.27	32.20	5.60	0.48	6.95	1.27	1.38	1.57
1.14	1.54	0.30	n.d.	0.65	n.d.	6.74	0.70	n.d.	n.d.	n.d.	
1.96	1.51	0.46	n.d.	0.18	57.74	7.14	1.29	5.85	0.87	1.40	1.32
2.50	1.19	0.31	n.d.	0.30	31.74	5.32	0.71	5.05	0.92	1.16	1.14
2.18	1.45	0.51	12.20	0.15	56.77	6.34	1.29	4.76	1.02	1.41	1.08
1.06	0.79	0.39	14.71	0.21	60.68	5.70	0.32	1.67	0.85	0.47	0.38
0.42	1.19	0.13	0.34	0.17			0.29	7.36	3.37	4.20	1.66

mechanism may work inside the Kameni Islands where the acid leaches the calc-alkalic rocks. It has been shown by experiments by SCHORIN (1972) that under exposure of andesites and dacites to sulfuric acid of pH 1 at 100 °C iron and manganese are extracted preferentially. Aluminum and silica are dissolved later and the silica especially slowly.

Iron and manganese can be transported to the vents where the solution debouches into the sea. The form of the migrating iron and manganese compounds will be a sulfate or a hydrogencarbonate complex, and the solution will be oversaturated in respect to carbondioxide. In the moment of the emergence of the solution into the sea the pH is increased drastically. Thus all compounds which are very sensitive to pH changes will precipitate in the form of hydroxides or carbonates. Other compounds as silica, the solubility of which is less pH-dependent in an acidic solution (KRAUSKOPF, 1967), are practically not precipitated in this way. For silica the decrease of temperature and change of ionic strength will be a stronger precipitating factor. Thus, gel precipitation will occur from the supersaturated solution and the gel itself is flocculated by iron hydroxide gels (HARDER and FLEHMING, 1970). Aluminum is rather insoluble as a hydroxide (solubility product 10^{-32}), but much more soluble than the co-

Table 2. Quantitative chemical determinations of core GK

	depth below sediment-water interface	description	water below 105°C	pH	H_2O^+ %	CO_2 %	Fe_2O_3 %	MnO %
GK 1	0 - 10 cm	brown gel	67.3	5.9	6.00	13.10	46.2	0.08
GK 2	10 - 19 cm	yellow-grey gel	59.3	6.1	3.56	23.30	44.5	0.10
GK 3	19 - 27 cm	yellow-grey-brown gel		5.8	3.20	23.16	45.4	0.09
GK 4	27 - 30 cm	grey gel	56.5	5.6	4.81	25.4	42.9	0.09
GK 5	30 - 38 cm	dark grey gel		5.7	4.70	15.40	44.2	0.10
GK 6	38 - 44 cm	grey gel	69.3	5.5	6.26	4.54	26.2	0.07
GK 7	44 - 52 cm	greyish yellow gel with some sand	69.1	5.6	6.85	0.75	23.6	0.17
GK 8	52 - 60 cm	grey gel	73.8	5.7	6.60	<0.10	27.2	0.19
GK 9	60 - 70 cm	grey gel	74.3	5.6	7.45	1.62	22.4	0.18
GK 10	70 - 80 cm	grey gel with organic residue	69.7	5.9	6.80	0.16	28.5	0.17
GK 11	80 - 90 cm	grey gel, highly decomposed pumice and much organic residue	71.2	5.8	2.02	0.19	31.8	0.06
GK 12	90 - 100 cm	grey gel with highly decomposed pumice	70.7	6.0	4.26	<0.10	25.0	0.07
GK 13	100 - 110 cm	grey gel	75.7	5.6	6.56	0.19	25.0	0.09
GK 14	110 - 120 cm	grey gel with white schlieren	74.4	5.6	5.79	1.19	21.0	0.14
GK 15	120 - 130 cm	grey gel with very little pumice	75.3	5.6	7.02	<0.10	33.1	0.03
GK 16	130 - 140 cm	grey gel with some pumice	71.3	5.6	5.00	<0.10	36.2	0.02
GK 17	140 - 150 cm	grey gel with some pumice	65.5	5.6	4.89	<0.10	38.0	0.05
GK 18	150 - 160 cm	pure grey gel	70.3	5.8	4.98	0.42	28.5	0.07
GK 19	160 - 170 cm	grey gel with sand and pumice	65.5	5.8	6.45	0.72	30.3	0.04

239

	MgO %	TiO$_2$ %	P$_2$O$_5$ %	SiO$_2$ %	Na$_2$O %	K$_2$O %	Cl$^-$ %	SO$_3$ %	org.C %	Cu ppm	Pb ppm	minus oxygen for Cl
5	2.19	0.17	0.57	23.65	5.25	0.36	4.50	0.86	0.82	12	<1	1.02
2	2.07	0.12	0.45	20.34	3.77	0.482	3.24	0.69	0.67	28	2	0.73
+	2.19	0.13	0.52	21.12	2.82	0.349	3.07	0.56	0.56	9	<1	0.69
5	1.96	0.18	0.53	21.29	3.64	0.325	4.02	1.07	n.d.	16	<1	0.91
2	2.17	0.23	0.48	24.10	3.10	0.301	4.24	1.54	0.74	15	<1	0.95
9	1.65	0.30	0.31	39.10	5.65	0.820	6.73	2.17	1.37	35	5	1.52
3	1.25	0.29	0.20	42.29	5.38	0.663	4.48	2.79	2.35	2	4	1.10
2	0.39	0.29	0.19	43.39	5.38	0.325	6.13	3.17	3.16	1	1.5	1.38
5	0.78	0.11	0.16	43.44	4.65	0.277	6.46	3.66	2.66	<0.5	3.5	1.46
7	0.30	0.19	0.29	26.98	5.19	0.28	5.55	3.20	2.62	21	5	1.25
5	0.28	0.15	0.25	34.09	4.45	0.39	4.86	3.35	1.93	n.d.	n.d.	1.10
5	1.04	0.24	0.22	41.60	5.73	0.615	8.46	2.74	0.86	0.5	n.d.	1.91
0	1.15	0.12	0.19	40.72	5.32	0.277	6.01	3.61	1.96	5	2	1.36
8	1.06	0.30	0.25	43.87	4.85	0.735	6.30	2.89	1.32	2	<1	1.42
7	0.73	0.15	0.25	35.14	3.97	0.253	5.45	3.28	2.26	4	<1	1.23
7	0.20	0.16	0.26	27.52	5.59	0.43	4.54	3.11	1.81	4	2	1.02
9	0.65	0.30	0.31	29.07	4.78	0.313	7.18	2.68	1.63	6	<1	1.62
6	0.84	0.17	0.25	32.43	4.58	0.481	5.40	3.10	0.26	3	1	1.22
5	1.02	0.25	0.24	33.87	3.37	0.723	4.30	1.28	0.58	11	5	0.97

Table 3. Quantitative chemical determinations of core NK9

	depth below sediment–water interface	description	water below 105°C	pH	H_2O^+ %	CO_2 %	Fe_2O_3 %	Mn %
9/1	0 – 3 cm	with black FeS	71.7	6.3	6.47	12.77	37.9	0.0
9/2	3 – 19 cm		64.5	5.8	4.30	22.36	42.8	0.0
9/3a	19 – 21 cm	reddish–brown gel	58.2	5.7	7.05	10.56	39.8	0.0
9/3	21 – 25 cm	green–yellow– brown gel	67.2	5.7	4.72	24.09	40.3	0.0
9/4	25 – 30 cm	grey gel	68.2	5.5	3.90	18.86	39.3	0.0
9/5	30 – 34.5 cm	almost black gel	63.4	5.3	5.92	4.44	19.4	0.0
9/6	34.5 – 42.5 cm	dark green and white gel	76.2	5.9	6.04	0.88	18.4	0.0
9/7	42.5 – 48 cm	light green– white gel	81.3	5.8	8.26	0.68	10.00	n.
9/8	48 – 55 cm	dark grey gel	75.2	5.8	5.98	12.32	24.4	0.0

55 cm   total length of core

existing ferric hydroxide (solubility product 10^{-38}). But since $Al(OH)_3$ is less soluble than $Fe(OH)_2$ (solubility product 10^{-15}), it will be the compound which is precipitated prior to $Fe(OH)_2$. $Al(OH)_3$-rich layers should occur in deeper parts of the sediments not sampled so far. Although it is shown by the experiments of SCHORIN (1972) that silica is dissolved considerably in hot acid solutions, the silica content of the solution itself cannot be the only source for all the silica which is contained in the sediments of the bay. From several of the cores it has been learnt that frequently pumice is caught in the iron sediment and that this volcanic glass is decomposed the more, the longer (deeper) it lies in the slightly acid gel. Only by this way the TiO_2 values can be explained which are found in the core, since in acidic leaching of volcanic rocks TiO_2 always remains in the residue, i.e. it is practically not transported. New pumice analyses which present an impression on the field of pumice chemistry are reported by PUCHELT and SCHOCK (1972). From the data obtained by SCHORIN (1972) and the observations of disintegration of rocks by solfatari gases it is logic to derive the iron sedimented in the bays of the Kameni Islands from rock-leaching. This is also supported by the trace element data which were obtained for several samples. Most of the silica is introduced into the sediment by in situ decomposing pumice, and only lower amounts of silica are brought by the thermal solutions. The ratio of silica/iron oxides is not constant for all sediment layers, but it is often close to that of chamosite.

CaO %	MgO %	TiO_2 %	Al_2O_3 %	P_2O_5 %	SiO_2 %	Na_2O &	K_2O %	Cl^- &	SO_3 %	org.C %	minus oxygen for Cl
1.91	1.14	0.28	<0.20	0.37	24.2	4.92	0.560	5.27	n.d.	1.50	1.18
2.69	1.24	0.23	1.98	0.38	20.20	3.64	0.355	3.97	0.78	0.53	0.89
2.02	1.10	0.28	2.57	0.38	22.37	3.64	0.493	3.31	0.87	5.53	0.75
2.46	1.30	0.21	0.30	0.50	26.47	4.05	0.379	4.80	0.87	0.64	1.08
2.49	1.18	0.22	2.23	0.38	23.38	4.32	0.482	4.40	0.94	0.94	0.99
1.68	1.06	0.38	6.96	0.22	43.18	5.18	1.132	3.80	1.37	4.37	0.86
0.63	1.47	0.17	4.02	0.20	55.74	5.65	0.639	6.45	0.81	1.38	1.45
n.d.	1.29	0.14	3.90	0.10	65.39	7.60	0.53	9.38	n.d.	1.68	2.11
0.37	0.62	0.11	1.89	0.15	41.95	5.25	0.337	6.10	4.22	2.03	1.38

The idealized conditions for the fields of existence for the different iron compounds are given in Fig. 10 (cf. also CURTIS and SPEARS, 1968). The Eh- and pH-values could not be measured exactly, since no equipment was available for in-situ work. Eh- and pH-values measured on samples which have been exposed to the air for only a short time can be considerably in error, as described by GARRELS and CHRIST (1965).

The results of this investigation differ somewhat from those of BUTUZOVA (1969) for a bay on Nea Kameni. There, sediments of less thickness occur and H_2S of volcanic origin is reported for some bays. BUTUZOVA (1969) found authigenic vivianite crystals in her sediments which was not the case for the Palaea Kameni bay. The mechanism for the iron sediment formation proposed by her differs considerably, as in her bay always the lowest layers of a sediment consists of ferrous sulfide.

In our investigations, no crystallized substance could be observed by X-ray-diffraction, but a few delicate authigenic siderite crystals are to be seen under the microscope. At the same time in areas of higher silica content some feldspar crystals were found which obviously are relicts from pumice decomposition.

Table 4. Quantitative chemical determinations of core K 12 (partial)

	depth below sediment-water interface	pH	Fe$_2$O$_3$ %	SiO$_2$ %	TiO$_2$ %	CaO %	MnO %	Na$_2$O %	CO$_2$ %	P$_2$O$_5$ %	Rb ppm	Sr ppm
K 12/1	0 - 2 cm	n.d.	32.75	21.0	0.25	n.d.	0.05	9.44	n.d.	0.43		
K 12/2	2 - 6.5 cm	6.8	35.0	23.3	0.26	n.d.	0.00	7.01	n.d.	0.44		
K 12/3	6.5 - 12.5 cm	6.6	31.25	21.1	0.23	n.d.	0.05	10.78	n.d.	0.43	11	110
K 12/4	12.5 - 17 cm	6.5	35.75	27.9	0.44	2.80	0.11	3.77	15.5	0.46	18	109
K 12/5a	17 - 21.5 cm	6.1	38.75	21.3	0.33	2.80	0.13	3.37	19.95	0.39	8	76
K 12/5b	soft pumice of layer 5	n.d.	7.8	n.d.	0.38	1.96	0.08	6.74	n.d.	0.14		
K 12/6	21.5 - 25.5 cm	6.1	32.5	27.1	0.33	2.64	0.11	4.36	18.4	0.35	4	73
K 12/7	25.5 - 27 cm	6.1	33.5	20.7	0.22	3.10	0.11	4.59	23.5	0.31	12	80
K 12/8	27 - 34 cm	5.9	31.95	20.0	0.25	2.80	n.d.	5.14	22.1	0.42	10	95
K 12/9a	34 - 38.5 cm	5.7	24.2		0.15	2.12	0.04	6.07	0.37	0.15		
K 12/9b	38.5 - 40.5 cm (amorphous SiO$_2$ rich gel)	n.d.	12.0	48.0	0.22	1.18	0.06	4.48	n.d.	0.28	21	71

Table 5. Composition of pore solutions from
core GK

	Fe mg/l	Cl⁻ g/l	Ca mg/l	Mn mg/l
GK 2+3	41	23.0	682	0,6
GK 4-6	53.4	22.6	626	2,4
GK 7	14.8	22.8	560	1,3
GK 8	11.4	22.6	n.d.	
GK 9	14.6	23.0	n.d.	
GK 10	14.6	22.5	n.d.	
GK 11	11.6	22.5	542	
GK 12	9.0	23.0	690	
GK 13	13.0	23.0	530	1,4
GK 14	17.4	22.8	548	o,9
GK 15	20.0	22.6	542	0,8
GK 16	27.0	22.8	542	0,8
GK 17	37.0	23.0	542	5,6
GK 18	30.8	22.7	530	0,6
GK 19	14.6	22.4	545	< 0,5

The resources of iron in this Palaea Kameni bay with 2400 m³ sediment containing 480 m³ of water-free iron sediment can be calculated. With a specific gravity of three this would correspond to ≈ 1500 t of iron-ore with perhaps an average content of 33 % ≈ 500 t Fe_2O_3. This is not very impressive an amount, but it is still a valuable example of iron-ore formation with the full story of geochemical information and physico-chemical conditions of sedimentation. The available information puts this bay in the row with iron-rich sediments (cf. JAMES, 1966) as an important recent example for submarine exhalative sedimentary iron-ore formation.

Acknowledgements

The work has been supported by a grant of the Deutsche Forschungsgemeinschaft. The field investigations were much helped by Mr. Peter M. Nomikos and his men. In the analytical work Mrs. Hedy Hahn was engaged with much interest. H. W. Hubberten took several of the photographs and was of much assistance with the drawing work. I want to thank all these persons and the Deutsche Forschungsgemeinschaft for their support.

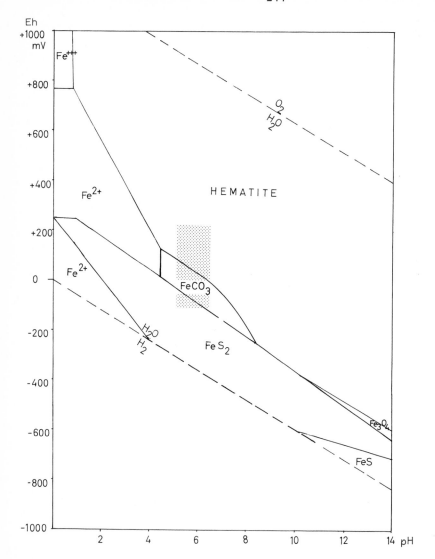

Fig. 10. Eh-pH-range of ore-sediments in bay of Palaea Kameni

$$p_{CO_2} = 1 \text{ atm.} \qquad \sum S \approx 10^{-2} \text{ molar} \qquad \left[Fe^{2+}\right] \approx 10^{-3} \text{ molar}$$

Bibliography

BEHREND, F.: Eisen und Schwefel fördernde Gasquellen auf den Kameni Inseln. In: H. RECK, ed., Santorin. Der Werdegang eines Inselvulkans und sein Ausbruch 1925-1928. Band II, Kapitel XII. Dietrich Reimer/Andrews & Steiner, Berlin (1936).

BONATTI, E., HONNOREZ, J., and JOENSUU, O.: Submarine iron deposits from the Mediterranean Sea. Paper read at the VIII International Sedimentological Congress, September 1971, in Heidelberg, Germany (1971).

BUTUZOVA, G. Yu.: Iron ore sediments of the fumarole field of Santorin Volcano, their composition and origin. Dokl. Akad. Nauk SSSR 168, p. 1400-1402 (1966).

— Recent volcano-sedimentary iron-ore process in Santorin Volcano-caldera (Aegean Sea) and its effect on the geochemistry of sediments. Academy of Sciences of the USSR, Geol. Inst. Transactions, v. 194 (1969).

CURTIS, C. D., and SPEARS, D. A.: The formation of sedimentary iron minerals. Econ. Geology 63, p. 257-270 (1968).

GARRELS, R. M., and CHRIST, Ch. L.: Solutions, minerals, and equilibria. Harper & Row Publ., New York (1965).

HARDER, H.: Rezente submarine vulkanische Eisenausscheidungen von Santorin, Griechenland. Fortschr. Miner. 38, p. 187-189 (1960).

— Kohlensäuerlinge als eine Eisenquelle der sedimentären Eisenerze. In: G. C. AMSTUTZ, ed., Sedimentology and Ore Genesis. Vol. 2. Elsevier, Amsterdam, 184 p. (1964a).

— Untersuchung rezenter vulkanischer Eisenausscheidung zur Erklärung der Erze vom Lahn-Dill-Typus. Ber. Geol. Ges. DDR 9, p. 439-623 (1964b).

— und FLEHMIG, W.: Quarzsynthese bei tiefen Temperaturen. Geochim. Cosmochim. Acta 34, p. 295-305 (1970).

JAMES, H. L.: Chemistry of the iron-rich sedimentary rocks. In: Data of Geochemistry. Chap. W. Geol. Surv. Prof. Paper 440-W (1966).

KRAUSKOPF, K. B.: Introduction to geochemistry. McGraw-Hill, New York (1967).

LIPPERT, H. J.: Bericht über eine Studienfahrt nach Santorin in der südlichen Ägäis. Zeitschr. Dt. Geol. Gesellsch. 105 (1953).

PUCHELT, H., HOEFS, J., and NIELSEN, H.: Sulfur isotope investigations of the Aegean volcanoes. Acta of the International Scientific Congress on the Volcano of Thera, p. 303-310 (1971).

— and SCHOCK, H. H.: Geochemische Untersuchungen an Santorinbimsen. Zeitschr. Dt. Geol. Gesellsch., in print (1972).

— KROSE, G., and MIOSGA, G.: Infrared investigation of the volcanic Santorini Island, Greece. In preparation (1972a).

RECK, H.: Santorin. Der Werdegang eines Inselvulkans und sein Ausbruch 1925-1928. Dietrich Reimer/Andrews & Steiner, Berlin (1936).

SCHORIN, H.: Experimentelle Untersuchungen zur fumarolischen Zersetzung von Gesteinen und Mineralen und ihre Anwendung auf die Veränderung von Laven und Tuffen auf Santorin/Griechenland. Dissertation Univ. Tübingen (1972).

Address of the author:

Abteilung Geochemie,
Mineralogisches Institut der
Universität Tübingen
74 Tübingen / Germany

Ore Deposits and Continental Weathering: A Contribution to the Problem of Geochemical Inheritance of Heavy Metal Contents of Basement Areas and of Sedimentary Basins

J. C. Samama

Abstracts

Les concentrations de métaux lourds (Pb, Zn, Cu, U) contenues dans les formations détritiques continentales (ou à affinité continentale) se prêtent bien à l'étude des relations géochimiques entre socle et couverture.

De l'étude des altérations actuelles ou anciennes, il apparait que, suivant le processus majeur de l'altération d'une aire continentale, un même élément lourd peut être soit éliminé vers les bassins sédimentaires, soit enrichi et provisoirement stocké sur le continent. Ce modèle conduit à l'analyse de la correspondance observée entre la nature (donc la géochimie) des altérations et le chimisme des concentrations métallifères contenues dans les séries détritiques qui résultent du décapage des aires altérées : seuls les éléments lourds stockés lors de l'altération se retrouvent concentrés dans ces molasses.

L'altération continentale semble ainsi déterminer l'existence et la nature des concentrations métallifères dans les séries détritiques contemporaines et immédiatement postérieures.

Ce modèle d'héritage de zone altérée à couverture détritique complète considérablement le modèle précédent d'héritage de socle à couverture en rendant compte, en particulier, du caractère monométallique accusé de ces gisements.

Heavy metals (Pb, Zn, Cu, U) concentrations which occur in detric continental (or subcontinental) environments are found to be suitable for studying basement-sedimentary basin relationships.

If heavy metals of these deposits originate from basements (geochemical sources), then the influence of the weathering affecting a basement prior to erosion must be an important phenomenon to which this paper is devoted.

From the study of recent or old weathering profiles, it appears that each heavy element can either be leached out towards a basin or enriched and provisionally accumulated within weathered covers according to the major weathering process acting upon continents.

Geochemical analysis of three regional examples leads to the conclusion that only the heavy elements accumulated during a weathering period are found in economic deposits in the resulting sandstone formations, and that geochemical evolutions or zonations may derive from climatological evolutions or zonations.

This model of geochemical inheritance from weathered basements to detric basins improve considerably the previous one of inheritance from basements to detric basins, explaining, for instance, the well marked mono-(bi-) metallic character of the sandstone type deposits.

1. Introduction

During the last twenty years, it has become more and more apparent that continental weathering and erosion processes influence the nature of sedimentary supply and, of course, the nature and the geochemistry of the sediments themselves. Several models for the relationship between continental erosion and sedimentation have been presented, but the biorhexistasic model of H. ERHART (1955) is probably the most fitting one for ore deposit problems. In this theory, ERHART distinguishes two main types of continental evolution in tropical or equatorial regions. In the first one, under equilibrium conditions (so-called "biostasic" period), the vegetation and the weathered mantle act as a "separative filter" fixing a residual phase (Al and Fe hydroxides, for instance), but allowing leaching and migration of a "soluble" phase (Na, Ca, Mg, etc) towards a chemical basin. Disruption of such an equilibrium (so-called "rhexistasic" period) induces continental erosion of the weathered cover and consequently sedimentation of a clastic residual phase.

This model has been elaborated for the supergene separation of major elements, but it may be interesting to introduce this concept in geology of ore deposits to which this paper is devoted. The attempt to introduce such a model in the geochemistry of trace elements and consequently in the genesis of stratabound ore deposits is not new (1), but the previous results are not so clear because of great differences in methods and scales of the studies, but chiefly because of a very often wrong conception of trace elements behaviour during weathering.

The purpose of this paper is to give an up-to-date idea of this important problem and to point out a solution built on paleoclimatology, on recent data about trace element behaviour during weathering and, of course, on geology of stratabound ore deposits.

---

(1) See for instance: BERNARD, 1962, 1964; BERNARD and FOGLIERINI, 1963; BERNARD and SAMAMA, 1968; CÁRRIE et al. , 1966; DEVIGNE and NICOLINI, 1963; ERHART, 1961, 1966, 1967; FOGLIERINI and BERNARD, 1967; FUCHS, 1969a and b; FUCHS and PINAUD, 1969; ROUTHIER, 1969; SAMAMA, 1969; TAMBURINI and URAS, 1965; VALERA, 1966; VIOLO, 1966, etc.

(2) The reasons for such restrictions are evident: investigating for the possible basement influence on stratabound deposits, we must choose examples in which the heavy metals could be considered as deriving from the basements.

The choice of terrigenous continental formation is imposed by methodological evidences. In this case, a continental origin of heavy metals agrees much better with sedimentological results than a margine origin.

The choice of examples for which exhalative influence does not appear to be of great importance, is imposed by the same reasons. That does not mean heavy metals are not of volcanic origin, but just they occurred on continental areas prior to mineralized sandstone deposition. From this point of view, we would consider some sandstone type deposits such as the cupriferous red beds of Yunnan, the copper source of which is thought to be a 400-700 ppm Cu basalt (LI HSI CHI et al., 1964) or such as the uranium deposits of Niger, the uranium source of which would be a 100 ppm U volcanic tuff (BIGOTTE and OBELLIANNE, 1968).

In this way and restricting the problem to the simplest case, we have chosen the example of heavy metal concentrations (Pb, Zn, Cu and U) in clastic continental (or subcontinental) formations, where the exhalative volcanism does not seem to be of primary importance (2). In this simple case, a sandstone type deposit can be considered as deriving from a trace rich basement (geochemical inheritance) which has been weathered during a long period. The erosion affecting this rich weathered cover yields selectively enriched products to the sedimentary basin in the border of which heavy metals such as Pb, Zn and Cu would be concentrated.

Our purpose is not to develop here this general idea, but to try to establish the role of the weathering processes in the geochemistry and the genesis of stratabound deposits.

2. Heavy Metals and Terrigenous Clastic Phases

Among heavy metal deposits in detritic continental formations, the group of placer deposits is well characterized by the origin, the manner of concentration and the nature of minerals: when there is no deep diagenesis or metamorphism, they are always chemically and physically resistant in the region and epoch conditions prevailing during weathering and transportation.

In constrast to this group, numerous economic concentrations of lead, zinc, copper, vanadium, uranium are found in detritic formations of continental (or subcontinental) facies: they are known as "sandstone type" or sometimes as "red bed type" (s. 1). This group is characterized by the association of a terrigenous residual phase (siliceous rudites or arenites, often rich in feldspars) cemented by a chemical phase, unstable in surface conditions, which contains heavy metals, as sulfides or oxides.

In this case, according to H. ERHART (1966), heavy metals would be leached out from a partly weathered basement just when the terrigenous phase is eroded from the same basement. Nevertheless, this model does not explain one of the main characteristics of the sandstone type deposits: whereas basements contain many elements in the same amount, only one or two heavy metals are concentrated in an economic deposit, even if paleogeographic conditions required for their sedimentary concentration are similar, as for Pb, Zn and Cu (SAMAMA, 1969). This characteristic can only be explained by continental supplies selectively enriched in the concerned elements, in contradiction with ERHART's theory (3). In fact, in the same spirit, it is possible to build detailed geochemical models of selective concentrations. For instance, in the Lower Triassic border of the French Massif Central (Largentière district) and its Pb-Zn-Ag associated occurrences, a succession of moderate weathering processes acting upon the granitic and metamorphic Hercynian basement has been established (SAMAMA, 1969). In a first period, plagioclases and ferromagnesian minerals are destroyed and Cu, U and partly Zn are leached out, whereas Pb and Ba are concentrated in resistates with K-feldspars. After erosion of the weathered cover and its pediment-like sedimentation along the border of the Hercynian cratonic area, the weathering goes on and the leaching of K-feldspars by percolating waters yields important amounts of Si, Ba and Pb. This selectively enriched water carries down these elements to the border of the sulphatic basin where internal chemical sedimentation processes concentrate them as the cement of the detritic phase.

---

(3) According to H. ERHART, heavy metals do migrate with soluble elements, so that a clear separation between the different metal traces is highly improbable.

So, in this model, weathering processes are considered as an important factor of enrichment and as the main geochemical agent of heavy metals separation. However first of all, in order to establish and to generalize this model, it is necessary to improve and to throw more light on the behaviour of trace elements during the principal weathering processes.

3. Behaviour of Heavy Metals during Weathering

Behaviour of heavy trace metals during weathering may be either deduced from theoretical considerations or established from experimental works and from recent or old profiles. But, in any case, it is necessary to classify the main types of weathering processes.

3.1. The main weathering processes

According to G. PEDRO (1966, 1968), the definition of the four main weathering processes acting on silico-aluminous rocks would rely on the relative speed of elimination of rock components from the weathered mantle; these main processes are podzolization, bisiallitization, monosiallitization and allitization (table 1). Except in mountainous regions, the areas of potential weathering are established according to the climatic conditions from cold regions (podzolization) to equatorial ones (allitization). The behaviour of iron could be introduced in the same sketch, being or not individualized as oxides or hydroxides (ferruginization - rubefaction). Of course, mineralogical neoformations in the weathered cover can characterize these processes.

From this general sketch of weathering processes, it is now possible to examine the data about the behaviour of trace elements during weathering (4).

3.2 The behaviour of Pb, Zn, Cu and U traces in different weathering processes

Behaviour of trace elements during weathering depends clearly upon partly empiric considerations about weathering of ore deposits and geochemical prospecting experiences: generally speaking, one thinks in terms of "differential mobility" (for instance, zinc is more "mobile" than lead). These considerations are more or less wrong, being established on very special cases (ore deposits) in which heavy metals generally occur in special minerals and not in the lattice or along microfissures of minerals. For instance, tin, as cassiterite, may be concentrated in placer deposits which involve a great stability and a very low geochemical mobility; but as a matter of fact, traces of tin, which are common in biotite, have an intermediate mobility (TARDY, 1969). Independantly of these considerations, V.M. GOLDSCHMIDT (1934) has proposed a general classification of metals behaviour in conditions prevailing near the surface: the higher the ionic potential of a heavy metal, the lower its mobility.

A very good application of this theoretical consideration has been proposed by M. GORDON and K.J. MURATA (1952) about trace amounts of heavy metals occurring in Arkansas bauxites. All trace contents are in good agreement with Goldschmidt's rule except Pb and mainly Cu which is concentrated eight times instead of being depleted. This example sets a limit to Goldschmidt's rule: when a weathering profile

---

(4) As the climatic environment of sandstone type deposits is very far from podzolization conditions (STRAKHOV, 1962) it is possible not to take into account this process in the following considerations.

Table 1 – Geochemical and crystallochemical characteristics of the main weathering processes (according to G. PEDRO, 1968).
q(SiO$_2$) and similar formulas indicate the speed of leaching of silica and other elements.

Weathering processes	Allitization	Monosiallitization	Bisiallitization	Podzolization
Geochemical characteristics of the evolution	$q(Al_2O_3)$ < $q(SiO_2)$			$q(Al_2O_3) > q(SiO_2)$
	$q(SiO_2) \geqslant q(bases)$	$0,64\ q(bases) \leqslant q(SiO_2) < q(bases)$	$q(SiO_2) < 0,64\ q(bases)$	
Mineralogical characteristics of the eluvial horizon	Al hydroxides	Phyllosilicates 1/1	Phyllosilicates 2/1	Free silica

is rich in special compounds or minerals (for instance Al and Fe hydroxides considered as "typomorphic" by A. I. PEREL'MAN, 1967) the behaviour of trace elements is largely controlled by these minerals or compounds.

So, the concept of mobility of an element, when established from mineralogical or theoretical considerations does not permit a good judgement of the element behaviour at the scale of a profile, and of course, at that of a continent: a "mobile" element, liberated from a crystal can be trapped and fixed in slightly different conditions (another horizon or another station). For these reasons, the concept of mobile (or immobile) element, which is characteristic at the crystal level, must be distinguished from the concept of migration (or residual) element which characterizes the regional or continental level.

In this paper it is considered that

- elimination of an element (migrating element) corresponds to the leaching out of the region (or the continent) towards a sedimentary basin.

- concentration of an element (residual element) corresponds to its fixation and concentration within the weathered cover at the scale of the region (or the continent) as long as the weathering profile would not be modified or eroded.

- indifference of an element corresponds to its partial leaching so that it is neither concentrated nor depleted with respect to its value in the fresh parent rock.

The most suitable way to study the tendency of trace element behaviour at a regional scale consists of an examination of weathering profiles, either through analysis of trace elements within the profiles, or through analysis of percolating waters (TARDY, 1969). Of course, in the first case, recent or old (so-called regoliths) weathering profiles might be taken under consideration.

From published data and unpublished results, it is possible to analyse the tendency of Pb, Zn, Cu, U behaviour in the different weathering processes (fig. 1).

Under allitization processes, uranium is concentrated in recent profiles (ADAMS and RICHARDSON, 1960; PATTERSON, 1967; de WEISSE, 1970) as well as in older ones (FUCHS and PINAUD, 1969; HECHT and TOMIC, 1957; PLIER and ADAMS, 1962). Copper behaviour is quite similar, being concentrated in recent profiles (PINTA and OLLAT, 1961; RAMBAUD and PINTA, 1970; ZEISSINK, 1969) as well as in older ones (COMBES, 1969; FUCHS and PINAUD, 1969; GORDON and MURATA, 1952; TISSOT, 1956). Under the same weathering conditions, lead is either indifferent or partly eliminated (COMBES, 1969; GORDON and MURATA, 1952; PINTA and OLLAT, 1961; ZEISSINK, 1969). Zinc behaviour is more difficult to establish, being either clearly eliminated or indifferent or even concentrated (ZEISSINK, 1969) (5) during allitization processes.

Under monosiallitization (kaolinization) processes, uranium behaviour is not so well known; however, according to Y. FUCHS (1969a), to Y. FUCHS and C. PINAUD (196 and to personal results, this element appears to be concentrated, but less than in th

---

(5) In this peculiar example, the high zinc concentration (up to ten times) may be related to zinc bearing resistant minerals.

Fig. 1. The weathering processes and the corresponding behaviour of heavy metal traces.
The weathering zones, being characterized by their major chemical and mineralogical processes are presented as a half theoretical succession according to the latitude in the northern hemisphere. The behaviour of trace elements is represented by (+) or (-) symbols according to their residual (+) or migrating (-) characters during weathering.
Numbers 1 to 8 are references to the occurrences of the second example (fig. 2) and letters A and B to those of the third one (fig. 3)

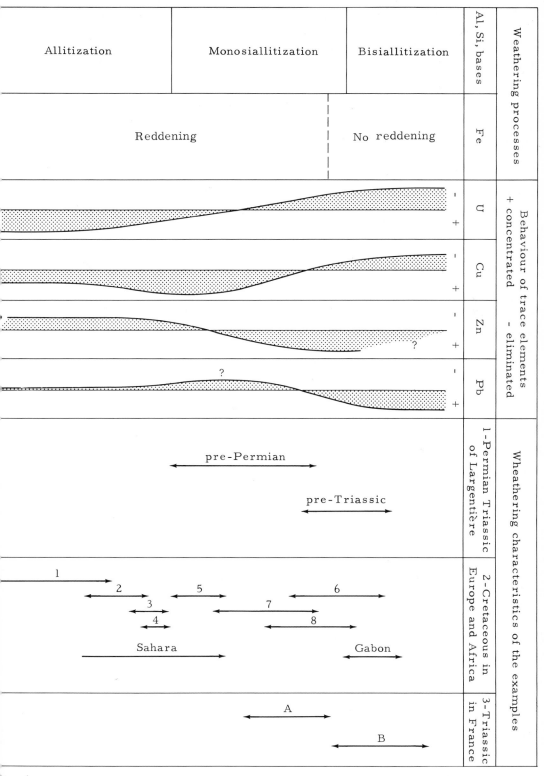

Fig. 1

previous case. Under the same weathering conditions, copper is clearly concentrated even in greater proportion (6) than previously, whereas zinc is often concentrated (COLLINS and LOUREIRO, 1971; FUCHS, 1969a; GUILLOU, oral com.; NEUZIL and KUZWART, 1964) and lead is slightly eliminated or indifferent (NEUZIL and KUZWART, 1964).

Under bisiallitization processes, uranium is nearly always eliminated like copper, but to a lesser extent (SAMAMA, 1969), whereas lead and zinc are more or less concentrated (FUCHS, 1969a; SAMAMA, 1969).

Of course, some examples appear to be in contradiction with these statistical results which are, however, considered to represent the global trend of trace element behaviour. But in spite of the significance of these results, it is not yet proved that such a geochemical phenomenon has any genetic influence on economic sandstone type deposits.

On the basis of these results, it is considered that geochemical continental concentrations of an element in a specific weathering zone (i. e. a specific climatic zone) as well as major residual mineral associations would be destroyed by erosion. Of course, it is now necessary to look for some possible relations between well characterized relicts of a weathered basement and geochemistry of the concentrations occurring within the terrigenous formations deriving from erosion, transport and sedimentation of the same weathered cover.

4. Regional Examples of Relations between Weathering of a Basement and Geochemistry of a Sedimentary Basin

Trace element abundance in basements as well as in sedimentary basins is not well known, except in a few cases, so that any attempt to draw up an adequate geochemical balance sheet becomes a delicate problem. However, it would not be impossible to make progress with incomplete or imprecise data, either with differential geochemical balance (to say with the comparison of relative abundance of two or three trace elements) or with suitable hypotheses on heavy metal repartition in a broad region. Accordingly, three regional examples will be presented here.

4.1. Differential balance sheet of elements with a constant basement: Permian and Triassic formations of the Massif Central border (France)

The sedimentological work carried out on the Permian and Triassic formations in the south-eastern border of the French Massif Central (Largentière area) lead to the conclusion that terrigenous sediments accumulated in these two successive basins derive from erosion of the same geological units of the Hercynian basement (SAMAMA, 1969). In fact, Permian sediments seem to derive from a slightly more restricted area and, of course, from a higher level of the basement. However, geochemical mapping of the Hercynian basement (BERNARD and SAMAMA, 1968; 1970) leads to the conclusion that the fresh Hercynian rocks from which Permian and Triassic sediments derive, have a very similar trace element content; in first approximation, trace contents of exposed fresh rocks of the basement have been constant during Permian and Triassic.

---

(6) According to Y. FUCHS (1969a), in Pre-Liassic profiles of the French Massif Central, copper should be 5 to 35 times higher than in fresh rocks.

However, Permian and Triassic terrigenous formations are drastically different from a geochemical point of view:

- In Permian beds, occurrences of Cu and U are found, and the zinc background is higher than that of lead (Zn/Pb = 1,5 to 2).
- In Lower Triassic beds, Pb is largely dominant over Zn and still more over Cu (Pb/Zn = 5. 5; Pb/Cu = 2000), uranium occurrences not being known.

As the primary trace element content in Hercynian rocks is the same, as the conditions of transport and sedimentation are very similar, this drastic geochemical change must be interpreted in terms of drastic change of the "geochemical landscape" (PEREL'MAN, 1967) built on the Hercynian basement.

If we compare now the two corresponding weathering profiles and if we study them in the light of the general sketch of the weathering processes, it appears that pre-Permian weathering profiles belong to the monosiallitization type and are consequently favourable to Cu and partially to U accumulations, whereas the pre-Triassic ones belong to the bisiallitization-monosiallitization type and are as such favourable to Pb and partially to Zn accumulations (fig. 1). This observed parallelism between geochemical landscapes (weathering) and geochemistry (ore deposits) of the following continental formations can be interpreted as follows: the weathering processes have controlled geochemistry of the terrigenous formations, changing with time in the same place.

In conclusion, it seems that the trace elements of the basement have become separated during weathering. Migrating elements (Pb and partly U for the pre-Permian weathering, U and Cu for the pre-Triassic one) have been largely eliminated and are no more found in the corresponding terrigenous sediments. Residual elements (Cu and partly U and Zn for the pre-Permian weathering, Pb and partly Zn for the pre-Triassic one) have been fixed and then redistributed and concentrated during erosion - transport - sedimentation - diagenesis processes. This first example strongly suggests the geochemistry of the cover to be under the dependance of continental weathering processes, the influence of which is stronger than that of trace metal contents of the fresh parent rocks.

4.2. Geochemical zonation related to zonal weathering processes. First example: U, Cu, Pb in European Cretaceous formations.

For a better view of the "geochemical landscape" influence, let us choose a well known example of zonation of weathering processes related to a zonal climate.

In this connection, the works of M. KUZWART and J. KONTA (1968) about Cretaceous weathering crusts in Europe are found to be suitable. According to these authors, the distribution of kaolin, laterite and bauxite deposits of Lower and Upper Cretaceous shows a zonal distribution of weathering processes (fig. 2) The development of bauxite-kaolin deposits is characteristic of allitization and partly monosiallitization processes in equatorial areas located north of the limits drawn by M. KUZWART and J. KONTA. South of these limits, the weathering conditions are different, less intense, and characterize the monosiallitization and the limit of mono-bisiallitization processes.

On the same map of zonal weathering processes, heavy metals concentrations occurring in the terrigenous formations of the top of Lower Cretaceous or of the basis of Upper Cretaceous are also shown (fig. 2). In the northern area (allitization - monosiallitization area), the main heavy metal occurrences are the uraniferous ones in the laguno-continental Cenomanian formations (Bohemian deposits). Uranium is also known in allochthonous bauxites of Austrian Upper Cretaceous (HECHT and TOMIC, 1957). Copper occurrences are known either in the same type of bauxites in the Alps (TISSOT, 1956) or in southern France (Languedoc) (COMBES, 1969) or in typical red bed deposits in Spanish Wealdian (Horiguela, Province of Burgos). In the southern area (transition between monosiallitization and bisiallitization), the detritic formation following the weathering crusts and deriving from them are characterized by Pb, Z and Cu deposits. The main deposits occur either in Infracenomanian of Morocco or i Aptian-Albanian of Algeria and Tunisia. In the first case, according to J. CAIA (196 a large group of Pb, Zn, Cu occurrences appears in the first detritic beds following the general regression of the Upper Jurassic. The relative contents of Pb, Zn and C change from one point to another, but Pb seems to be the major element. In the sec-

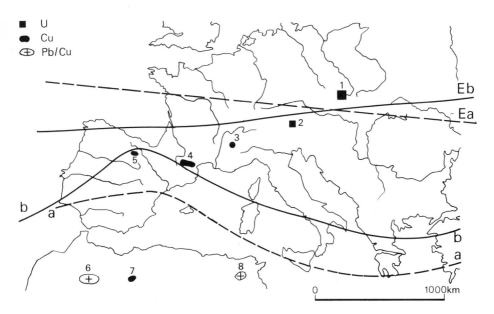

Fig. 2. Distribution of the Lower and Upper Cretaceous weathering areas and of the following ore deposits in Europe and North Africa.

a) southern limit of Lower Cretaceous occurrences of bauxite, laterite and kaolin;
b) southern limit of Upper Cretaceous occurrences. Ea and Eb, corresponding positions of Equator (according to M. KUZWART and J. KONTA, 1968).

1. Uranium deposits in Cenomanian of Bohemia; 2. Austrian uranium occurrence; 3. and 4. Copper occurrences from the Alps and Languedoc; 5. Copper occurrences of Hortiguela; 6. Lead, zinc, copper deposits from Northern Atlas (Morocco); 7. Copper deposit of Ain Sefra (Algeria); 8. Lead-zinc-copper deposit of Koudiat Safra (Tunisia).

The geochemical trend along a northeast-southwest direction (U to Pb/ Cu) can be explained by a weathering zonation, in good agreement with the geochemical model of trace elements behaviour during weathering (fig. 1).

At this continental level, the great superiority of the weathering profiles over common paleoclimatological data is due to the better knowledge of the former and to the fact that the fossil weathering profile moved with the continental blocks. For instance the rotation of the southern limits of bauxite kaolin deposits in Spain is perhaps an effect of post-Cretaceous rotation of the Iberic block

ond case, according to A. CAILLEUX and J.Y. THIEBAULT (1960-1962), Cu is more abundant than Zn, Pb or U occurrences being not known. In the third case, Pb is more abundant than Zn which is itself more abundant than Cu (7).

It would be possible to follow the weathering zonation and the corresponding mineralizations in Africa by considering, for instance, the very deep weathering crust of Central Sahara (CONRAD and CONRAD, 1967) and the U-Cu deposits of Niger (IMREH and NICOLINI, 1962; BIGOTTE and OBELLIANNE, 1968) as well as the weathering profiles without rubefaction of Gabon and the Pb-Zn deposit of Kroussou (CARRIE et al. , 1966).

So, on one hand, the observed geochemical zonation in the Cretaceous detritic beds is nearly impossible to explain by geochemical characteristics of basements (in that case, pre-Cretaceous formations); on the other hand, the geochemical zonation can be interpreted as resulting from the behaviour of trace elements during the corresponding zonal weathering (fig. 1); but heavy metals accumulated within local weathering profiles occur in the following detritic beds as geochemical or economic concentrations.

Table 2 - Pre-Triassic weathering types and geochemistry of heavy metal occurrences in the Lower Triassic
Numbers 1 to 12 refer to fig. 3.

Area	Weathering type	Name	Geochemical characteristics
A	Deep reddening and kaolinization (monosiallitization process)	1 Maubach	$Zn \geqslant Pb \gg Cu$
		2 Saint Avold	$Pb \geqslant Cu$
		3 Le Cerisier	$Cu > Pb \gg Zn$
		4 Le Luc	Cu
		5 Cap Garonne	$Cu \gg Pb - Zn$
		6 Lodève	Cu
		7 Massif de l'Arize	Cu
		8 Plana de Monros	Cu - U
B	Light reddening and little kaolinization (bisiallitization process)	9 Largentière	$Pb > Zn$
		10 Mas de l'Air	$Pb \gg Zn$
		11 Saint Sébastien	$Pb \gg Zn$
		12 La Plagne	Pb

According to ARRIBAS (1966) for 8; BERNARD (1958) for 10; BERNARD and FOGLIERINI (1963) for 7; BERNARD and SAMAMA (1970) for 9; BERTRANEU and DESTOMBES (1963) for 7; COURNUT (1966) for 4; GUILLEMIN (1952) for 5; CARALP (1908) for 8; GRANDCLAUDE (1965) for 1; de LAUNAY (1913) for 1, 2, 3 and 10; LAUNEY (1966) for 2; NICOLINI (1970) for 3 and 6; ROGEL (1963) for 12.

---

(7) The mineralized rocks of this deposit (Koudiat Safra) belong to detritic beds which seem to be Aptian-Albian (SAHLI, 1971, oral com.), even if they were previously considered as Tertiary (SAINFELD, 1952; ROUVIER, 1967) so that the connexion of this deposit with the general zonation is uncertain.

4.3. Geochemical zonation related to zonal weathering processes. Second example:
Lower Triassic deposits in France

At the continental scale of the previous example, the weathering zonation was very
important and evident. But, if a clear physiographic limit (southern limit of bauxite -
kaoline - laterite occurrences) was drawn in figure 2, it is obvious that, in fact,
weathering zones present some transitional types so that it may be interesting to investigate the same kind of geochemical processes on a more detailed scale; Pb-Zn-
Cu occurrences within the detritic beds of the Lower Triassic were found to be suitable for this purpose.

In this third example, two well marked areas could be distinguished by their weathering profiles. In the first one (A on fig. 3 and table 2), reddening is well marked
and kaolinite abundant (monosiallitization process); in the second one (B on fig. 3 and
table 2), reddening is nearly unknown or weakly perceptible, and kaolinite is generally not the main weathering product (bisiallitization process).

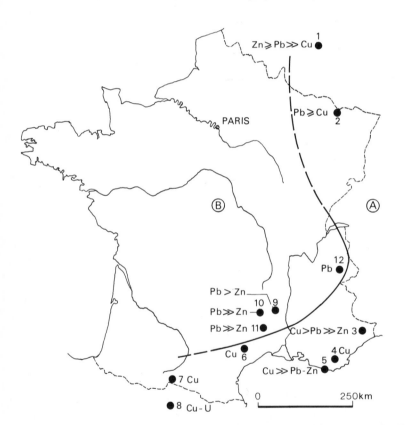

Fig. 3. Map of pre-Triassic weathering facies and of sandstone type deposits of
the Lower Triassic.
The weathering areas A and B have been established according to various descriptions (BERNARD, 1958; CHEVALIER et al. , 1970; EVEN and SAMAMA, 1969;
FUCHS, 1969a and references of table 2 and unpublished personal works).

A - Area of pre-Triassic monosiallitiziation process and of sandstone type deposits
with at least small amounts of copper; B - Area of bisiallitization process and of
sandstone type deposits without any copper

The heavy metal occurrences found in the Lower Triassic arkoses or conglomerates follow the same antagonistic trend. In the first area (A), Cu is always present among heavy metal occurrences, at least as by-product, sometimes as major element; the second one (B) is characterized by Pb-Zn contents with very low traces of Cu (Cu/Pb less than 10^{-2}). This geochemical zonation can be completed by small occurrences of U associated with Cu (8) (fig. 3).

In this third example, once more, a strict relationship between weathering zonation and geochemical characteristics of the following detritic beds is obvious; again, the basement peculiarities would not permit an explanation of the detritic cover characteristics, whereas the weathering processes effect does.

4.4 Conclusion about balance sheets and geochemical zonations

From the three examples briefly presented in this note, it appears that the geochemistry of sandstone type occurrences is under the dependance of the weathering process type developed on the continental areas.

The following correspondances can be pointed out:

- deep allitization process: U - (Cu)
- transition between allitization and monosiallitization processes: Cu - (U)
- monosiallitization process: Cu - Zn - (Pb)
- transition between mono- and bisiallitization processes: Pb - Zn - (Cu)
- bisiallitization process: Pb - (Zn).

This result is not absolutely new. U-Cu sandstone type deposits do occur in typical red bed formations characterized by their Fe-hydroxides and the frequency of wood fragments which indicates an equatorial to tropical climate corresponding to allitization or monosiallitization processes. Pb-Zn sandstone type deposits do occur in arkosic formations characterized by the lack of Fe hydroxides and the scarcity of wood fragments which indicates a drier climate (arid tropical climate) corresponding to bisiallitization processes. This well known regularity has the same significance as the previous one (U-Cu in allitization and Pb-Zn in bisiallitization processes).

For the description of the three presented examples, the physiographical conditions have been considered as being stable (or at least oscillating around a mean position) during the whole weathering period, that is to say, during the few hundred thousand or the few million years required for the described phenomenon (9). It is obvious that, the more we know about age, duration and type of weathering, the better the geochemical conclusions. Of course, as H. ERHART (1956) emphasizes, the development of a weathering mantle and its erosion-sedimentation must be considered as two distinct problems. This distinction is due to the nature of phenomena and to different physiographical conditions. In case of too fast a physiographical change in respect to the whole weathering period, the geochemical and mineralogical effects are generally confused and difficult to interprete. However, the physiographical conditions of the three presented examples were stable enough and the proposed interpretation seems to be worthy of acceptance.

---

(8) These occurrences are located near "Le Cerisier" (3 on fig. 3) according to F. KERVELLA (1965) and in the Spanish Pyrenees (8 on fig. 3) according to S. CARALP (1908) and to A. ARRIBAS (1966).

(9) These durations are much longer than those required for the development of a weathering cover or of a soil (a few thousand years).

5. Remarks about the Global Continental Behaviour of Heavy Metals

According to the geochemical data on weathering profiles, the same heavy metal has been considered as residual, indifferent or migrating, in contradiction with Erhart's model in which Cu, Pb, Zn and U are migrating elements, even when associated to sandstone formations. The general sketch of trace metal behaviour built on Pedro's zonal weathering sketch (fig. 1) is in good agreement with general or detailed data and allows a coherent interpretation of the geochemistry of sandstone type deposits. For this proposed interpretation, the different examples have been plotted in fig. 1 not as a point, but as an overlapping phenomenon due to the following reasons:

- lack of precision in local definitions of the weathering profiles,
- large extent of studied areas in comparison with climatic regional variations,
- superposition of weathering conditions slightly changing with time.

From a methodological point of view, the first cause of lack of precision can be eliminated by better studies of old profiles; the second can be reduced by more detailed investigations, but the third one is not surmountable: what are the characteristics and the role of weathering profiles found on a basement weathered during 200 million year as for instance the central part of the French Massif Central? To continue with this type of investigation or to try to apply these results, two main conditions are necessary: a good record of profiles and an approximative stability of physiographical conditions during the whole weathering period.

This sketch of the geochemical behaviour of some heavy metals is surely not perfect; it is sure that new data on sandstone type deposits and also on weathering profiles would lead to some modifications; much work has still to be done in this field (10).

6. Conclusions : Pedological Differentiation and Sandstone Type Deposits

According to N. M. STRAKHOV (1962), paleoclimatic control of stratabound deposits is well established but without any details about geochemical zonations or about characteristics of the deposits themselves. This new model, built on the continental behaviour of trace elements (elimination, concentration or indifference), seems to be much more suitable for the study of geochemical regularities of sandstone type deposits. In this model, the weathering cover is acting as a very efficient "separating filter" which, according to the type of weathering process, eliminates or concentrate a trace element. This separative filter function is the best explanation of the mono- (or bi-) metallic character of stratabound deposits (for instance the contrast between the association U-Cu and Pb-Zn, see fig. 1); such a phenomenon could be called the "pedological differentiation" of the heavy metal traces.

The role of weathering agents already evoked by A. BERNARD and F. FOGLIERINI (1963) seems now of primary importance for the deposits themselves as well as for a better understanding of the differences between mineralized and barren detritic environments. These environments (or molasses) when originated at the end or immediately after an orogeny and so deriving from continental areas, morphologically you and without any deep weathering are generally barren (exept some iron deposits of carboniferous molasses). Mineralized molasses are of peculiar types; either being

---

(10) All this study has been devoted to the problems of sandstone type deposits. Similar investigations on base metal deposits in black shales or in carbonate environments would surely be possible, even if more difficult.

restricted to immediately post-orogenic formations with oxidation characters (red beds according to F. J. PETTIJOHN, 1957) which indicate a decrease in the erosion rate and a concomitant development of weathering (MILLOT, 1967) or representing cratonic covers or borders which derive from a weathered basement but without any definite relation with an orogenic cycle.

However, whatever the weathering importance is supposed to be, it would be erroneous to restrict the ore deposit genesis to a weathering action, leaving aside the two other main aspects, i. e. syn-diagenetic concentration and geochemistry of the basement.

Of course, syn-diagenetic processes occurring in very peculiar and local sedimentary environments represent the major factor without which any economic deposit cannot exist. These sites and the related physico-chemical processes have been largely investigated for 15 years; it does not seem necessary to emphasize here this fundamental aspect of stratabound deposits.

The geochemistry of the basements from which mineralized formations derive has only been investigated in a few cases during the last 5 to 10 years (LI HSI CHI et al., 1964; BAKUN et al., 1966; MALYUGA et al., 1966; NARKELYUN et al., 1968; BERNARD and SAMAMA, 1968, 1970; SAMAMA, 1969); from these studies, it appears that, the richer a basement is (either in a diffused form or may be as pre-concentration), the richer are the following detritic formations; prospecting programs according to this regularity are already started.

The relations between these three geochemical processes and their respective contributions to the genesis of any stratabound deposit must be emphasized. It is easy to distinguish the syn-diagenetic processes (concentration in situ) from the two other processes (selectively enriched concentration). But it is more difficult to separate the influence of basement geochemistry from that of weathering, i. e. the two processes which give rise to the selectively enriched supply by a superposition of their own concentrating effects. So, a copper rich basement (on diffuse or concentrated forms) the background of which is 5 times higher than the clark, would induce by simple mechanical erosion only a sedimentary supply no more than 5 times the normal supply. If the weathering processes are of bisiallitization type, a great part of the heavy metal bulk would have been leached out before erosion (see fig. 1) and detritic sedimentation: the copper supply would be nearly normal. But, if the weathering process is of monosiallitization type inducing a weathered cover 3 to 5 times richer than the fresh rock, this cover would be 15 to 20 times higher than the clark; the copper supply to the sedimentary basin will be quite rich!

This very simple example shows how the weathering processes can accentuate (or attenuate) the geochemical characteristics of a basement: but a continental area, both rich in an element and weathered under conditions concentrating the same element, can give rise to a really important metalliferous supply to the basins and, therefore, induce in appropriate sedimentary sites, deposits of high grade and important tonnage.

So, the metal content of a basement, the weathering characteristics and the paleogeographic peculiarities of detritic environments are the three main genetic aspects of the sandstone type deposits. Among these three aspects, the second one, i. e. the weathering process and its selective concentration effect seem to open the door on a very broad field of genetic investigations and also to provide a new clue for prospecting for stratabound deposits.

Bibliography

ADAMS, J. A. S. , RICHARDSON, K. A. : Thorium, uranium and zirconium concentrations in bauxite. Econ. Geol. (New Haven), 55, p. 1653-1675 (1960).

ARRIBAS, A. : Mineralogia y metalogenia de los yacimientos espanoles de uranio. Los indicios cupro-uraniferos en el Trias de los Pirineos contrales. Est. Geo logicos (Madrid), 22, p. 31-46 (1966).

BAKUN, N. N. , VOLODIN, R. N. , KRENDELEV, F. P. : Genesis of Udokansk cupriferous sandstone deposit (Chitinsk oblast). Intern. Geol. Rev. (Wash.), 8, p. 455-466 (1966).

BERNARD, A. : Contribution à l' étude de la province métallifère sous-cévenole. Sci. Terre, (Nancy), 7, 3-4, p. 125-403 (1958).

— Notions de métallogénie sédimentaire. In: Coll. Gîtes stratiformes. Notes et Mém. Serv. géol. Maroc, (Rabat), 181, p. 267-282 (1962).

— Précipitation et adsorption sédimentaires. In: Developments in Sedimentology. Elsevier (Amsterdam), 2, p. 19-27 (1964).

— FOGLIERINI, F. : Aperçu sur le Trias métallifère en France. In: Coll. Trias de la France et des régions limitrophes. Mém. B. R. G. M. (Paris), 15, p. 635 -648 (1963).

— SAMAMA, J. C. : Première contribution à l' étude sédimentologique et géochimique du Trias ardéchois. Mém. Sci. Terre (Nancy), 12, 106 p. (1968).

— SAMAMA, J. C. : A propos du gisement de Largentière (Ardèche). Essai méthodologique sur la prospection des "Red Beds" plombo-zincifères. Sci. Terre (Nancy), 15, p. 209-264 (1970).

BERTANEU, J. , DESTOMBES, J. P. : Minéralisations du Trias des Pyrénées françaises. In: Coll. Trias France et régions limitrophes. Mém. B. R. G. M. (Paris 15, p. 706-709 (1963).

BIGOTTE, G. , OBELLIANNE, J. M. : Découverte de minéralisations uranifères au Niger. Mineral. Deposita (Berl.), 3, p. 317-333 (1968).

CAIA, J. : Les minéralisations plombo-cupro-zincifères stratiformes de la région des plis marginaux du Haut-Atlas oriental : un exemple de relations entre des minéralisation et une sédimentation détritique continentale. Notes et Mém. Serv. géol. Maroc (Rabat), 29, no. 213, p. 107-120 (1969).

CAILLEUX, A. , THIEBAULT, J. Y. : Les concentrations de cuivre stratiformes de l'Atlas saharien (Algérie). In: LOMBARD et NICOLINI, Gisements stratiformes de cuivre en Afrique, Paris, p. 33-42 (1960-1962).

CARALP, S. : Note sur les grès cuprifères d'uranium et vanadium de Montanuy (Aragon). Bull. Sec. géol. Fr. (Paris), 4, VIII, p. 480-481 (1908).

CARRIE, KNUP, NICOLINI, P. : Minéralisations plombo-zincifères du Crétacé gabonais. Symp. gisements de plomb-zinc en Afrique, Ann. Mines et Géol. (Tunis), 23, p. 295-323 (1966).

CHEVALIER, Y. , DEJOU, J. , LENEUF, N. : A propos d'une paléo-altération développée sur diorite quartzique dans le massif du Tanneron (Var). C. R. somm. Soc. géol. Fr. (Paris), p. 262-263 (1970).

COLLINS, J. J. , LOUREIRO, A. R. : A metamorphosed deposit of Precambrian super gene copper. Econ. Geol. (Lanc.), 66, p. 192-199 (1971).

COMBES, P. J. : Recherches sur la genèse des bauxites dans le Nord-Est de l'Espagne, le Languedoc et l'Ariège (France). Mém. Centre Et. Rech. Geol. Hydrogeol. (Montpel.), II-IV, 335 p (1969).

CONRAD, G. , CONRAD, J. : Les altérations à la base du Continental intercalaire Crétacé inférieur du Tidikelt occidental (Sahara central). Bull. Soc. géol. Fr. (Paris), 7, IX, p. 307-311 (1967).

COURNUT, A. : Contribution à l'étude sédimentologique et métallogénique du grès bigarré de la région du Luc en Provence (Var). Thèse 3ème cycle, Nancy, 83 p. (1966).

DEVIGNE, J. P. , NICOLINI, P. : Les minéralisations plombifères stratiformes de la région de Florac-Meyrueis (Lozère). Chron. Mines et Recherche Min. (Paris), 319, p. 152-174 (1963).

ERHART, H. : "Biostasie" et "Rhexistasie". Esquisse d'une théorie sur le rôle de la pédogenèse en tant que phénomène géologique. C.R. Acad. Sci. (Paris), 241, p. 1218-1220 (1955).

— La genèse des sols en tant que phénomène géologique. Masson & Co. , Paris, 90 p. (1956).

— Sur la genèse de certains gîtes miniers sédimentaires en rapport avec le phénomène de biorhexistasie et avec les mouvements tectoniques de faible amplitude. C.R. Acad. Sci. (Paris), 252, p. 2904-2906 (1961).

— Biorhexistasie, biostasies évolutives, hétérostasie. Importance de ces notions en gîtologie minière exogène. C.R. Acad. Sci. (Paris), 263, D, p. 1048-1051 (1966).

— Intervention de la biorhexistasie dans le genèse des gîtes miniers Crétacé-Eocène. In: Biogéographie du Crétacé-Eocène de la France méridionale. Imp. Nle. , Paris, p. 67-74 (1967).

EVEN, G. , SAMAMA, J. C. : Argiles de socle et argiles de couverture. Le problème des altération et des néoformations au contact entre le Trias et le socle hercynien du Vivarais. C.R. Acad. Sci. (Paris), 268, D, p. 3005-3008 (1969).

FOGLIERINI, F. , BERNARD, A. : L'histoire géologique d'un gisement stratiforme plombo-zincifère : Les Malines (Gard, France). Econ. Geol. (New Haven), Monograph. 3, p. 294-307 (1967).

FUCHS, Y. : Contribution à l'étude géologique, géochimique et métallogénique du détroit de Rodez. Thèse, Nancy, 257 p. (1969a).

— Quelques exemples de remobilisation dans le domaine épicontinental (Sud du Massif Central). Conv. sullo rimobilizzazione dei minerari metallici e non metallici, Cagliari, p. 161-183 (1969b).

FUCHS, Y. , PINAUD, C. : Sur l'existence d'un ravinement entre Autunien et Saxonien dans le détroit de Rodez et ses conséquences sur le comportement géochimique de certains éléments en traces (Cu, U). Bull. Soc. géol. Fr. (Paris), 7, XI, p. 459-463 (1969).

GOLDSCHMIDT, V. M. : Drei Vorträge über Geochemie. Geol. Fören. Förhandl. , 56, p. 385-427 (1934).

GORDON, M. , MURATA, K. J. : Minor elements in Arkansas bauxite. Econ. Geol. (Lanc.), 47, p. 169-179 (1952).

GRANDCLAUDE, P. : Rapport de stage à la mine de Maubach (Allemagne). Rapport inédit. E. N. S. G. (Nancy) (1965).

GUILLEMIN, C.: Etude minéralogique et métallogénique du gîte plombo-cuprifère du Cap Garonne (Var). Bull. Soc. fr. Minér. (Paris), 75, p. 70-160 (1952).

HECHT, F., TOMIC, E.: Uranforschung in Österreich. Österreich. Chemiker-Zeitung, 58, 19-20, p. 221-227 (1957).

IMREH, L., NICOLINI, P.: Les minéralisations cuprifères du "continental inter-calaire" d'Agadez (République du Niger). Bull. B. R. G. M. (Paris), 3, p. 51-108 (1962).

KERVELLA, F.: Gisements et indices sédimentaires divers. In: M. ROUBAULT. Les minerais uranifères français, P. U. F., Paris, 3, 2, p. 266-303 (1965).

KUZWART, M., KONTA, J.: Kaolin and laterite weathering crusts in Europe. Acta Univers. Carolinae (Prague), Geologica 1-2, p. 1-19 (1968).

LAUNAY, L. de: Traité de métallogénie : gîtes minéraux et métallifères. Paris et Liège, 858, 801 et 934 p. (1913).

LAUNEY, P.: Les grès triasiques plombifères de Saint Avold et les grès infrahet-tangiens à baryum, plomb, cuivre et zinc de la région de Figeac. Coll. B. R. G. M. Orléans, Déplacement des fluides dans les sédiments détritiques, p. 1-6 (1966).

LI HSI CHI, P'AN K'AI-WEN, YANG CH'ENG-FANG, TS'AI CHIEN-MING: Copper-bearing sandstone (shale) deposits in Yunnan. Intern. Geol. Rev. (Wash.), 10, p. 870-882 (1964).

MALYUGA, V. I., PROKURYAKOV, M. I., SOLOLOVA, T. N.: Distribution of exo-genic copper concentrations in the Ural region. Lith. and mineral ressources, Cons. Bur. (New York), 6, p. 765-774 (1966).

MILLOT, G.: Signification des études récentes sur les roches argileuses dans l'in-terprétation des faciès sédimentaires (y compris les séries rouges). Sediment-ology (Amsterdam), 8, p. 259-280 (1967).

NARKELYUN, L. F., YURGENSON, G. A.: Copper sources in the formation of de-posits of the cupriferous sandstone type. Lith. and Mineral Ressources, Cons. Bur. (New York), 6, p. 739-747 (1968).

NEUZIL, J., KUZWART, M.: Lateritic and kaolinitic weathering in Ghana. 22ème Congrès géol. intern. (New Delhi), 14, p. 188-212 (1964).

NICOLINI, P.: Gîtologie des concentrations stratiformes. Gauthier-Villars, Paris, 792 p. (1970).

PATTERSON, S. H.: Bauxite reserves and potential aluminium ressources of the world. U. S. Geol. Surv. Bull. 1228 (Wash.), 176 p. (1967).

PEDRO, G.: Essai sur la caractérisation géochimique des différents processus zonau résultant de l'altération des roches superficielles (cycle alumino-silicique). C. R. Acad. Sci. (Paris), 262, p. 1828-1831 (1966).

— Distribution des principaux types d'altération chimique à la surface du globe. Présentation d'une esquisse géographique. Rev. Géog. phys. et Géol. dyn. (Paris), 10, p. 457-470 (1968).

PEREL'MAN, A. I.: Geochemistry of epigenesis. Plenum Press, New York, 266 p. (1967).

PETTIJOHN, F. J.: Sedimentary rocks. Harper and Brothers, New York, 718 p. (195

PINTA, M., OLLAT, C.: Recherches physicochimiques des éléments traces dans les sols tropicaux. I. Etude de quelques sols du Dahomey. Geochim. et Cosmochim Acta (Oxf.), 25, p. 14-23 (1961).

PLIER, R. , ADAMS, J. A. S. : The distribution of thorium and uranium in a Pennsylvanian weathering profile. Geochim. Coscmochim. Acta (Oxf.), 26, p. 1137 -1146 (1962).

RAMBAUD, D. , PINTA, M. : Sur la répartition et la concentration de quelques éléments à l' état de traces dans les profils de sols latéritiques : leur rapport avec la granulométrie. C. R. Acad. Sci. (Paris), 270, D, p. 2426-2429 (1970).

ROGEL, P. : Exemple de gisement de galène du Trias des Alpes. In: Coll. Trias de la France et des régions limitrophes. Mém. B. R. G. M. (Paris), 15, p. 695-704 (1963).

ROUTHIER, P. : Essai critique sur les méthodes de la géologie (de l' objet à la genèse) Masson & Co. , Paris, 204 p. (1969).

ROUVIER, H. : Sulfures et régime hydrodynamique du milieu de sédimentation. Exemple tunisien : le minerai de plomb de Koudiat Safra. Mineral. Deposita (Berl.) 2, p. 38-43 (1967).

SAINFELD, P. : Les gîtes plombo-zincifères de Tunisie. Ann. Mines et Géol. (Tunis) 9, 286 p. (1952).

SAMAMA, J. C. : Contribution à l' étude des gisements de type Red Beds. Etude et interprétation de la géochimie et de la métallogénie du plomb en milieu continental. Cas du Trias ardéchois et du gisement de Largentière. Thèse (Nancy), 450 p. (1969).

STRAKHOV, N. M. : Principles of lithogenesis. Moscow, Trad. Cons. Bur. (New York) 1967-1970. 245, 609 and 577 p. (1962).

TAMBURINI, D. , URAS, I. : Contributo alla conoscenza delle mineralizzazioni del Cambrico sardo : la zona di Monte Flacca. Symp. A. M. S. (Cagliari), p. 293 -314 (1965).

TARDY, Y. : Géochimie des altérations. Etude des arènes et des eaux de quelques massifs cristallins d' Europe et d' Afrique. Mém. Serv. Carte géol. Als. Lorr. (Strasb.), 31, 199 p. (1969).

TISSOT, B. : Etude géologique des massifs du Grand Galibier et des Cerces. Trav. Lab. Géol. Grenoble, 32, p. 111-193 (1956).

VALERA, R. : I solfuri metallici della serie sedimentaria del Mte La Nave (Varese). Symp. intern. sui Giacentimenti Minerari delle Alpi (Trento), pres. A. 10, 21 p. (1966).

VIOLO, M. : Nuove indagini sperimentali sulla deposizione della galena. Nota III Symp. intern. sui Giacentimenti minerari delle Alpi, (Trento), Prest. 1-3, 18 p. (1966).

WEISSE, G. de: Bauxite sur un atoll du Pacifique. L' île de Rennel dans l' Archipel des Salomons. Mineral. Deposita (Berl.), 5, p. 181-183 (1970).

ZEISSINK, H. E. : The mineralogy and geochemistry of a nickeliferous laterite profile (Greenvale, Queensland, Australia). Mineral. Deposita (Berl.), 4, p. 132 -152 (1969).

Address of the author:

Ecole Nationale Supérieure
de Géologie Appliquée
54 Nancy / France

On the Origin of "Kies"-Ore and Pb-Zn Deposits in Sediments

T. N. Schadlun

Abstract

A brief summary of the data obtained by numerous investigators on the sulfur isotope composition from stratiform copper-zinc and lead-zinc deposits from the USSR occurring in sedimentary and volcanogenic-sedimentary rocks is presented in this paper. The problems of the genesis and the source of metals and sulfur are still open to discussion. The comparison of the character of isotope composition variations and certain features of the ore textures and structures suggest that the origin of ores took place during two periods - sedimentary or diagenetic and metasomatic or magmatogen-hydrothermal.

- : -

The origin of copper and lead-zinc sulfide deposits occurring in sediments and not associated with magmatic activity still remains a problem; objective criteria are needed to solve it. One of these criteria may perhaps be the sulfur isotopic composition of the sulfur in the sulfides.

In the past few years many extensive sulfur isotope studies of sulfide minerals from "kies" and lead-zinc ore deposits occurring within sedimentary and volcanic rocks have been carried out in the U.S.S.R. Numerous S-isotope analyses have revealed that the pyrite in some deposits (Filistschai, Kisil-Dere) contains biogenic sulfur, whereas the pyrite in other deposits contains predominantly magmatic sulfur.

Table I shows the δS^{34} values of the dominant (ore-forming) sulfide minerals from several stratiform-type copper, copper-zinc, and lead-zinc ore deposits. All of the data were obtained from the literature (BOGDANOV and GOLUBTSCHINA, 1969 and 1971; GRINENKO et al., 1969 and 1971; CHUKHROV, 1969; ANDREEV et al., 1970). In Table I the thin lines indicate the spread in δS^{34} values for each of the dominant sulfide minerals. The short, thick portions superimposed on each thin line indicate the spread of the most frequent δS^{34} values. In all, more than 400 isotopic analyses were reported.

From the data shown in Table I, it is evident that the sulfides from the typical "kies"-ore deposits (copper and copper-zinc ores) of the Urals (Levichi, Sibai), which occur in extrusive rocks, have a narrow spread in δS^{34} values ranging from +3 to -4 ‰. The polymetallic (Cu-Pb-Zn) "kies"-ore deposits, that occur within sandstone - shale strata and contain concretionary, diagenetic pyrite (Filistschai, Kisil-Dere in the Trans-Caucasus) are characterized by the fact that the pyrite in them shows a very large spread in δS^{34} values ranging from +30 to -22 ‰. The spread in δS^{34} values for the other sulfide minerals, on the other hand, is here slightly larger than for the same sulfide minerals from the "kies"-ores of the Urals. For sphalerite, e.g., the spread ranges from +6 to -6 ‰, for chalcopyrite from +6 to 0 ‰, for galena from +4 to -2.5 ‰, and for pyrrhotite from +6 to -2 ‰.

268

Table I - The sulfur-isotope composition of sulfides from various ore deposits

I   Deposit Levichi (GRINENKO et al. , 1969)
II   Deposit Sibai (GRINENKO, 1963)
III   Deposit Tekeli (ZAIRI)[1]
IV   Deposit Filistschai (GRINENKO et al. , 1971)
V   Deposit Mirgalimsai (BOGDANOV and GOLUBTSCHINA, 1971)
VI   Deposit Udokan (BOGDANOV and GOLUBTSCHINA, 1969)
VII Deposit Dzeskasgan (CHUKHROV, 1969)

     1 - pyrite; 2 - chalcopyrite; 3 - sphalerite; 4 - galena; 5 - pyrrhotite;
     6 - bornite; 7 - chalcocite

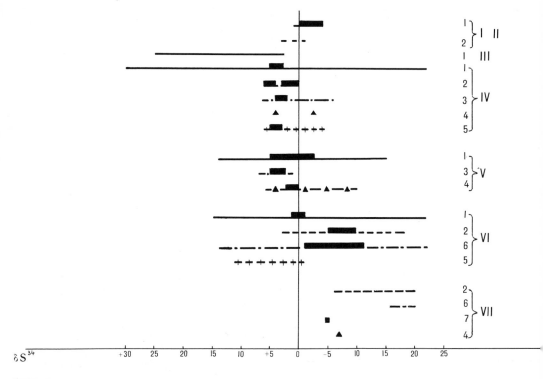

1) analyzed by ZAIRI in 1971.

On the other hand, pyrite from the Pb-Zn deposit Mirgalimsai in Kazakhstan - a Mississippi Valley-type deposit - was found to have a rather small spread in S-isotopic composition. Of interest here, however, is that the spread in the sulfur isotopic composition of pyrite with δS^{34} values ranges from +14 to -15 ‰, that of sphalerite from +7 to + 1 ‰, and that of galena from +6 to -10 ‰. Therefore, only the spread in the sulfur isotopic composition of galena is somewhat larger than for the above mentioned localities in the Caucasus.

The differences in the spread of the δS^{34} values of all the sulfides from the copper-sandstone-type deposits Dzeskasgan in Kazakhstan and Udokan in Transbaikalia are much larger. Not only pyrite, but chalcopyrite and bornite as well show a spread ranging from +15 to -22 ‰.

269

If, however, only the bulk of all samples - not the individual minerals - are considered, all deposits are typified by a smaller spread in δ S^{34} values. In this case the spread, e.g. for Filistschai and Mirgalimsai, in δ S^{34} values of the principal sulfide minerals (excepting pyrite) amounts to only 2 ‰; only bornite from the Udokan deposit shows a larger spread in δ S^{34} values, ranging to -10 ‰.

All available data reveal that the formation of pyrite in all sulfide deposits, which occur within sedimentary rocks (limestones, sandstones, and argillites), took place over a long period of time, that the process probably took on many forms and was repeated many times, further that sedimentary, diagenetic and biogenetic processes have played a significant role in its formation. As for the other sulfides, their sulfur may have been derived from biogenic pyrite, whereas the metals (Cu, Pb, Zn) probably have been largely derived from hydrothermal magmatogenic solutions and in part through extraction from the host rock. This means that the Cu, Pb, and Zn sulfides were largely formed epigenetically (against pyrite); however, they may in part be diagenetic or biogenic in origin, as for example in the ores of Mirgalimsai and Dzeskasgan.

As has been shown by many authors, e.g. relatively recently by CHUKHROV (1970) and CHUKHROV et al. (1971) and much earlier by AULT and KULP (1960), concretionary pyrite from several Paleozoic sediments shows a spread in δ S^{34} values ranging from +14 to +29 ‰, which corresponds exactly to the values obtained in bacterial sulfide reduction experiments, whereas pyrite from other localities shows a spread ranging from -27 to +44 ‰.

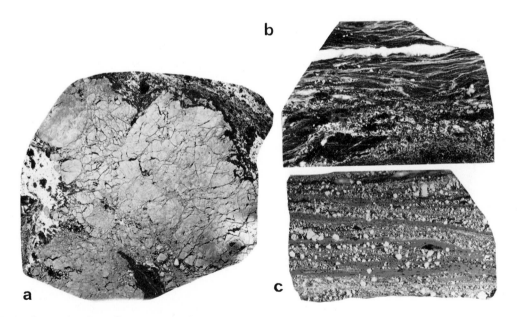

Fig. 1. Deposit Tekeli, Kazakhstan
a - Finely granular fractured pyrite aggregate (with δ S^{34} = +25 ‰), which formed during the first sedimentary cycle.
b - Pyrite-bearing banded shale. The white band is recrystallized calcite (top). Below this band some galena and sphalerite occurs.
c - Banded ore. Gray = sphalerite with some galena. White = recrystallized and newly formed, magmatogenic pyrite (with δ S^{34} = +3 ‰).
Polished sections, natural size. Reflected light

Other criteria pointing to the sedimentary origin of iron disulfides are the texture and structure (fabric) of the aggregates, for example the framboidal or globular forms, closely associated with very thinly layered or finely granular pyrite aggregates. In the very pyrite-rich ores of the Pb-Zn deposits of Tekeli in Kzakhstan, which are similar to the Rammelsberg ores, such pyrite forms independent layers or lenses within the Paleozoic shales (Fig. 1 - 4). In the galena-sphalerite layers, however, such pyrite occurs only as relicts. Several isotope analyses of this relic pyrite show a spread in δS^{34} values ranging from +3 to +25 ‰, which permits us to assume a sedimentary or diagenetic origin for this pyrite.

Fig. 2. Deposit Tekeli, Kazakhstan

a - Finely layered sedimentary pyrite ore with thin intercalations of carbonaceous-siliceous shale.

b - Banded ore. White = pyrite. Black = shale with graphitic coaly substance.

c - Banded ore. Light gray = boundinaged crushed pyrite aggregate. Dark gray = sphalerite. White = galena. Black = shale.

Polished sections, natural size. Reflected light

Similar pyrites are characteristic for many "kies"-ore deposits or Pb-Zn deposits, as for example for Filistschai, U.S.S.R., and Rammelsberg, Germany. At the same time, however, we find within the same ores granular, crystalline or colloform later generation pyrite, which is more closely associated with the other sulfides and has δS^{34} values close to that of the standard ($\pm$ 0.5 ‰ to a maximum of $\pm$4 - 8 ‰), which, if regarded for itself, would tend to point to a magmatogenic origin of the sulfur.

From the foregoing discussion it becomes apparent that the sulfur in most of the above cited deposits is biogenic or diagenetic in origin, whereas the metals may have been magmatic in origin.[1] The same conclusion was reached at an earlier date for

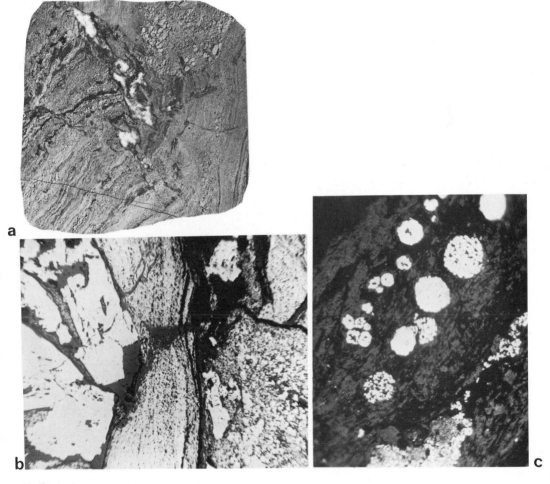

Fig. 3. Deposit Tekeli, Kazakhstan
a - Folded thinly layered pyrite ore. White = recrystallized calcite.
   Polished section, natural size. Reflected light.
b - Partially thinly layered, partially recrystallized pyrite. Magnification 40 x.
c - Pyrite balls (globules, framboids) in graphitic coaly substance. Magn. 760 x

---

1) CHUKHROV (1969) made the same assumption for the stratiform-type Cu and Pb-Zn deposits.

the Tekeli deposit (SCHADLUN, 1959), principally on the basis of a structural-tex-
tural study of the ore minerals. For the polymetallic "kies"-ore deposits located in
the southern half of the "Greater" Caucasus, a sedimentary-hydrothermal origin has
been proposed by several authors (a.o. by SMIRNOV, 1967) based on an analysis of
geological criteria. Other authors (GRINENKO et al., 1971) take the standpoint that
the ores are magmatic-hydrothermal-metasomatic in origin and that only a small
portion of the pyrite is of a sedimentary-diagenetic origin based on the preserved
structural and textural forms of the pyrite relicts which occur in disseminations and
concretions within the ore, analogous to the relicts of shale (host rock) in the ore.

It may be concluded that the majority of ore deposits discussed here were formed in
two cycles, the first one being sedimentary or diagenetic and the second one being
magmatic-hydrothermal in nature.

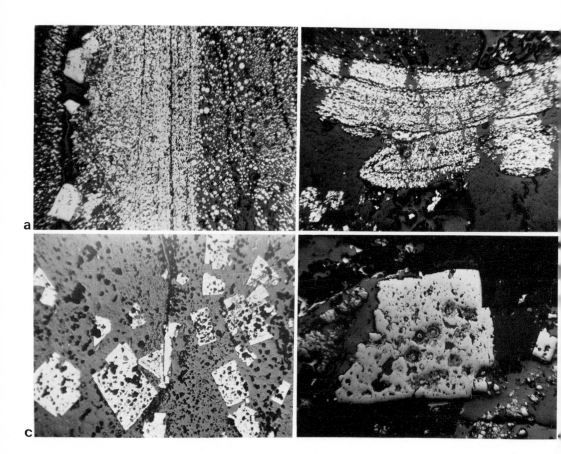

Fig. 4. Deposit Tekeli, Kazakhstan
a - Pyrite ore with globules (right), thinly layered (center) and prophyroblastic
    pyrite (left). Magnification 40 x.
b - Fragment of thinly layered pyrite in part replaced by quartz (dark gray). Magni-
    fication 40 x.
c - Sphalerite band (gray) with pyrite porphyroblasts (white). Black = pyrite globules
    etched with HCl. Magnification 85 x.
d - Partially fractures pyrite porphyroblast with enclosed pyrite globules (framboids)
    in part replaced by sphalerite. Magnification 85 x

Bibliography

ANDREEV, U. P. , BESSANOVA, I. I. , SAVIN, S. W. , TSCHERNIZIN, W. B. : Copper-
pyrrhotite deposit Kisil-Dere in Dagestan. Geol. Rudn. Mestor. N 5, p. 16-32
(1970) (russ.).

AULT, W. , KULP, J. : Sulfur isotopes and ore deposits. Econ. Geol. 55, p. 73-100
(1960).

BOGDANOV, J. W. , GOLUBTSCHINA, M. N. : Isotopic composition of sulfur of the
sulfides in the stratiform copper deposits of the Olekmo-Witim Mountains.
Geol. Rudn. Mestor. N3, p. 3-17(1969) (russ.).

BOGDANOV, J. W. , GOLUBTSCHINA, M. N. : Isotopic composition of sulfur of the
sulfides in the deposit Mirgalimsai. Geol. Rudn. Mestor. N3, p. 61-70
(1971) (russ.).

CHUKHROV, F. W. : Sulfur isotope composition and genesis problems of the ores
from Dzeskasgan and Udokan. Geol. Rudn. Mestor. N 3, p. 18-25 (1969)
(russ.).

—   On the problem of the isotope fractionation of sulfur during lithifaction. Litol.
Poles. Iskop. N 2, p. 76-88 (1970) (russ.).

CHUKHROV, F. W. , ERMILOVA, L. P. : On the isotopic composition of concretions.
Litol. Poles. Iskop. N 3, p. 36-46 (1971) (russ.).

GRINENKO, L. N. : Sulfur isotope composition of the sulfides in the deposit Sibai
(S. Urals). Geol. Rudn. Mestor. N 4, p. 86-100 (1963) (russ.).

GRINENKO, L. N. , GRINENKO, W. A. , ZAGRYAZHSKAYA, G. D. , STOLYAROV,
J. M. : Isotopic composition of sulfur from sulfides and sulfates of the kies-ore
deposit Levichi and the question of their origin. Geol. Rudn. Mestor. N 3,
p. 26-39 (1969) (russ.).

GRINENKO, L. N. , ZLOTNIK-CHOTKEWITSCH, A. G. , ZAIRI, N. M. : Sulfur isotopes of the kies-ore polymetallic deposit Filistschai in the Caucasus. Geol.
Rudn. Mestor. N 1, p. 62-76 (1971) (russ.).

SCHADLUN, T. N. : Some regularities in the metamorphism of pyrite-rich lead-zinc
ores in the deposit Tekeli. Geol. Rudn. Mestor. N 5, p. 39-56 (1959) (russ.).

SMIRNOV, W. I. : The relationship between the sedimentary and the hydrothermal
processes in the formation of the kies-ores in the Jurassic flyschoids of the
Great Caucasus. Dikladi Akad. Nauk, SSSR, 177, N 1, p. 179-181(1967) (russ.).

Address of the author:

IGEM Ac. Sc. USSR
Staromonetny per. 35
109017 Moscow J-17 / USSR

Annotation: This paper was translated from the German original by E. Schot.

Sedimentology of a Paleoplacer: The Gold-bearing Tarkwaian of Ghana

G. Sestini

Abstract

The Precambrian conglomerates of the Tarkwaian Series are alluvial fan deposits with braided stream channels. The variations of thickness, facies and crossbedding direction in the conglomerates indicate four coalescing fans with dispersal from the east and southeast. The conglomerate layers tend to decrease in number and become thinner downcurrent, and it can be shown that erosion and reworking were active in the upstream parts of the fans. The distribution of gold is clearly related to sedimentary features; its grain size distribution,for instance, is very similar to that of placer gold. In general, gold is more abundant in conglomerates that are thinner, well packed, and contain larger pebbles, and a matrix rich in hematite. In these conglomerates the quartz and hematite, and possibly also gold grains of the matrix are in approximate hydraulic equilibrium with each other, but not with the pebbles. One mechanism of gold concentration was, therefore, the trapping into open-work gravel; this gravel was formed in channels and at the margins of point-bars. The distribution of ore-shoots and borehole values in relation to dispersal shows that gold was deposited mainly in two situations: one - two main channels in each fan, and in relatively up-fan areas where reworking was active.

Introduction

Gold occurs in Precambrian conglomerates of ages ranging 1900 to 2600 million years in South Africa, Ghana, Brazil, Canada; as well as in India, Gabon and Finland (KREN-DLEV et al. , 1966), and in various parts of the Soviet Union (SMIRNOV, 1969, LEVIN et al. , 1970). The stratiform distribution of gold in conglomerates is suggestive of a syngenetic origin, and for the deposits most studied (Witwatersrand, Tarkwaian, Blind River) the weight of evidence favours a paleoplacer interpretation. In the Witwatersrand, the association of gold with sulphides, and the fact that microscopic gold occupies cavities and is vein-like and replacive, naturally suggests an epigenetic origin (DAVIDSON, 1965). It has been shown, however, that recrystallization due to metamorphism has produced 'pseudohydrothermal' (SCHIDLOWSKI, 1968) relationships between the minerals, but that the distance of movement has been negligible (LIEBEN-BERG, 1955; RAMDOHR, 1958), in no way affecting the occurrence of gold in terms of exploration and mining. On the other hand, numerous studies of the areal relationship between gold and sedimentary features (REINECKE, 1927; PIROW, 1920; SHARPE, 1949; HARGRAVES, 1962; PRETORIUS, 1966; ARMSTRONG, 1968; SIMS, 1969; MINTER, 1970), as well as studies of mineralogy and grain size (KOEN, 1961; COTZEE, 1965; VILJOEN, 1968; SCHIDLOWSKI, 1968) have clearly demonstrated that the primary genetic factors of gold accumulation are sedimentary. In this situation, a sedimentological study of the deposits becomes an essential tool of exploration.

The Banket of the Tarkwaian of Ghana presents a favourable situation for the study of relationships between mineralization and sedimentary features: Gold occurs in small discrete grains, is very low in silver as it is in known placer deposits, and is not associated with sulphides. It was recognized early that the oreshoots followed the mean direction of crossbedding, and the majoritiy opinion has favoured a sedimentary origin (JUNNER et al., 1942; BERGNE, 1944).

Essentially, the approach to the study of paleoplacers lies in the formulation of models establishing the relationships between the accumulation of heavy minerals, the dynamic conditions of transport and the direction of sediment dispersal. Apart from source factors, like rate of supply, which are often difficult to assess, the relationships are a function mainly of the environment of deposition. The position and geometry of the concentrations will vary according to whether transportation occurred in a floodplain with meandering streams, in a glacial outwash plain, on coastal beaches, or on the shallow continental shelf. Consequently, research must be directed towards identifying the environment of deposition and the direction of dispersal, and the factors controlling the areal distribution and local concentration. In an alluvial fan, for instance, the ideal places ofr concentration are channels and the median portions of the fan (PRETORIUS, 1966). The occurrence of economic heavy minerals will have to be correlated with those sedimentary features that reflect the dynamics of transportation, like grain size and sorting, packing, types of crossbedding, etc.; dispersal, such as crossbedding directions, areal variations of grain size, composition and thickness; and the environment of deposition: texture and sedimentary structures, geometry and facies changes of the formations.

The present work was carried out in the winters 1969-70 and 1970-71 mainly in the Tarkwa area, with the aim of determining which sedimentary factors control the distribution of gold in the conglomerates. The investigation has dealt with the microscopic and mesoscopic occurrence of gold, the geometry and origin of the oreshoots, the environment of deposition and its relevance to the regional distribution of gold. A study was also made of the Pleistocene gold placers of the Ofin River, in order to compare the occurrence of gold there with that in the Tarkwaian (SESTINI, 1971b).

The research was financed by the Research Institute of African Geology of the University of Leeds, under the support of Anglo-American Corporation. The author is particularly indebted to the Minister of Lands and Mineral Resources of Ghana, and the managers of the A. B. A. Mine and the Dunkwa Goldfields, for making the work possible and facilitating its progress in various ways. The field assistance of Mr. J. Kwesie is also warmly acknowledged, and Drs. P. Garrett and P. W. G. Tanner are thanked for their critical revision of the manuscript.

The Tarkwaian

The Tarkwaian rests upon and is folded with the Birrimian Series, which in Ghana has a minimum age of 1915 - 2110 million years (KOLBE, 1967; PRIEM, 1967). The Tarkwaian has not been dated yet with precision; a minimum age of 1645 million years was reported by HOLMES and CAHEN (1957). Both series display the same tectonic vergence, a decreasing intensity of folding from NW to SE and bear the same degree of metamorphism (middle greenschist to middle almandine-amphibolite facies). A marked angular unconformity has been assumed between them (JUNNER et al., 1942; SERVICE, 1943), but in most parts it is an overstep, in parts probably only an erosional disconformity, the greater apparent deformation of the Birrimian may be due to its less competent nature (pelitic flysch and volcanics facies). The Tarkwaian has in general a molassic facies, with fluviatile deposits formed in elongated internal basins, and probably represents the final depositional stage of the Eburnean geosyn-clinal cycle (older than 1850 ± 250 million years) (TAGINI, 1966, 1967).

The conglomerates and quartzites of the Tarkwaian Series form two major outcrops in Ghana. One band is in the south, extending for 250 km from Tarkwa to Konongo in a NE-SW direction, the other is a narrow 120 km strip in west-central Ghana across the Black Volta at Bui (Fig. 1). The main outcrop in the Ivory Coast is in the Bon-doukou region.

In the Tarkwa-Konongo outcrop gold is found in the conglomerates of the eastern mar-gin, in two areas: in the SW from Tarkwa to Damang (45 km), and in the NE from Ban-ka to Ntronang (20 km). In the Tarkwaian of Bui and Bondoukou gold is present, but in very minor amounts (MESCHERIAKOV and YAKCHIN, 1964; SONNENDRUCKER, 1958).

In the Tarkwa region, the Tarkwaian has a thickness of probably up to 2400 m; it has been divided into four formations (JUNNER et al. , 1942) - upwards, Kawere Conglo-merates, Banket, Tarkwa Phyllites and Huni Sandstones. These are in effect facies with notable vertical and lateral variability (VAN ES, 1964; SESTINI, 1971a).

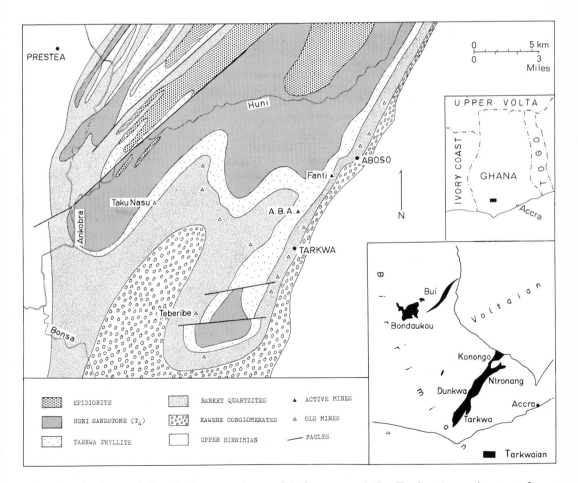

Fig. 1. Geology of the Tarkwa region and index map of the Tarkwaian outcrops of Ghana and the Ivory Coast

The Tarkwaian arenites and mudstones are low grade metamorphic rocks; the common minerals are sericite, chlorite, tourmaline, zoisite, calcite, ankerite and chlor itoid. Locally there are higher grade facies with biotite, garnet, staurolite and kyanite (VAN ES, 1964; HUNT, 1964). Psammoblastic and lepidoblastic textures are common under the microscope and the pebbles and matrix of the Banket conglomerates are occasionally strongly recrystallized, but the associated tectonism has not appreciably affected the sedimentary features of the rocks: stretched quartz pebbles are uncommon and crossbedding is clearly visible and undeformed.

The four formations are cut by sills and dykes of metamorphosed quartz porphyry and dolerites, the latter usually cutting the former. Larger sills of gabbroic rocks, now epidorites, occur in the Huni Sandstones.

The structure of the Tarkwa region is characterized by a series of synclines and anticlines with axial planes striking N 30º E and fold axes plunging to the NE. Except in the areas of the Huni Sandstone outcrop the folds are tight and asymmetric, with development of steep to overturned northwest limbs, often accompanied by thrust faulting (Fig. 1).

Geology of the Banket

a) Stratigraphy

The main feature of the Banket in the Tarkwa region is the occurrence of a conglomerate zone consisting of at least three psephitic bands in the lower part of the formation. Otherwise, the formation is composed mainly of fine to medium grained sericitic quartzites, frequently crossbedded. There are occasional lenses of small pebble conglomerates, but phyllite intercalations are rare and very thin.

The hanging-wall quartzites are light pinkish-grey, feldspathic and contain epidote, zoisite, chlorite and carbonate; the quartzites in the conglomerate zone are characterized by hematitic seams and contain sericite, tourmaline, zircon, rutile, garnet and leucoxene, but little feldspar. The footwall quartzites are dark grey, feldspathic and argillaceous and do not contain much hematite.

JUNNER et al. (1942) named four 'reefs' or conglomerate bands, downwards: the breccia reef, the middle reef, the basal reef and the sub-basal reef. This succession is typical in the western and southern parts of the area, whereas in the east and north (Tarkwa, A.B.A. and Fanti mines) the sub-basal reef is absent and the middle and basal reefs are called, respectively, west reef and main reef. HIRST (1938) and BERGNE (1944) stressed the lenticularity of the reefs and did not consider them as traceable units. The correlation of borehole logs shows, however, that at least three units are persistent, the breccia reef, middle reef and a unit of 'basal' conglomerates', - even if the boundaries of the first two are in some cases uncertain and subjective, and the subdivision of the last one is not always possible, nor perhaps significant.

The breccia reef is so named after the abundant tabular, angular pebbles of schist and phyllite, measuring 2 mm to 20 mm. These fragments have previously been considered to be derived from the Birrimian, but a number could well be metamorphosed flakes of Tarkwaian mudstone or shale. The breccia reef is generally a zone of 2 - 10 m thickness, made of breccia and gritty breccia set between quartzites and gritty quartzites, but it is up to 15 - 20 m thick in places and associated with quartzite or grit containing thin breccia bands, quartzite with scattered schist fragments or quartz pebbles, and conglomerate beds.

279

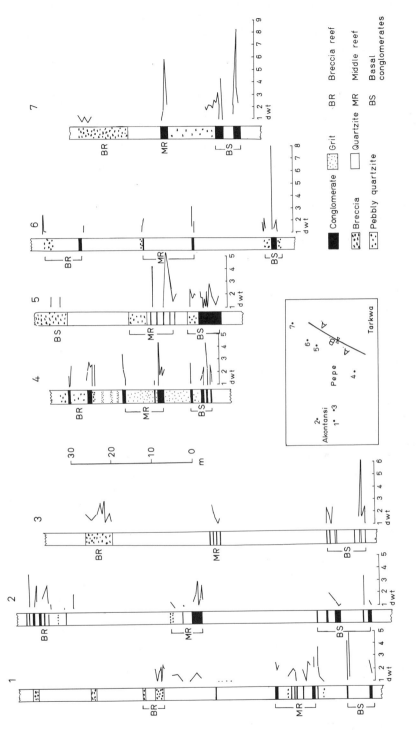

Fig. 2. Selected stratigraphic columns of the conglomerate zone showing the vertical variations of gold content. Boreholes of the Akontansi (1, 2, 3) and A.B.A. areas (4, 5, 6, 7)

The middle reef is 1.5 to 15 m thick on the west side of the Tarkwa syncline, but to the east it rapidly becomes thinner and it wedges out or merges into the basal conglomerates. Frequently, it includes two beds of conglomerate and in places up to five. Where the underlying quartzites contain pebbles or breccia beds its separation from the basal conglomerates is somewhat arbitrary, and the middle reef cannot be considered as a separate depositional unit. It is characterized by abundant hematite seams with small-scale trough crossbedding, and by quartz pebbles smaller than those in the basal conglomerates. Angular schist fragments are common, and often the conglomerate beds pass laterally into grit, breccia or pebbly quartzite.

In the basal conglomerates the number of beds varies from two to ten, being more abundant in the west than in the east. The individual conglomerate beds are discontinuous lenses which vary in number over a short distance and pass laterally into grit, breccia-grit, breccia and pebbly quartzite. The lenses are 600 - 1000 m long, in the direction of the current, and 100 - 150 m wide. The variability of the conglomerate beds probably reflects lateral and possibly also upward channel shifting, but this cannot be deduced objectively, because of the uneven quality of the borehole log descriptions.

The occurrence of gold in the conglomerates bands provides a means of correlation, in general supporting the lithological correlations. Most gold is in the basal conglomerates, where it appears abruptly and, west of the axis of the Tarkwa syncline, is concentrated mainly in the beds of the middle - upper part. It occurs again, in progressively decreasing quantities, in the middle and breccia reefs.

Considering the random sampling represented by assays on borehole samples, the correlation of vertical variations of gold values between boreholes is quite remarkable (Fig. 2). In the basal conglomerate, there are two or three peaks, and two seem to occur in the middle reef, one corresponding to conglomerate beds in the lower part the other to conglomerate, breccia, or to pebble bands.

In the breccia reef, the values are quite low and variable, but they do occur in all the boreholes assayed. Gold is present also in the quartzites between the reefs, but only a few have been sampled.

b) Thickness and geometry

The overall thickness of the Banket varies between 120 and 600 m, being greater south and west of Tarkwa. The conglomerate zone increases considerably in thickness, from 30 - 45 m to 60- 75 m, westward from the western margin of the outcrop. In the Akontansi - Kotraverchy area, the thickness increases to the north from 60 to 100 m (Fig. 3). The increases in thickness are due to the increase in the total thickness and the number of individual beds in the basal conglomerates and the middle reef. The breccia reef thickens towards the east, rather than the west.

Along the eastern margin of the outcrop, parallel with the strike, all the units of the conglomerate zone pinch and swell. Such thickening and thinning in a NE-SW direction is also noticeable in the basal conglomerates along the axis and the west limb of the Tarkwa syncline.

c) Texture and sedimentary structures

In the basal conglomerates, there are lenticular units up to several feet thick entirely made of pebbles and apparently without crossbedding, and /or crossbedded units

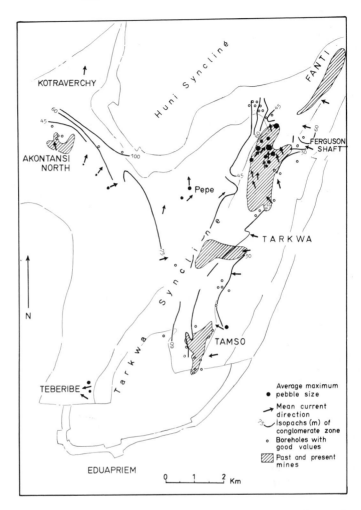

Fig. 3. Map showing mean current directions, thickness of the conglomerate zone, average maximum pebble size and the known areas of gold occurrences in the Tarkwa region (BG = Bogoso Road area)

of quartzite with pebbles at the base (Fig. 4). The crossbedded units vary in width from one meter to about 10 m; they are 25 - 60 cm thick and 10 - 15 m long. Trough crossbedding is common, but the troughs are rather shallow. In the A. B. A. Mine, most trough axes lie between N 40° E and N 70° W. Some units have trough axes directed northward, but eastward dipping foresets; they could represent stream point-bars characterized by lateral, rather than downstream accretion.

The conglomerate beds often alternate with quartzite showing small-scale trough crossbedding, and abundant hematite marking the foresets. The conglomerate is generally poorly sorted. The average size of the pebbles is 5 - 8 cm, and the largest are generally 10 - 15 cm across, with a maximum recorded length of 18 cm.

The largest pebbles tend to be well rounded, whereas those in the 35 - 60 mm range are not so rounded, corners, edges and concave surfaces being still in evidence.

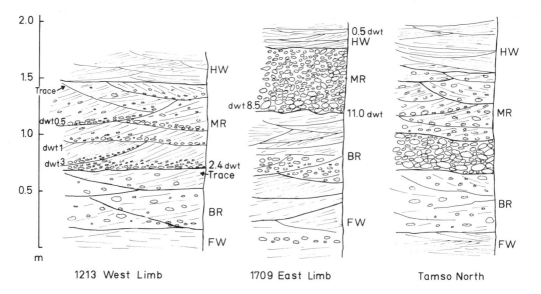

Fig. 4. Three types of occurrence of the main reef in the A.B.A. Mine. The numbers indicate gold content in pennyweight

Elongated to tabular forms are common, especially among the larger ones, but an imbrication in the plane normal to bedding is not obvious because of the high percentage of isometric pebbles.

Over 90 % of the pebbles are made of quartz. The rest are of Birrimian phyllite and schist, hornstone, chert and gondite (JUNNER et al., 1942). The matrix is mostly of quartz grains with sericite, hematite and magnetite; accessory heavy minerals are tourmaline, zircon, rutile, garnet, chloritoid, epidote, pyrite; RHOZKOV (1967) reported also some bornite and chalcopyrite.

d) Dispersal

(i) Crossbedding. The directions of crossbedding in the A.B.A. Mine are shown in Fig. 3 and 5. In the west limb of the syncline, they are generally eastward (NE, NNE) except that they turn to the NNW in the north. In the axial part of the syncline, the mean direction of crossbedding is to the north and northwest. These changes are related to two main directions of dispersal in the Tarkwa area (Fig. 3). On the eastern outcrop margin, the currents flowed to the west, with a certain radial pattern (WSW to NW) between Efuenta and Abbontiakoon; a radial pattern is possible also further south, near Tamso.

In the west (Pepe, Mantraim and Akontansi) dispersal was to the north and east-northeast, with a definite fan-like arrangement between Mantraim and Akontansi north. There is unfortunately no information for the axial part of the syncline in the south; possibly the currents flowed to the north, after a confluence of streams from the east and the southwest (but the original width of this area was probably 5 km greater).

(ii) Pebble size variation. The areal variations of pebble size in the main reef over parts of the west limb of the Tarkwa syncline are shown in Fig. 5. The isopleths indicate the average size of the largest pebbles in the conglomerate (arithmetic mean

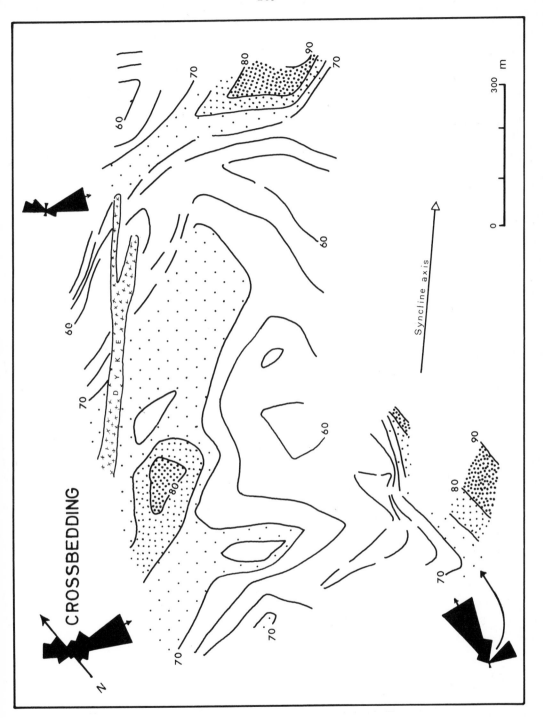

Fig. 5. Variations of the average size of the 50 largest pebbles in a part of the A.B.A. Mine. Isopleths in mm

of the longest apparent diametre of 50 pebbles). Measurements were made on a grid with about 100 m sides. The size of the largest pebbles is a parameter easy to measure and it is more indicative of the power of the transporting current than the modal size (PETTIJOHN, 1957).

The distribution of the largest pebbles in the main reef is characterized by elongated strips with large pebbles (average 75 - 85 mm), alternating with strips of smaller one (50 - 70 mm). The strips are elongated E-W and WSW-ENE, which is in agreement with the mean direction of crossbedding. There are pockets of larger and smaller pebbles (up to 90 mm, and below 50 mm). There is also a definite tendency for the size of the largest pebbles to decrease eastward, but in the axial part of the syncline, at least where it is accessible in the mine, the pebbles are again larger and the orientation is north-south.

Only a few, scattered measurements of pebble size could be made on the surface. The statement of previous authors that the pebbles are smaller near the eastern margin of outcrop and that they increase in size westward, could not be tested. In the south, however, there is in the basal conglomerates a definite decrease westward in the direction of the current, from 70 mm at Tamso to 50 mm at Teberibe, in the distance downdip of 6 km. The rate of decrease of 3mm/km being rather less than is found in the lower parts of modern alluvial fans (BLISSENBACH, 1952; BLUCK, 1964), but comparable with that found in the Witwatersrand conglomerates (STEYN, 1963; SIMS 1969).

Elsewhere, there are trends of size decrease, but they would give directions of dispersal opposite to those indicated by crossbedding (Fig. 5). It is evident that the decrease of pebble size can be suggestive of sediment dispersal only if measurements are made on a dense two-dimensional network and can be checked with other current indicators (POTTER and PETTIJOHN, 1963).

(iii) Thickness variations. Only in a few areas, the density of boreholes is sufficient to bring out variations of thickness that are elongated and may be related to dispersal On the Pepe anticline, there is a strip oriented NE-SW, i.e. in the direction of the current, in which the basal conglomerates consist of more numerous and thinner bed than either to the NW and SE; in the Bogoso road area, a similar strip is oriented north-south. These strips could represent channels in which the conglomerate has been thinned by reworking. Again in the Bogoso road area, the conglomerate zone as a whole is thicker (60 m) along a NE-SW strip (Fig. 3). There are no current measurements there, but the orientation of the strip agrees with that of crossbedding in the A. B. A. Mine, a little to the south.

e) Environment of deposition

The Banket formation represents a fluviatile series. The fluviatile character of the conglomerates and quartzites is indicated by the following features:

1) Lenticularity and great variability of the number of conglomerate beds, with transitions from conglomerate to breccia and pebbly quartzite.

2) Trough crossbedding in the conglomerates and the quartzites.

3) Upward-fining cycles, from conglomerate to fine-grained quartzite, accompanied by a decrease in the size of the crossbedded units.

4) Poor sorting and abundance of sand-supported gravel.

5) Abundance of elongated-equidimensional pebbles, many of which lie with their long axes normal to the direction of the current, as in modern stream gravel; occasionally, there is upcurrent imbrication.

6) Relatively low roundness, indicating a short distance of transport.

The next question is whether the Banket conglomerates were the floodplain deposits of a meandering stream, or of one or more braided streams. It is a problem of some importance to exploration, because the frequency and geometry of the ore-shoots would depend on the type and evolution of the stream channels.

The gold-bearing bands are continuous over a distance of at least 25 km across the direction of the current; a meandering belt could not have been so wide, with sediments that are so coarse and poorly sorted. The crossbedding does not indicate much channel variability which could be attributed to meandering; on the contrary, the channel deposits seen in the A.B.A. Mine tend to be long, straight lenses with fairly sharp margins.

There is a notable absence of very fine-grained deposits and of features suggestive of floodplain interchannel areas, such as lamination and mudcracks. The interchannel areas instead are made of relatively coarse arenites, often containing scattered pebbles, or breccia and conglomerate bands. The only possible evidence of pelites could be the schist fragments of the breccia. If some are Tarkwaian, they need not indicate older beds that were being eroded upstream, but flakes of mud reworked from deposits in abandoned channels. Silt and mud are known to be laid down in temporarily abandoned parts of braided streams (WILLIAMS and RUST, 1969). The poor sorting is indicative of rapid deposition from overloaded aggrading channels. Braided stream channels develop on surfaces with a relatively high gradient, and/or from streams that are overloaded (LEOPOLD et al., 1964).

The Tarkwa area was, therefore, not a plain with a low gradient, but a piedmont surface formed by various coalescing alluvial fans, generally sloping to the west, with a northerly slope in the southwest of the region. This hypothesis is supported also by:

1. The concentration of gold near the eastern margin of outcrop with a rapid decrease westward.
2. The rapid westward increase of the thickness of the conglomerate zone in the direction of the current, accompanied by pinching and swelling in a NE-SW direction.
3. The radiating pattern of current directions. This is very similar to that found in alluvial fans; especially those at the margin of the Great Valley of California and the Death Valley of California and Nevada (BULL, 1964; DENNY, 1965).

The suggestion of a piedmont surface of fans is not meant, however, to imply an arid climate, by the analogy with the fans of these semi-arid regions. The Banket, in fact, does not contain mud-flows, which are characteristic of such fans (ALLEN, 1965). The piedmont surface may have been more like that at the southern foot of the Alps or the Himalayas (GEDDES, 1960).

The abrupt appearance of gold-bearing conglomerates and the great increase of hematite in the Banket must reflect a general event affecting the source area of the clastics. A climatic event is possible, with changes in the rate of erosion. Changes in the source area are indicated also by the composition of the Tarkwaian arenites; the quartzites of the footwall are feldspathic and argillaceous (and the underlying Kawere arenites are often arkosic), whereas the quartzites of the conglomerate zone are quite mature.

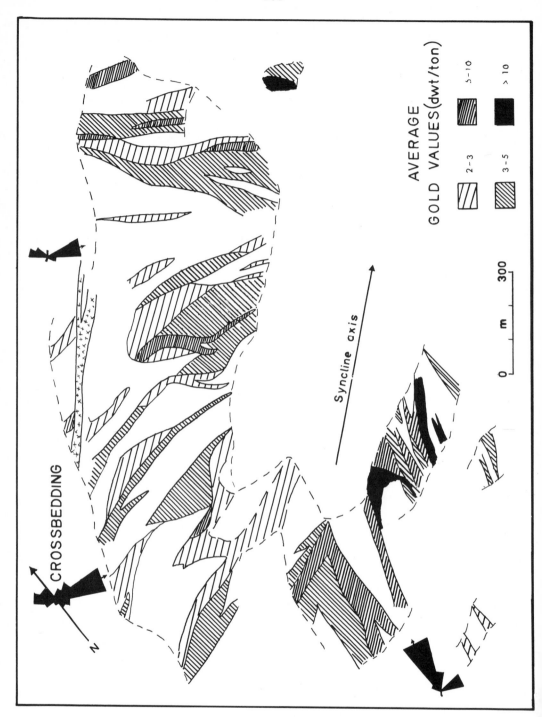

Fig. 6. Areal distribution of gold in the main reef in parts of the Tarkwa syncline

Occurrence and Distribution of Gold

Horizontal distribution (ore shoots):

In order to determine the areal distribution of gold in relation to sedimentary features, the average assay values in pennyweight per ton (dwt/ton)[1] were plotted for the entire thickness of the main reef in parts of the A. B. A. Mine. The values were obtained by averaging the gold content of channel samples situated 1. 5 m (five feet) apart, in a number of selected drives. The results are shown in Fig. 6. In the west limb of the Tarkwa syncline, the areas of equal average values are elongated narrow strips and lenses mainly with an east-west orientation. On approaching the axis of the syncline, the orientation gradually changes to northeast, and on the southeast limb of the syncline it is north-south.

The zones of values over 5 dwt/ton (most of which are part of pay zones) are very narrow, lenticular streaks occurring 200 - 500 m apart; they are 50 - 150 m wide, 200 - 500 m long. The lenses with average values of over 3 dwt/ton often border with zones of very low values, and transitions from 5 - 10 dwt/ton to 1 - 2 dwt/ton and less are generally abrupt. This situation is interesting, because it is common also in modern placer deposits (LINDGREN, 1911, p. 66; SESTINI, 1971b).

The orientation of the zones of equal gold content follows closely the average direction of crossbedding, and the pattern of the lenses is reminiscent of braided stream channels, again much like the pattern seen in modern placers (Fig. 7).

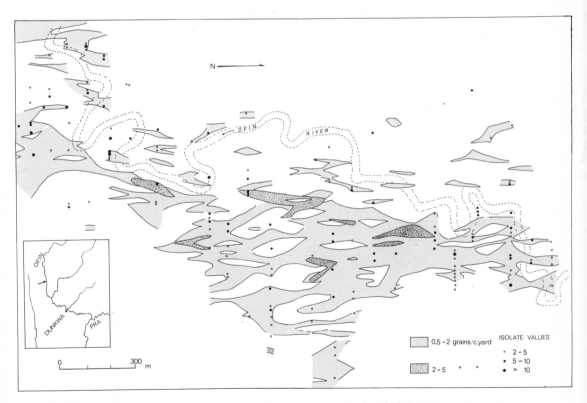

Fig. 7. Distribution of gold in the Pleistocene gravel of the Ofin River placers, 35 km northwest of Dunkwa

1) One pennyweight per ton = 1. 53 g / metric ton.

There is a rough correlation of the areas of high gold values with those containing large pebbles (Fig. 5). In both maps, the orientation of the isopleths is similar to that of the mean crossbedding, but the zones of equal gold content often cut across the pebble isopleths. This is certainly due to the much smaller number of measurement points for pebble size, as compared with the close sampling for gold assays. Nonetheless, there is a generally positive correlation between the average gold values and the size of the largest pebbles at the places where measurements were made (Fig. 8). Values over 5 dwt are not found where the average pebble size is less than 70 mm, and where the largest pebbles are below 60 mm, the average gold content of the conglomerate is usually less than 3 dwt.

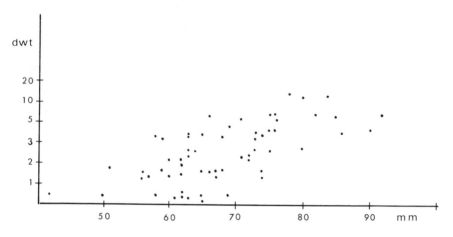

Fig. 8. Correlation between the average gold values and the average largest-pebble size in the main reef, A. B. A. Mine

The elongation of gold streaks in mines of the Tarkwa area had been noted previously. It was said to be E-W to WSW-ENE at Aboso; N-S at Mantraim (JUNNER et al., 1942). In the Fanti Mine, there are elongated lenses with a NE direction, and on the east side of the Tarkwa syncline (old part of the A. B. A. Mine), the zones over 180 inch/pennyweight form an anastomosed pattern with a predominant NW direction (JEFFERY, 196..) such a direction corresponds to that of crossbedding, which in the area indicates currents to the northwest.

Vertical distribution in the main reef:

Gold occurs mainly in the lower part of the crossbedded conglomerate units and in the lower 25 - 50 cm of the main reef. In the ore-shoots of the A. B. A. Mine, the lower 20 cm of the main reef contain two to six times more gold than the entire reef. For instance, in the main reef tract shown in Fig. 9, 79 % of the values over 10 dwt occur in the basal 40 cm of the reef; all values over 15 dwt are in the lower 20 cm of the reef. In the middle-upper parts, there are some scattered values from 3 to 6 dwt but in the vast majority values are below 2 dwt.

Generally where the values are high, the main and west reefs tend to be thinner (Fig. 10), contain larger pebbles, and the quartzites between the conglomerate layers show abundant small-scale trough crossbedding and are very hematitic. On the other hand,

289

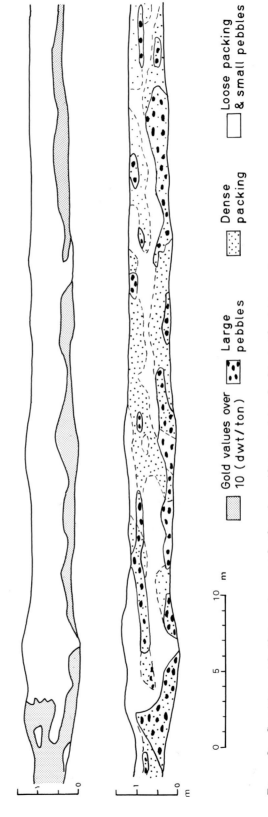

Fig. 9. Section of the main reef showing the vertical distribution of gold values (top), and of zones with large pebbles, and zones with dense packing (bottom) in the conglomerate

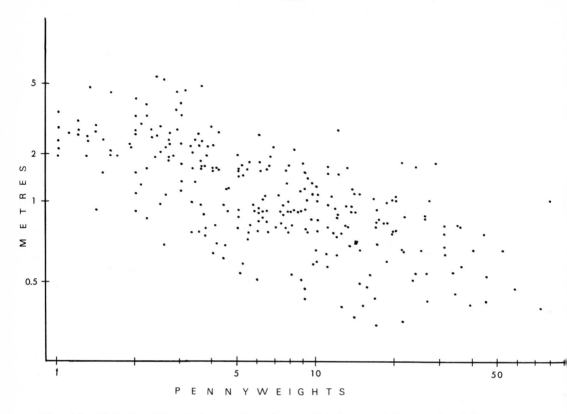

Fig. 10. Relationships between the average thickness of the reef and the average gold content in the main reef (A. B. A.) and west reef (Fanti)

where the average values are low, the highest single values are often not at the base. There is very little gold in the quartzites, and the vertical scatter of values is relate to random clusters or thin layers of pebbles.

Microscopic occurrence:

In polished section, the gold grains appear to be equidimensional and of irregular out line, with a diameter generally 10 - 15 microns; small crystals were reported by WHI TELAW (1929). The grains occur isolated or in clusters, in the vicinity of hematite or lodged within hematite areas; also in the quartz and sericite matrix and near the margins of quartz pebbles. Numerous grains of gold are seen only in the interstices of the conglomerates which contain over 20 % of hematite.

Much of the hematite is euhedral and often porphyroblastic, especially in the quartz- ites and in conglomerates with a large amount of matrix. Recrystallization of hema- tite has obviously taken place, but this has occurred in situ, because the mineral is concentrated mainly along the crosslaminations and is accompanied by water-worn grains of rutile and zircon. The same can be said of pyrite.

In the matrix of the main reef conglomerate, hematite ranges from 2 to 60 % and it occurs as isolated grains with random distribution, in patches, in 1 - 2 mm layers and thin stingers. In some of the smaller patches and in the stringers, the hematite is crystalline and has the same dimensional orientation as the metamorphic fabric. In most of the larger patches and layers, hematite occurs as distinct oval grains which have a crystalline texture. They show point contacts with detrital quartz and are not porphyroblastic.

The hematite is, therefore, clearly detrital and gold, the quantity of which is in direct proportion to that of hematite, must have been deposited with the heavy mineral concentrations of the sand.

It may be argued that the Banket gold is much finer than gold in modern placers and that the isometric grains contrast with the generally flaky shape of placer gold (RAMDOHR, 1965; LINDGREN, 1933). In regard to size, JUNNER et al. (1942) showed that the grain size distribution of Tarkwaian gold compares quite well with that of the gold panned from Ghanaian rivers, the latter containing a large proportion of very fine grains. The histograms and cumulative curves of Fig. 11, based on grain size data in JUNNER et al. (1942), indicate a very similar grain-size distribution: fairly sorted, but polymodal and coarse skewed, with two characteristic breaks at 0.12 - 0.18 mm and 0.08 - 0.06 mm. The cumulative curves are very similar to one given by BOGGS et al. (1970) for gold in the Sixes River, Oregon. The grain size distribution of alluvial gold has a higher mean: 0.1 mm in the Ghanaian rivers, up to 0.8 mm in the Sixes and its tributaries.

The finer grain size of the Tarkwaian gold can be attributed to various factors, among them: 1. finer-grained gold at the source; 2. sorting during transport (the coarser gold was deposited upstream of the present Tarkwaian margin, in rocks now eroded); 3. fragmentation during compaction and dynamic metamorphism (RHOZKOV, 1967, showed that some gold is aligned with the metamorphic fabric); 4. the methods of analysis; the latter were very systematic in the Sixes River study, whereas in the Tarkwaian they were based on odd samples of unspecified location, which disregards the fact that most of the coarsest gold is concentrated near the base of the conglomerates.

Nature and Origin of the Ore Shoots

The problem

The size of pebbles in a conglomerate is generally taken as an indication of the velocity of the current (PETTIJOHN, 1957). Since the zones of maximum pebble size and the gold streaks follow the direction of crossbedding, the zones of high gold concentration could be interpreted as stream channels in a braided river system.

Concentrations of gold could be formed in three principal ways.

1) The conglomerate could represent a channel bottom deposit (lag gravel) which was reworked by the current free of sand; later a traction load of sand with heavy minerals was transported over it and the heavier particles were trapped in the openwork gravel, gold gradually moving to the bottom. This mechanism was proposed by MINTER (1970) on the basis of the positive correlation between high gold values and high sorting, large pebble size and dense packing in the thin conglomerate of the Ventersdorp Contact Reef of the Witwatersrand; a hypothesis supported by the results of flume experiments (MINTER and TOENS, 1970). In this case, the pebbles and their matrix would represent separate depositional events; the quartz grains and the heavy mineral sand need not be in hydraulic equilibrium with the pebbles.

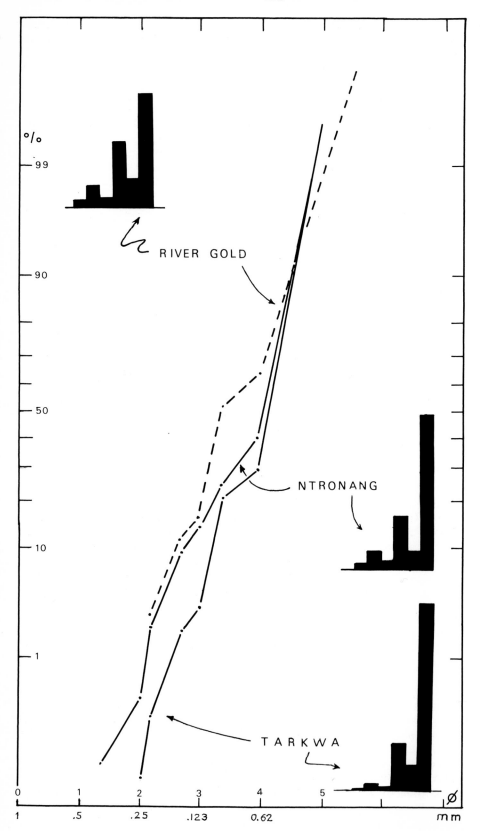

Fig. 11

2) The more auriferous conglomerate represents braided stream bars, and particularly their margins, where the coarser and heavier particles are preferentially deposited. This happens in the lateral bars of fairly straight streams, and especially in
the bars of braided channels, which tend to grow laterally and downstream (LEO-
POLD and WOLMAN, 1957; DOEGLAS, 1962). In a number of modern placers gold
occurs at the edges or in the upper parts of bars (PRINDLE, 1904; EAKIN, 1919;
HILL, 1916; SMIRNOV, 1965). In the Sixes River, BOGGS and BALDWIN (1970)
showed that the largest concentrations of gold (and the largest gold particles) are
in gravel made of large pebbles (mean diameter over 10 cm) which contains less
than 40 % sand, on the backside of low-relief bars, and at the front, stream-side
edge of high-relief bars.
In this case gold is deposited with the gravel, and its concentration is related to
channel geometry and the dynamic conditions of the different parts of the bars.
These conditions may be indicated by the thickness, distribution and packing of the
conglomerates.

3) The concentration derives from the reworking of channel or point bar deposits, and
particularly of gravelly sand containing dispersed gold. In the Ofin River (SESTINI,
1971b) the Pleistocene gravel overlies a scoured bedrock surface, and has itself an
eroded surface, with evidence of several stages of incision; the gravel is overlain
by Recent silts and clays. The considerable scatter of gold values and the absence
of consistent pay streaks is probably due to an initial deposition of gold in the gravel of the shallower (not the deeper!) scours in the bedrock, and to a later lateral
and vertical reworking with a weak concentration in the more recent scours of the
gravel surface. Reworking as an active mechanism of gold re-distribution has been
suggested for parts of the Witwatersrand conglomerates (SIMS, 1969).

In order to test the relative importance of one or the other of these modes of emplace-
ment, the distribution of gold in the ore shoots was examined in relation to reef thickness, and to pebble size, sorting and packing, and to the grain size of the matrix.

General features of an ore-shoot

A transversal crossection of an ore-shoot (ABA Mine 24 level East Limb) was examined in detail over a distance of 310 m (Fig. 12). The average gold values of 10 to
30 dwt/ton within the pay zone decrease very rapidly to less than 2 dwt at both ends
of the ore-shoot. These boundaries separate parts of the reef of different characters.
In the ore-shoot the main reef is thinner, below one meter, and it contains 2 to 4 continuous thin bands of conglomerate, made of generally well packed large pebbles (75
to 90 mm maximum average size). The reef presents very long and thin crossbedded
units and its base is channelled and crossbedding indicates a current in the direction
of the ore-shoot elongation. Outside the ore-shoot, the reef is thicker, about 70 % is
crossbedded quartzite with scattered pebbles, and/or with pebbles concentrated at the
base of the crossbedded sets. The pebbles are smaller (60 - 70 mm). In the ore-shoot,
the conglomerate is pebble-supported, outside it is predominantly sand-supported.

Reef thickness

It seems to be general in the A.B.A. and Fanti Mines that where the gold values are
higher, the reef is thinner (Fig. 10). There is, however, a great deal of variability,
and in three-dimensions the correlation is much less evident (Fig. 13). In a small

Fig. 11. Grain-size distribution of gold from the Tarkwa and Ntronang mines, and
of alluvial gold from Tarkwaian rivers (based on data in JUNNER et al., 1942, p.67)

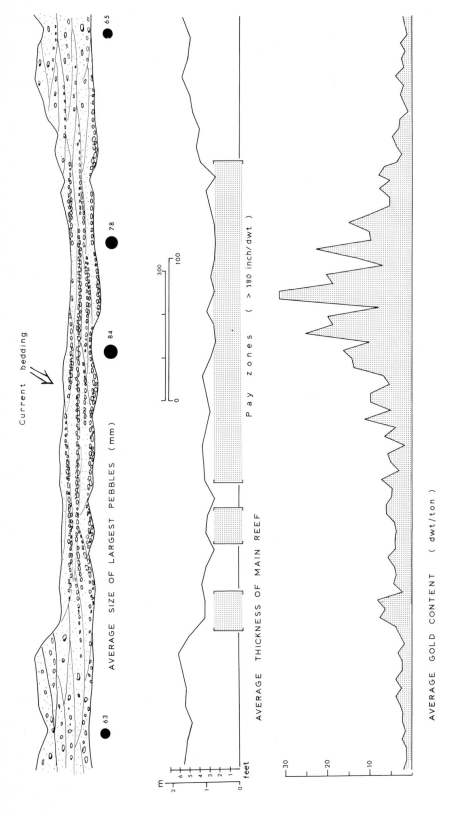

Fig. 12. Crossection of an ore-shoot in the A. B. A. Mine: variations of gold content, reef thickness and sedimentary features

295

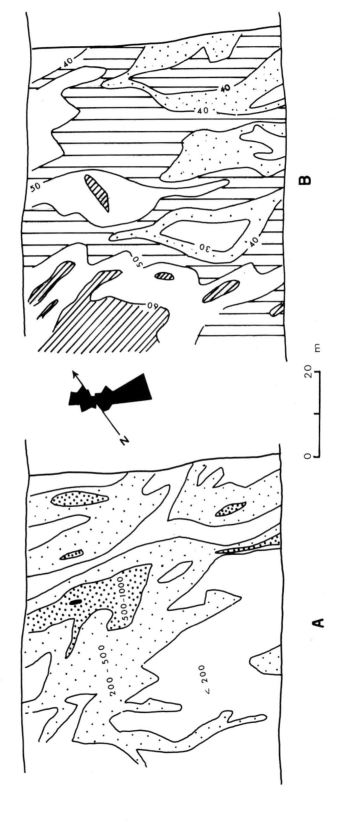

Fig. 13. Areal distribution of gold values (A) and average thickness of the reef (B - isopleths in inches) in a small section of the main reef (A.B.A.)

section of the main reef now stoped out, both the gold values and thickness isopleths are elongated in the direction of the current - the gold isopleths following the mean direction, the isopachs the modal direction of crossbedding. Most gold appears to be in reef of moderate thickness, there is little or no gold where the thickness of the reef is below 100 m or above 175 cm. This lack of correlation may not be too relevant, because reef is not equivalent of conglomerate: It can be made of one, but often two or three pebble bands (Figs. 4 and 9). It is interesting that in several parts of the area of Fig. 13 the high values tend to lie at the margins of the thicker lenses.

Pebble size, sorting and packing

The relationships between amounts of gold and these properties were investigated in two ways. First, samples were collected from areas of conglomerate 25 x 30 cm in which the number and diameter of quartz pebbles were recorded, and the amount of matrix and degree of packing were estimated by comparison with prepared charts of known pebble/matrix ratios. The samples were taken randomly from conglomerate tracts of high and low values. The results showed no correlation between gold content and the average size and sorting of the pebbles in the small area of sampling. But the

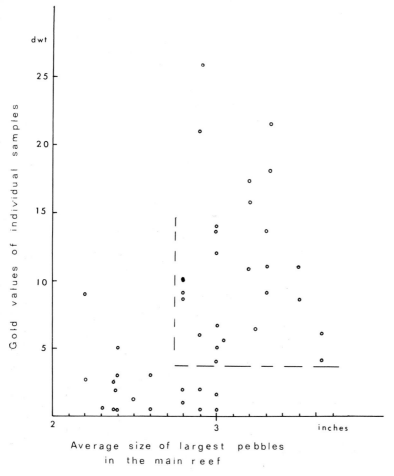

Fig. 14. Gold content of small samples (20 x 30 cm) of conglomerate in relation to the average largest-pebble size of the main reef in the vicinity (2 x 4 m areas)

samples derived from the reef tracts in which the average size of the largest pebbles was high (70 - 90 mm), contained more gold than those from areas of smaller pebbles (Fig. 14). This would indicate that the amount of gold is more directly related to the dynamics of deposition of the entire reef area (i. e. in the scale of meters), than to that of the sampled area (scale of decimeters).

As a local factor, packing seems to be mor relevant than size and sorting: in 30 samples with matrix ranging 15 - 35 % (long and point contacts between pebbles) the arithmetic mean of gold values was 7. 1 dwt; in 34 samples with matrix between 40 and 60 % (pebbles not in contact with each other), the mean value was 4. 8 dwt. This finding, itself not too conclusive statistically, has been confirmed by the second study of the distribution of gold values in relation to fabric. In the routine sampling for assays, the size of the pebbles is normally recorded in terms of 'big, medium, small', and the packing in terms of compact versus loose pebbles. These estimates are subjective, but personal testing by the author indicated that they can be used statistically, at least as distinctions between extremes: conglomerate with, or without, pebbles over 75 mm (largest average size), and with matrix definitely below or above 40 %.

In the main reef tract shown in Fig. 13, most of the conglomerate of the lower 40 cm is well packed, and somewhat less than 2/3 of it contains also large pebbles. In the packed conglomerate, 76 % of the values are over 5 dwt, whereas in the conglomerate that is poorly packed and devoid of large pebbles, the majority of values is below 5 dwt (Fig. 15). Pebble size appears to be important, but not as much as packing,

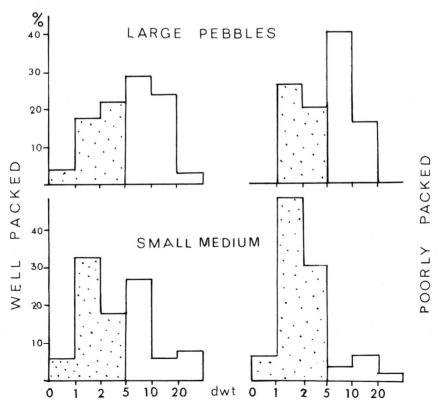

Fig. 15. Statistical distribution of gold values in relation to packing and pebble size, in the main reef section of Fig. 9

although there are rather more high values in the conglomerate with large pebbles, whether it is packed or not, than in the packed conglomerate with medium and small pebbles. In general, where there are large pebbles, the conglomerate is low in matrix.

Grain size of the conglomerate matrix

The matrix of the main reef conglomerate was examined in four samples each, from areas of known high and low gold content and respectively with larger and smaller pebbles. There is definitely more hematite in the conglomerate with large pebbles and high packing; where the pebbles are in close contact and the matrix amounts to only 20 - 30 %, over 50 % of the matrix often is hematite. The mean size of the hematite grains in the conglomerate with large pebbles (70 to 85 mm average maximum size) was 0.63 mm, that of quartz grains 1.38 mm; in the conglomerate with smaller pebbles (55 to 63 mm), the means for hematite and quartz were, respectively, 0.41 and 0.84. Thus the two minerals appear to be in approximate hydraulic equilibrium with equivalent diameters not far from the value of one phi found by RITTENHOUSE (1943).

In regard to the size of gold, there is too little significant information. For a mean diameter of 0.05 mm (Fig. 11) the equivalent size of hematite, calculated from the settling velocities of spheres (TOURTELOT, 1968), would be only 0.2 mm. But in the polished sections examined, most gold grains appeared to be larger, 0.08-0.2 mm i.e. possibly in hydraulic equilibrium.

Conclusions

The presence of conglomerate is obviously very important for the sedimentary accumulation of gold. There is no dynamic reason why gold should not have concentrated also in the quartzites, considering its grain size and that of hematite, and the fact that hematite is often abundant also in the quartzites; gold could easily have been transported with the hematitic sand.

There is a definite positive correlation between the higher gold values and the high degree of packing of the conglomerate. This and the lack of hydraulic equilibrium between the pebbles and the quartz-hematite, and possibly also gold, matrix, indeed suggest that pebbles and sand were not deposited at the same time. The random distribution of hematite in the matrix of packed conglomerate is a further evidence of a filtering of sand into an openwork gravel.

Thin openwork gravel is likely to form at the bottom of channels where the current is faster; the thin conglomerate layers with large pebbles in some ore-shoots (Fig. 12) may indeed be channel-lag gravel. On the other hand, not everywhere are high and moderately high values associated with thin sheets of gravel; coarse and well packed gravel can be irregularly distributed (Fig. 9). The occurrence of high values at the margins of relatively thicker conglomerate (Fig. 13) could indicate channel and point bars; and in general patches of openwork gravel could form by reworking, due to the lateral shifting of channels.

Gold and hematite could also be concentrated at the margins and in the upper parts of stream bars by the winnowing of sand during floods.

Areal Distribution of Gold

Knowledge of the areal distribution of gold in the Tarkwa region is very generalized, because the information available is patchy and incomplete outside the old and present mines and prospects. North of Fanti and Abosso, gold occurs only near the eastern margin of outcrop with, at intervals, areas of economic interest; to the west, nothing is known beyond 2.5 - 3 km from the margin, because the conglomerates dip under the large Huni syncline. It would appear from borehole data that the amount of gold decreases very rapidly to the west.

South of Fanti, various anticlinal structures extend the conglomerate outcrops to at least 15 km further west, and the conglomerates have been prospected in various parts (Figs. 1, 3). In these areas, high gold values and definite ore-shoots occur only in the eastern and central parts of the Tarkwa syncline; from what is known in the west, the values are moderate at Akontansi north, Kotraverchy and Teberibe; quite low at Taku Nasu, between Akontansi and Tarkwa, and in the south at Eduapriem, Detchikrom.

When the distribution of gold is considered in relation to the current directions and to the thickness variations of the conglomerates, a number of significant patterns become apparent (Figs. 3, 16).

1. The dispersal of crossbedding between Tarkwa and Ferguson Shaft agrees very well with the fan-like area of the main reef in which the values are over 180 inch/pennyweight.
2. The high values in the Tarkwa syncline seem to correspond to an area of convergence of current directions, with a flow first to the northeast, then to the north. Boreholes have indicated high values in the Bogoso road area; this lies on the continuation of the same channel system, probably with tributaries also from the Ferguson Shaft area.
3. Further south, the occurrence of gold decreases rapidly in the direction of the current, except along the line from Tamso to Teberibe. Given the large pebble size at Tamso and the persistence of a westward dispersal across the syncline, it is likely that the high values near Tamso and the promising ones to the west (borehole E 10 and Teberibe) lie on a main stream channel.
   Another stream important for gold concentration should extend from Akontansi north to Kotraverchy. The area between Tarkwa and Akontansi appears to be marginal with respect to both the Tarkwa and Akontansi fans, a possible explanation of its low gold content.
4. In relation to conglomerate thickness it is clear from Fig. 16 that the gold concentrations on the east side of the Tarkwa syncline correspond to areas in which the basal conglomerates and the middle reef are very thin (less than 5 m), whereas to the west, where there is little gold, the conglomerates are thicker and contain mor schist fragments and layers of breccia.
   In parts of the Bogoso Road area, there are high values even where the basal conglomerates are over 10 m thick; such values, however, are not scattered throughout the band, but are at the base of thin beds in the middle-upper part.

The conclusions that can be drawn from these observations are that gold is concentrated along certain main directions of stream flow, particularly where there is a confluence of various channels; and preferably it occurs in thinner conglomerate containing larger pebbles and less sand. It would be worthwhile to speculate, where in the fans these conditions are more favourably developed, and how they are related to fan geometry and facies, - along the lines of the model discussed by PRETORIUS (1966) for the conglomerate fans of the Witwatersrand. In the Tarkwa region, strati-

300

Fig. 16

graphy and dispersal are complicated by the fact that there are several interfering fans, and unfortunately the amount of information presently available is too discontinuous and not detailed enough. One feature probably related to fan geometry does appear to be important: the role of erosion and reworking in certain parts of the fan.

The rapid upcurrent thinning of the basal conglomerates and the middle reef could be an original depositional feature; but it could also arise from a later partial erosion in the upstream areas. In modern alluvial fans, the greatest thickness generally is towards the apex, but this is also the area in which erosion starts, once the fan surface there and the profile of the feeding stream behind,have reached graded conditions (BLISSENBACH, 1954). Erosion could account for the frequent absence of the middle reef near the eastern margin of outcrop, and reworking could have been the cause of the thinner and better sorted (i. e. less sand and no schist fragments) more auriferous basal conglomerates. The stratigraphic distribution of gold also suggests erosion and reworking: It occurs in the lowermost conglomerate in the east, but in the west it is generally found in the upper part of the basal conglomerates. The gold in the latter could have been derived from the reworking of the middle reef and of the upper part of the basal conglomerates in the east. The same phenomenon could also account for the apparent westward increase of gold in the middle reef. Another consequence of uplift and erosion in the upstream parts of the fans would have been the downcurrent migration of the zone of optimal conditions for gold deposition: they obtained earlier in the east than in the west, that is in a lower stratigraphic position.

Conclusions

The Tarkwaian Banket was formed on a piedmont surface of alluvial fans with dispersal by braided stream channels, mainly from the east. The absence of mudflows suggests that the fans formed under a relatively wet climate, not in semi-arid conditions; the slope of the surface probably was not as steep as that of modern semi-arid fans.

The larger quantities of gold are found where the conglomerate layers are relatively thin, contain less matrix, and a large amount of hematite. This type of conglomerate appears to represent channel-lag gravel, or reworked channel-bar and point-bar gravel. Reworking moved the sand away, produced a higher degree of packing, and concentrated the heavy minerals; or it produced an open framework gravel, into which gold was later trapped.

The position of such a conglomerate in relation to the geometry and dispersal in the fans, together with the possibility of substantial erosion in upstream areas, suggest that reworking was especially active in two zones. In an upper zone, it affected large parts of the conglomerates, whereas in the middle-upper zone of the fans it affected only some of the stream channels.

The confluence of channels in the central part of the Tarkwa syncline, with a flow to the north, makes that area and the one immediately west of Fanti one of the most

Fig. 16. Distribution of gold in relation to dispersal and the thickness of the basal conglomerates in the area between Tarkwa, Pepe and Ferguson Shaft (BG = Bogoso Road area). 1. Boreholes 2. Borehole values of 4-8 dwt 3. Borehole values of 8-16 dwt 4. Areas over 180 inch/pennyweight 5. Isopachs (meters) 6. Mean direction of crossbedding

promising for exploration. The western and southern areas are unfavourable from this point of view, possibly with the exception of the channels between Akontansi and Kotraverchy, and Tamso and Teberibe.

Bibliography

ALLEN, J. R. L.: A review of the origin and characteristics of recent alluvial sediments. Sedimentology, 5, p. 91-191 (1965).

ARMSTRONG, G. C.: Sedimentological control of gold mineralization in the Kimberley Reefs of the East Rand Goldfield. Univ. Witwatersrand Econ. Res. Unit, Inf. Circ. , 47, 24 p. (1968).

BELL, J. P.: The gold deposits of the Potaro River downstream of Tumatumari, British Guyana. Proceed. 5th inter-Guyana Conf. , Georgetown, 1961, p. 262 -272 (1959).

BERGNE, J. A. C.: West African Banket: some surveying, geological and other features. Trans. Inst. Min. Metall. , 53, p. 253-278 (1944).

BLISSENBACH, E.: Geology of alluvial fans in semi-arid regions. Bull. Geol. Soc. Am. , 65, p. 175-190 (1954).

BOGGS, S. Jr. , BALDWIN, E. M.: Distribution of placer gold in the Sixes River, Southwestern Oregon. U. S. G. S. Bull. 1312-I

BLUCK, B. J.: Sedimentation of an alluvial fan in southern Nevada. J. Sedim. Petrol. 64, p. 395-400 (1964).

BULL, W. B.: Geomorphology of segmented alluvial fans in western Fresno County, California. U. S. Geol. Surv. Profess. Papers, 437 A, 71 p. (1964).

COETZEE, F.: Distribution and grain-size of gold, uraninite, pyrite and certain other heavy minerals in gold-bearing reefs of the Witwatersrand basin. Trans. Geol. Soc. S. Africa, 68, p. 61-68 (1965).

DAVIDSON, C. F.: The mode of origin of banket ore-bodies. Trans. Inst. Min. Metall. 74, p. 319-338 (1965).

DENNY, C. S.: Alluvial fans in the Death Valley region, California, Nevada. U. S. Geol. Surv. Profess. Papers, 466, 62 p. (1965).

DOEGLAS, D. J.: The structure of sedimentary deposits of braided rivers. Sedimentology, 1, p. 167-190 (1962).

EAKIN, H. M.: The Porcupine gold placer district, Alaska. U. S. Geol. Survey, Bull. 699, 29 p. (1919).

GEDDES, A.: The alluvial morphology of the Indo-Gangetic plains. Trans. Inst. of British Geographers, 33, p. 253-276 (1960).

HARGRAVES, R. B.: Crossbedding and ripple marks in the Main-Bird quartzites in the East Rand area. Trans. Geol. Soc. S. Africa, 65, p. 263-279 (1962).

HILL, J. M.: Gold of the Snake River. U. S. Geol. Survey, Bull. 620, p. 271-294 (1916).

HIRST, T.: The geology of the Tarkwa goldfield and adjacent country. Gold Coast Geol. Survey, Bull. , 10, 24p. (1938).

HOLMES, A. , CAHEN, L. : African geochronology. Colon. geol. min. resources (Great Britain), 5, no. 3 (1955).

HUNT, B.N. F. : Sheet 20, Prestea NE. Rept. Director Geol. Survey Ghana 1961/62. p. 27 (1964).

JEFFERY, W. G. : The structural and economic geology of the area between Abbontiakoon and Fanti South mines, Tarkwa goldfield, Ghana. Ghana State Gold Mining Corp. , Unpubl. rept. (1962).

JUNNER, N. R. : Gold in the Gold Coast. Gold Coast Geol. Survey, Mem. 4 (1935).

JUNNER, N.R. , HIRST, T. , SERVICE, H. : The Tarkwa goldfield. Mem. Gold Coast Geol. Survey, 6, 75 p. (1942).

KOEN, G. M. : The genetic significance of the size distribution of uraninite in Witwatersrand Bankets. Trans. Geol. Soc. S.Africa, 64, p. 23-54 (1961).

KOLBE, P. , PINSON, W. H. , Jr. , SAUL, J. M. , MILLER, E. W. : Rb-Sr study on country rocks of the Busumtwi Crater, Ghana. Geochim. et Cosmochim.Acta, 31, p. 869-875 (1967).

KRENDELEV, F. P. , DMITRIYEV, A. N. , ZHURAVLEV, Yu. I. : Comparison of the geologic structure of Precambrian conglomerate deposits of foreign localities, using discrete mathematical formulas. Acad. Sci. U. S. S. R. , Dokl. , Earth Sci. Sect. , 173, p. 65-68 (1967).

LEOPOLD, L. B. , WOLMAN, M. G. : River channel patterns: braided, meandering and straight. U. S. Geol. Survey, Profess. Papers, 282 B, p. 39-85 (1957).

LEOPOLD, L. B. , WOLMAN, M. G. , MILLER, J. P. : Fluival processes in geomorphology. Freeman, San Francisco, 522 p. (1964).

LEVIN, V. I. , MIKHAYLOV, V. A. , NUZHOV, S. V. : Stratigraphic position and lithogenetic characteristics of gold-bearing conglomerates from the Proterozoic in the Davanya-Khugdinsk graben, Aldan Plateau. In: Sostayaniye i zadachi Sovetskoy litologii, Vses Litol. Soveshch, 8th Dokl. , 2, p. 213-219 (1970).

LIEBENBERG, W. R. : The occurrence and origin of gold and radioactive minerals in the Witwatersrand System, the Dominion Reef, the Ventersdorp Contact Reef and the Black Reef. Trans Geol. Soc. S.Africa, 58, p. 101-223 (1955).

LINDGREN, W. : The Tertiary gravels of the Sierra Nevada of California. U. S. Geol. Survey, Profess. Paper 73, 226 p. (1911).

— Mineral deposits. McGraw-Hill, New York, p. 214-251 (1933).

LUSTIG, L. K. : Clastic sedimentation in the Deep Springs Valley, California. U. S. Geol. Survey, Profess. Papers, 352 F, p. 131-192 (1965).

MESHCHERIAKOV, S. , YAKZHIN, A. : Report of work carried out in parts of the gold deposits of Dokrupe and Yoyo during 1962-63. Geol. Survey Ghana, Unpubl. Rept. , 106 p. (1964).

MINTER, W. E. L. : Gold distribution related to the sedimentology of a Precambrian Witwatersrand conglomerate, South Africa, as outlined by moving average analysis. Econ. Geol. , 65, p. 963-969 (1970).

MINTER, W. E. L. , TOENS, P. D. : Experimental simulation of gold deposition in gravel beds. Trans. Geol. Soc. S.Africa, 73, p. 89-98 (1970).

PETTIJOHN, F. J. : Sedimentary Rocks. Harpers, New York, 718 p. (1957).

PIROW, H. : Distribution of pebbles in the Rand Banket and other features of the rock Trans. Geol. Soc. S. Africa, 23, p. 64-97 (1920).

PLUMLEY, W. J. : Black Hills terrace gravels: a study in sediment transport. J. Geol. , 56, p. 526-577 (1948).

POTTER, P. E. , PETTIJOHN, J. F. : Paleocurrents and basin analysis. Springer, Berlin- Göttingen-Heidelberg, 296 p. (1963).

PRETORIUS, D. A. : Conceptual geologic models in the exploration for gold mineralization in the Witwatersrand basin. Univ. Witwatersrand Geol. Research Unit, Inf. Circular 17, 86 p. ; "Symposium on mathematical statistics and computer application to ore evaluation", Spec. Publ. , J. S. Africa Inst. Min. Metall. p. 225-266 (1966).

PRIEM, H. N. A. : Isotopic ages determinations on a biotite granodiorite and a biotite-hornblende diorite in the coastal area west of Accra, Ghana. Geol. en Mijnb. , 46, p. 206-210 (1967).

PRINDLE, L. M. : Gold placers of the Fairbanks district, Alaska. U. S. Geol. Survey Bull. 225, p. 64-73 (1904).

RAMDOHR, P. : New observations on the ores of the Witwatersrand in South Africa and their genetic significance. Trans. Geol. Soc. S. Africa, 61 (Annexure), 50 p. (1958).

— Rheingold als Seifenmineral. Jh. Geol. L. A. Baden-Württemberg, 7, p. 87-95 (1965).

REINECKE, L. : The location of payable ore bodies in the gold-bearing reefs of the Witwatersrand. Trans. Geol. Soc. S. Africa, 30, p. 89-119 (1927).

RITTENHOUSE, G. : Transportation and deposition of heavy minerals. Bull. Geol. Soc. Am. , 54, p. 1725-1780 (1943).

ROZHKOV, I. S. : The gold-bearing conglomerates of the Precambrian deposits of Tarkwa. Geol. Geofiz. Akad. Nauk SSSR, Sib. Otd. , 1, p. 60-74 (1967).

SCHIDLOWSKI, M. : The gold fraction of the Witwatersrand conglomerates from the Orange Free State Goldfield (South Africa). Mineral. Deposita, 3, p. 344-363 (1968).

SERVICE, H. : The geology of the Nsuta manganese ore deposits. Mem. Gold Coast Geol. Survey, 3, 32 p. (1943).

SESTINI, G. : Palaeocurrents and exploration for gold in the Tarkwaian of Ghana. 15th ann. Rept. Res. Inst. Afr. Geol. , Univ. Leeds, p. 23-25 (1971a).

— Occurrence and age of the Ofin River gold placers, Ghana. 15th ann. Rept.Res. Inst. Afr. Geol. , Univ. Leeds, p. 25-28 (1971b).

SHARPE, J. W. : The economic auriferous bankets of the Upper Witwatersrand beds and their relationship to sedimentation features. Trans. Geol. Soc. S. Africa, 52, p. 265-300 (1949).

SIMS, J. F. M. : The stratigraphy and palaeocurrent history of the Upper Division of the Witwatersrand System on President Steyn Mine and adjacent areas in the Orange Free State Goldfields, with special reference to the origin of the auriferous reefs (Bird Reefs Group). Ph. D. Thesis, Univ. Witwatersrand, Johannesburg (1969).

SMIRNOV, V. I. (Ed.): Geologiiya rossypei (Geology of placers). Akad. Nauk SSSR, Otdel. Nauk. o Zemle, Moscow, 400 p. (1965).

— Problem of the occurrence of metals in ancient conglomerates in the territory of the Soviet Union. Nauka, Moscow, 192 p. (1969).

SONNENDRUCKER, P.: L'or du Banket de Bondoukou. BUMIFOM, Abidjan, 11 p. (1958).

STEYN, L. S.: The sedimentology and gold distribution pattern of the Livingstone reefs on the West. M. Sc. Thesis, Univ. Witwatersrand, Johannesburg (1963).

TAGINI, B.: Esquisse géotectonique de la Côte d'Ivoire. Soc. Dev. Min. Côte d'Ivoire, Abidjan, 107 bis, 100 p. (1966).

TOURTELOT, H. A.: Hydraulic equivalence of grains of quartz and heavier minerals and implications for the study of placers. U. S. Geol. Survey, Profess. Papers, 594 F, 13 p. (1955).

VILJOEN, R. P.: The quantitative mineralogical properties of the Main Reef and Main Reef Leader of the Witwatersrand System. Univ. Witwatersrand, Econ. Res. Unit., Inf. Circ., 40, 63 p. (1968).

VAN ES, E.: Sheets 49 and 50, Dunkwa NW and NE. Rept. Direct. Geol. Survey Ghana for 1963-64, p. 18-23 (1968).

WILLIAMS, P. F., RUST, B. R.: The sedimentology of a braided river. J. Sed. Petrol., 39, p. 649-679 (1969).

WHITELAW, O. A. L.: The geological and mining features of the Tarkwa-Abosso goldfield. Mem. Gold Coast Geol. Survey, 1, 46 p. (1929).

Address of the author:

Dr. G. Sestini,
Department of Earth Sciences
University of Leeds, England

presently with:
UNESCO
Coastal Erosion Project
c/o P. O. Box 982, Cairo /Egypt

Size and Shape of Gold and Platinum Grains[1]

Harry A. Tourtelot and Leonard B. Riley

Abstract

Size analyses were made of two samples of gold concentrates, one sample of gold-bearing placer gravel, and two samples of platinum-bearing concentrates. The precious metals were separated from the sized fractions by use of an elutriator. Grain-shape analyses were made from grain dimensions measured microscopically. Breadth and length of each grain were measured by ocular micrometer intercepts; grain thickness was determined from the difference between readings of the fine focusing vernier when focused on the top of the grain and on the grain substrate. These four figures were translated into grain dimensions by a computer program that also calculated various shape factors and ratios between dimensions and mechanically plotted the results in different ways. The Corey shape factor that was plotted against the breadth of the grains, the dimension that controls response to sieving, conveyed the most useful information.

Relations between median sizes of black sand and precious metals, taking the effect of shape factors into account, suggest that the gold concentrate from one sluice accumulated under conditions of hydraulic equivalence and that another did not. Sharp breaks at the find ends of the precious metal cumulative curves seem to indicate the smallest size of effective accumulation for each sluice. The size is smaller for the material believed to have been accumulated under conditions of hydraulic equivalence. The size-distribution curves of a gold-bearing gravel indicate that all the gold could be concentrated in 35 percent of the gravel by sieving at a critical size.

The two gold samples reported have average Corey shape factors of less than 0.4 and the two platinum samples have factors greater than 0.4, indicating that the platinum is more spherical than gold. On the shape factor versus size plots, the grains become more spherical as the size decreases although there are some highly spherical, coarse grains of gold in one sample. One of the platinum samples shows a decrease in shape factor as the grain size decreases. The shape data suggest two populations of grains in one of the gold and in one of the platinum samples, though there are two possible methods of interpretation that identify two slightly different pairs of populations.

Sedimentologic data on detrital precious metal grains and host sediments can be useful in designing recovery processes and in interpreting nature and genesis of placer deposits.

Introduction

Sedimentologic data can contribute significantly to the understanding of the genesis of placer deposits of gold and platinum and to the search for additional deposits in

---

1) Publication authorized by the Director, U.S. Geological Survey.

ways similar to that demonstrated by data on compositional fineness of gold as pointed out by FISHER (1945, p. 562-563).

Recovery processes can be better designed and controlled if the particle-size distributions of the precious metals and the host sediments can be determined. The sedimentological behaviour of detrital grains of precious metals is not precisely understood; rules of thumb that have developed since the days of Jason and the Golden Fleece are still used and have been economically successful in past mining ventures. However, workable placer deposits are becoming more difficult to find. Many future placer operations, to be successful, may have to depend on recovery processes more effective than have been used to date, and such processes may have to be applied to particle sizes much finer than were considered recoverable in the past. Sedimentologic data and concepts apparently already are contributing to further developments in the gold fields of South Africa (PRETORIUS, 1966; VILJOEN, 1968).

This paper presents data on the size and shape of gold and platinum grains. Single samples from isolated deposits obviously are not a reasonable basis for drawing conclusions of general application. The data are interesting in themselves, however, and the authors suggest ways in which such data can be used to understand the nature of placer deposits and to clarify the problems of recovery of detrital grains of precious metals from such deposits.

Samples studied

Exploration samples of gravels and some known gold-bearing deposits were collected in Alaska, primarily for investigating the amounts of very fine grained gold that might be present (TOURTELOT, GANTNIER, and TERNES, 1971). Few of these samples contained sufficient coarse gold for obtaining useful sedimentologic data. Samples of concentrates also were obtained from placer operators, to whom we express our appreciation. We also are grateful to the Goodnews Bay Mining Co. and the International Mining Corp. for supplying samples of platinum concentrates from Alaska and Colombia, respectively.

Analytical methods

Size analyses. Mechanical size analyses were made of all samples using sieves from 8 mm to 0.061 mm that were spaced at quarter-Wentworth grade intervals in the sizes below 1 mm. Size fractions coarser than 1 mm were panned to recover any precious metals present. Precious metals were recovered from size fractions finer than 1 mm in a water elutriator similar to that described by FROST (1959). The narrow size range of the fractions that were elutriated facilitated separation of precious metals from other minerals by this gravity method. The precious-metal separates then were cleaned by hand picking under the microscope. The weights of the original size fractions and the weights of the separated precious metals were the bases for the size-distributions presented here.

The ratio between whole sample weight and weight of precious metals commonly is at least several hundred. The cumulative curves for the precious metals thus have a different statistical basis than the curves for the samples as a whole.

Shape descriptions. Shape descriptions for samples were made from microscope measurements of grains of gold and platinum selected from each of several size fractions. The coarsest size fractions contained only a few grains and all were measured. Several hundred grains were available from each of the finer size fractions, and a random sample of about 20 grains was taken from each by a micro-cone and quarter procedure under the microscope.

The first dimension determined for each grain was the breadth, which is the dimension that chiefly controls the response of the grain to sieving. The breadth was recorded only as the distance intercepted on the ocular micrometer. The grain was then rotated 90° and the length recorded in the same way as the breadth. Thickness was determined by focusing sharply on the substrate on which the grain lay and by recording the reading of the microscope fine-focus vernier. The microscope was then focused on the uppermost part of the upper surface of the grain and the vernier reading recorded. A simple computer program translated the ocular micrometer intercepts to millimeters, taking into account the magnification of the objective which was used in the measurement. The program also determined the difference between the vernier readings and converted the difference to millimeters according to the calibration curves that had been established for the vernier. Duplicate determinations of thickness indicate that thicknesses obtained in this way have a precision of ± 6 microns at a magnification of about 200 x. A similar method of determining grain thickness has been described by JOHANNSEN (1918) and more recently by WADSWORTH (1971).

The real effect of the measurements of breadth, length, and thickness is to enclose the grain in a rectangular parallelepiped. The squares and rectangles that enclose the breadth and length of near-spherical to elliptical grains are larger than the actual areas of the grains. This leads to the resulting factor describing the shape of the grains being somewhat smaller than is appropriate for the true areas of the grains. Conversely, for lunate and sigmoidal grains, and other kinds of complex shapes, the rectangles based on the grain breadths that would control responses to sieving do not enclose the grains and are smaller in area than the grains. The shape factor thus is made somewhat larger than the true areas would call for.

The computer program also calculated the Corey shape factor (COREY, 1949), the maximum projection sphericity (FOLK, 1957; SNEED and FOLK, 1958), and the ratios used by HAGERMAN (1936) and by MOSS (1962). Computer plotting techniques allowed rapid exploration of the amount of information given by each of these approaches to shape description. When plotted against the grain breadth, the Corey and Folk shape or sphericity factors give similar patterns that show differences in these parameters with different sieve sizes. Plots following the methods of Hagerman and Moss were not very informative.

The Corey shape factor was chosen rather than the maximum projection sphericity because the Corey factor is widely used in engineering problems of sedimentation (Inter-Agency Committee on Water Resources, 1957). The factor is the square root of the thickness squared, divided by the length times the breadth:

$$\left(\text{Corey shape factor} = \sqrt{\frac{T^2}{L \times B}}\,\right)$$

Small Corey shape factors indicate flaky grains and large factors indicate more spherical grains.

As pointed out by SNEED and FOLK (1958, p. 123), factors such as the Corey factor and the maximum projection sphericity factor designate grains of considerably different form that have the same numerical value for the factor. This is because the area of the grain (length times breadth) appears in the denominator of both shape factors, and the same area can be given by very different lengths and breadths. A grain with dimensions of 10 by 10 has the same area as another grain with dimensions of 4 by 25 and, if thicknesses are equal, the same shape factors, although the hydraulic behaviour of the grains would be very different. No method was found to illustrate this range and still plot the data against the sieve size of the grains. The following table, however, indicates the percentage of grains that are elongated to the extent that the length is more than twice the breadth.

Peters Creek gold	12 percent
Jack Wade gold	31 percent
Goodnews Bay platinum	0 percent
Nariño platinum	7 percent

Thus, only the Jack Wade gold contains a significant number of elongate grains.

Results

Size distributions. Both the Peters Creek and the Jack Wade Creek samples (fig. 1) are gold concentrates from Alaska. The black sand consists of magnetite and other minerals of relatively high specific gravity.

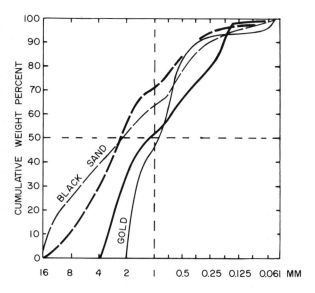

Fig. 1. Cumulative curves for black sand and gold in concentrates from Peters Creek, Alaska (heavy lines) and Jack Wade Creek, Alaska (light lines)

The black sand of the Peters Creek sample is remarkably poorly sorted. The gold also is rather poorly sorted with the mode being 1 to 2 mm and amounting to 33 percent of the gold. The median size of the gold is 0.7 mm. Considering that the aver-Corey shape factor for this sample is 0.39 (see fig. 5), the median size of 0.7 mm is about right to conclude that the gold is in hydraulic equilibrium with the black sand. It is assumed that the generally equidimensional grains of magnetite and associated minerals have shape factors approaching that of a sphere. The general parallelism between the cumulative curves of the gold and black sand also indicates that the two materials are in virtual hydraulic equilibrium.

The break in the gold cumulative curve at about 0.125 mm may be the result of the size distribution of the gold in the gravel put through the sluice. On the other hand, the break may indicate that the sluice operated in such a way that gold finer than 0.125 mm was not efficiently recovered.

The black sand coarser than 1 mm from the Jack Wade Creek sample (fig. 1) is very nearly perfectly unsorted and the black sand finer than 1 mm is almost as unsorted. The gold seems to be better sorted with a modal value of 37 percent in the 1- to 2-mm size fraction. The sharp truncation of the coarse end of the cumulative curve for gold suggests that a portion of the coarser gold had been removed before the sample was given to us. The median size of the black sand is about 2.4 mm and a high sphericity can be assumed. The median size of spherical gold in hydraulic equilibrium with 2.4-mm black sand would be about 0.7 mm. The actual median size of the gold is 1.0 mm. The average Corey shape factor for the sample is 0.36 (see fig. 6) and grains with this shape factor that are in hydraulic equilibrium with 2.4-mm black sand would be a little larger than 1 mm. If, however, the coarse gold that probably was removed from the sample could be restored, the median size of the gold would be increased, indicating the probability that the sluice was not operated at hydraulic equilibrium. This is also suggested by the lack of parallelism between the two cumulative curves.

The break in the cumulative curve for gold at about 0.35 mm could reflect the size distribution of the gold in the original gravel. The break may indicate instead, however, the lower size limit of gold that the sluice can accumulate effectively. This lower limit of effective accumulation from a sluice that probably did not operate at hydraulic equilibrium is more than twice as large as the limit for the Peters Creek sluice that seems to have operated nearly at hydraulic equilibrium. Perhaps the larger size of effective accumulation is related to the seeming lack of hydraulic equilibrium. This interesting possibility could be explored experimentally.

If this sample were to be sieved at 2 mm, all the gold would be concentrated in about half of the black sand.

What are the relations between the size distribution of an original gravel and its contained gold? The Stonehouse Draw sample of gold-bearing gravel (fig. 2) provides data at least partly applicable to the question. The cumulative curve for the gravel is drawn truncated at the coarse end because some large pieces were excluded in sampling. Possibly these pieces amount to 10 - 20 percent of the sample. On this basis, the gravel is no better sorted than any of the sluice concentrates, and in fact is nearly perfectly unsorted. The closest thing to a mode is the material coarser than 8 mm, which amounts to about 20 percent of the sample. A secondary mode is the material finer than 0.061 mm, which amounts to about 12 percent of the sample. The fine fraction in bimodal gravels often is interpreted as the result of material settling in the porosity of an openwork, framework-supported gravel (PLUMLEY, 1948). Such openwork gravels, however, generally have much more sharply distinguished coarse and fine modes and a larger deficiency in the intervening sizes. In the field, this gravel looked as if everything - fine and coarse together - had been deposited in a single sedimentational event, and the size distribution does not suggest otherwise.

The gold is relatively well sorted, with 95 percent of it falling between 0.125 and 1.0 mm. The median size of the gold is about 0.4 mm compared to the 0.3 mm for gold spheres called for theoretically by the 3.00-mm median size of the quartz in the gravel. No shape study was made of the gold from this sample but visual examination suggests that an average Corey shape factor from 0.3 to 0.5 probably is ap-

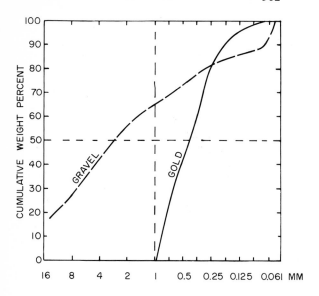

Fig. 2. Cumulative curves for gravel and gold from Stonehouse Draw, Alaska

plicable, and hydraulically equivalent gold with shape factors in this range would be larger than 0.4 mm. In addition, if the cumulative curve for the gravel could be re-drawn to account for the coarse material excluded in sampling, the median size of the gravel would be increased which would make the departure from hydraulic equi-librium even larger.

If particle-size distributions of gravel and of gold were known with confidence, such data could be valuable in planning the exploitation of a placer deposit. For example, in mining operations, if gold-bearing gravel from Stonehouse Draw were screened at 1 mm, 65 percent of the gravel could be immediately rejected; all the gold would be concentrated in only 35 percent of the gravel. The smaller volume of gravel, as well as its smaller particle size, should allow for more effective recovery with smal-ler amounts of water running through the gold-recovery sluice. More effective use of water could be an important factor in Alaska, and in other areas where water for sluicing is in short supply. The use of smaller amounts of water could also minimize the effects of placer mining on the environment.

The size distribution of the gold also provides data that might be useful to consider in the operation of sluices. If Stonehouse Draw gravel were to be sluiced under the conditions that produced the Jack Wade concentrate, only 65 percent of the Stonehouse Draw gold would be recovered.

Both gold and platinum are recovered from placer deposits at Goodnews Bay, Alaska. Size data are shown on figure 3. About half the platinum grains can be picked up with a hand magnet because of the ferromagnetic properties of some natural platinum al-loys (MERTIE, 1969, p. 7). The sample studied was a ball-mill product in which ev-erything was smaller than 0.25 mm. The cumulative curves for gold, magnetic plat-inum, and nonmagnetic platinum are in the upper right corner of the diagram. Be-cause the median sizes for all three materials are in the first size fraction, the ques-tion of hydraulic equivalence cannot be examined. The nonmagnetic platinum appar-ently has a higher specific gravity than the magnetic platinum inasmuch as the cumu-lative curve for nonmagnetic platinum lies toward the finer sizes.

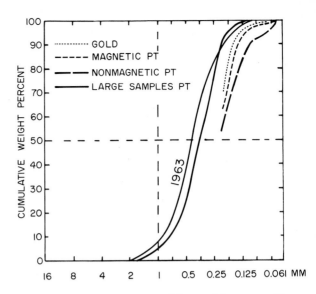

Fig. 3. Cumulative curves for large samples of platinum and for gold, magnetic platinum and nonmagnetic platinum in concentrate from Goodnews Bay, Alaska

The two complete cumulative curves labeled 1963 are based on data supplied by the Goodnews Bay Mining Co. to J. B. Mertie, Jr. , U. S. Geological Survey, who, with the permission of the company, made them available for this study. The data consisted of averages of the series of weekly dredge cleanups from each of two paystreaks worked that year. The median size for the coarsest curve is about 0. 45 mm, whereas that for the finest curve is about 0. 4 mm. The fine curve represents a paystreak recognized by the operators to contain finer grained platinum than other paystreaks. Both curves, however, indicate that only a little more than 10 percent of the material is finer than 0. 25 mm. The curves have no sharp breaks or deflection at the fine ends, suggesting that recovery processes are effective even at fine sizes.

These curves are unusual in the published literature and important because they are based on relatively large amounts of platinum, probably of the order of several hundreds of ounces.

Platinum is recovered as a by-product of gold dredging at Nariño, Colombia. The studied sample (fig. 4) was a special concentrate prepared by the International Mining Corp. to investigate the amount of very fine grained platinum present. The coarse fraction of the black sand consisted mostly of tramp metal from an unknown source and a complete cumulative curve has not been drawn. The finer part of the curve for the black sand is better sorted than is characteristic of the curves from gold sluice samples. The cumulative curve for the platinum is approximately parallel to the black sand curve and the platinum is about as poorly sorted as the Peters Creek gold (fig.1). The median size of the black sand is about 0. 3 mm. Spherical platinum grains in hydraulic equilibrium would be 0. 13 mm. The average Corey shape factor (fig. 8) is 0.42, which would account for the difference between the theoretical size (0.13 mm) and the actual median size (0. 175 mm) of the platinum. Thus the platinum is in hydraulic equilibrium with the black sand. The absence of a sharp break in the fine end of the cumulative curve suggests that the concentrating process operated effectively down to nearly the finest sizes present. Almost 10 percent of the platinum is finer than 0. 088 mm in sieve size.

314

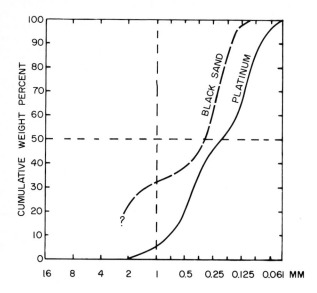

Fig. 4. Cumulative curves for black sand and platinum in a concentrate from
Nariño, Colombia

Shape distribution. The shape diagrams of figures 5 - 8 are made by plotting the Corey shape factor for each grain against the breadth of that grain. Also plotted are the
average factor for each of the size fractions from which grains were measured. These
averages are plotted at the midpoint of the nominal size fraction from which the grain
were taken. The points for some individual grains are displaced somewhat to the left
of this midpoint because the diagonal dimension of square sieve openings permits
flaky grains with a breadth greater than the nominal sieve opening to pass through.
A sieve opening of 0. 5 mm could pass along its diagonal a very flaky grain measuring 0. 7 mm in breadth.

The Peters Creek gold (fig. 5) seems to contain two populations with respect to shape
although there are two possible interpretations as to how the populations can be defined. The two size fractions coarser than 0. 5 mm have average shape factors ranging from 0. 42 to 0. 51. In contrast, the grains finer than 0. 5 mm have average shape
factors from 0. 30 to 0. 39 and the range of values around the average is smaller. The
shape factor for these fine grains increases systematically with the finest grains
having the largest shape factor.

Alternatively, it is possible that the grains between 0. 5 and 1. 0 mm represent two
populations because the range in shape factor is so large. The flaky grains of this
size fraction could represent the coarse end of a trend of increasing shape factor
accompanying decreasing grain size that would be continuous with the trend of the
grains finer than 0. 5 mm. The more spherical grains of those between 0. 5 and 1. 0
mm in breadth could then represent the fine end of another trend of increasing shape
factor with decreasing size - a trend that would begin with grains coarser than any
in the sample. Not enough information is available to suggest which of these interpretations is the more useful. The average shape factor for all grains is 0. 39.

The Jack Wade Creek gold (fig. 6) has quite a different pattern of shape factors. The
coarsest grains are flaky and have a shape factor of 0. 27. The shape factors then in-

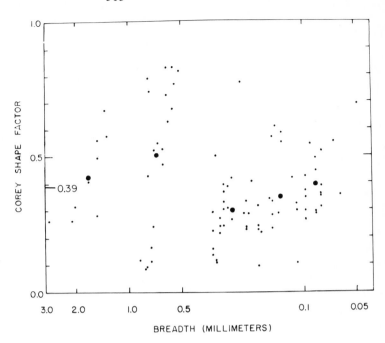

Fig. 5. Shape diagram for gold from Peters Creek, Alaska. Large symbols indicate average shape factors for size fractions. Average for all fractions shown on left side of diagram

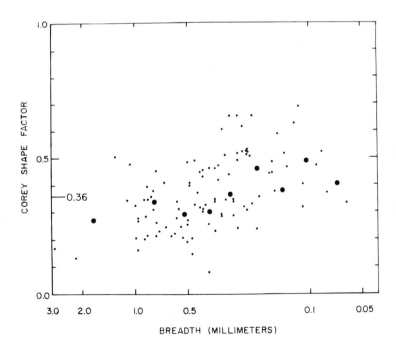

Fig. 6. Shape diagram for gold from Jack Wade Creek, Alaska. Large symbols indicate average shape factors for size fractions. Average for all fractions shown on left side of diagram

crease, with some irregularity, as the grain size becomes smaller, with values of 0.49 and 0.41 for the two smallest size fractions. Compared to the Peters Creek gold, the range of shape factors for each size fraction is relatively small. The average shape factor for the sample is 0.36.

These data do not agree with the general concept that gold becomes more flaky as it becomes finer grained.

The Goodnews Bay platinum (fig. 7), both magnetic and nonmagnetic, has larger shape factors than those of the two gold samples. The largest size fraction of the nonmagnetic platinum has an average shape factor of 0.72, but the two smaller size fractions have average shape factors of about 0.50. The average shape factor for all the magnetic platinum, however, is 0.57 as compared to 0.59 for the nonmagnetic platinum. The average shape factors for the magnetic platinum range from 0.49 to 0.58. The decrease in shape factor from the coarse grains of nonmagnetic platinum to the fine is chiefly responsible for the general trend of the shape factors for all grains taken together to decrease as the grains become smaller. The grains with the large shape factors are those that approach euhedral single crystals. Metallic platinum and most of its alloys crystallize in the cubic system so that grains tend to be equidimensional.

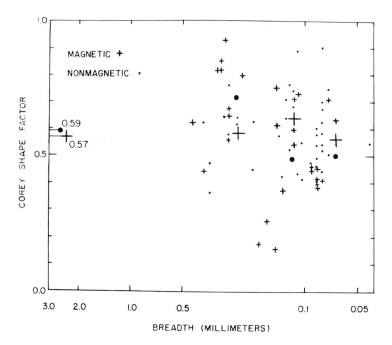

Fig. 7. Shape diagram for magnetic and nonmagnetic platinum from Goodnews Bay, Alaska. Large symbols indicate average shape factors for size fractions. Average for all fractions shown on left side of diagram

The pattern of shape factors for the Narino platinum (fig. 8) seems also to suggest two populations. The grains coarser than about 0.5 mm have average shape factors ranging only from 0.28 to 0.34 in contrast to the finer grains that have shape factors ranging from 0.48 to 0.56, with the largest shape factor applying to the smallest grains. The range of shape factors for the grains between 1.0 and 2.0 mm is very large, and it is possible that the flaky grains from this size fraction represent one population and the more spherical grains another, as has been suggested for the Peters Creek gold. There would then be one trend from the coarser flaky grains with increasing shape factors as the grains became smaller. The other trend would consist of grains for which the shape factor did not change with grain size.

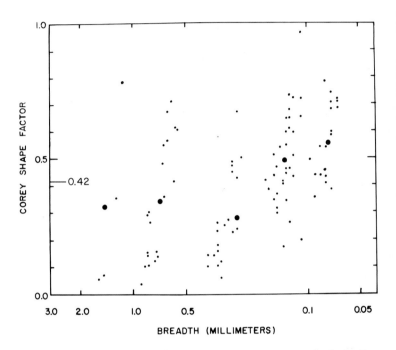

Fig. 8. Shape diagram for platinum from Nariño, Colombia. Large symbols indicate average shape factors for size fractions. Average for all fractions shown on left side of diagram

Discussion and summary

We emphasize that none of the samples studied provide data that do more than describe the samples themselves. No firm conclusions can be drawn regarding the nature of the deposits or the processes by which the concentrates were obtained. Nevertheless, the data allow certain interpretations to be made. We have not hesitated to make such interpretations in the hope that others will examine ways in which sedimentologic data can be applied to the problems of placer deposits.

The principal suggestions are:

1. Size distributions of sluice recovery products, both precious metals and black sand, can indicate the degree to which the sluice is operating to recover these products in hydraulic equilibrium with each other.

2. Size distributions of precious metal grains can indicate the minimum size that is effectively recovered in a hydraulic-gravity separation process. Data from two samples suggest that the minimum size of effective recovery is related to

the degree of hydraulic equilibrium that is attained between heavy precious metals and black sand.

3. Size-distribution studies of precious metal particles and their host gravel (or black sands) can indicate a critical size at which it would be useful to sieve to concentrate the metals in a relatively small portion of the host material.

4. Shape studies can aid in considering recovery processes designed to take into account either high or low sphericities of precious metal grains, which might lead to improved recoveries.

5. Shape studies of precious metal grains can identify different populations of grains with different sedimentational histories, which could have important implications as to the origin of the deposits.

References

COREY, A. T. : Influence of shape on the fall velocity of sand grains. Colorado A & M College (now Colorado State University), Fort Collins, unpub. Master's thesis, 102 p. (1949).

FISHER, N. H. : The fineness of gold, with special reference to the Morobe goldfield, New Guinea. Econ. Geol. , 40 (8), p. 537-563 (1945).

FOLK, R. L. : Petrology of sedimentary rocks. Hemphill's, Austin, Texas, 114 p. (1957).

FROST, I. C. : Elutriating tube for the specific gravity separation of minerals. Am. Mineralogist, 44 (7 & 8), p. 886-890 (1959).

HAGERMAN, T. H. : Granulometric studies in northern Argentine. Geografiska Annaler, 18 (2), p. 125-213 (1936).

INTER-AGENCY COMMITTEE ON WATER RESOURCES: A study of methods used in measurement and analysis of sediment loads in streams - Rept. 12, Some fundamentals of particle size analysis. Washington, U. S. Govt. Printing Office, 55 p. (1957; 1958).

JOHANNSEN, A. : Manual of petrographic methods. 2d ed. McGraw-Hill, Inc. , New York, 649 p. (1918).

MERTIE, J. B. , Jr. : Economic geology of the platinum metals. U. S. Geol. Surv. Prof. Paper 630, 120 p. (1969).

MOSS, A. J. : The physical nature of common sandy and pebbly deposits. Pt. I. Am. Jour. Sci. , 260 (5), p. 337-373 (1962).

PLUMELY, W. J. : Black Hills terrace gravels: a study in sediment transport. Jour. Geol. , 56 (6), p. 526-577 (1948).

PRETORIUS, D. A. : Conceptual geological models in the exploration for gold mineralization in the Witwatersrand basin. Econ. Geol. Research Unit IC No. 33, Univ. Witwatersrand, Johannesburg, 38 p. (1966).

SNEED, E. D. , FOLK, R. L. : Pebbles in the lower Colorado River, Texas - a study in particle morphogenesis. Jour. Geol. , 66 (2), p. 114-150 (1958).

TOURTELOT, H. A. , GANTNIER, R. F. , TERNES, E. B. : Silt-size gold. Geol. Soc. America Abstract with Programs 3 (6), p. 417 (1971).

VILJOEN, R. P. : The quantitative mineralogical properties of the main reef and main reef leader of the Witwatersrand system. Econ. Geol. Research Unit IC No. 41, Univ. Witwatersrand, Johannesburg, 58 p. (1968).

WADSWORTH, W. B. : Measurement of $\underline{c}$ intercepts in loose sand grains by optical height. Jour. Sed. Pet. , <u>41</u> (1), p. 30-37 (1971).

Address of the authors:

U. S. Geological Survey
Denver, Colorado / U. S. A.

Distribution of Certain Trace Elements in Marine Sediments Surrounding Vulcano Island (Italy)

J. N. Valette

Abstract

Vulcano is the southernmost of the islands of the Aeolian archipelago, situated to the south of the Tyrrhenian Sea and to the north of Sicily.

This volcanic island was affected by later phenomena, e.g. springs and fumaroles localised on the Fossa, the most recent part of the island. The analysis by quanto-metre (continuous arc) of the trace elements contained in the marine sediments found near the island and situated, above all, in the bay of Levant, permits the classifica-tion of these elements according to different categories:

- deep water concentrates,
- fumarolian concentrates,
- mixed concentrates.

As far as V, Cr, Sr, and Ba are concerned, the importance of the fumaroles is ap-preciable, but less so for the other elements. The relations between a) granulometry and trace elements were established with lutites and sands, and b) between the min-eralogy and the trace elements could be seen mainly to the south of the bay. Although these relations exist, they are nearly always hidden by the direct interference of fu-marolian phenomena which play the main part in the distribution of trace elements.

Introduction

Vulcano is the southernmost of the islands in the Aeolian archipelago, situated in the south of the Tyrrhenian Sea, to the north of Sicily and lies between 38°26' and 38°22' latitude North, 14°56' and 15°01' longitude East (Fig. 1). All the Aeolian islands are volcanic to different degrees (BERGEAT, 1899). At the present time only two zones on Vulcano are centres of later volcanic phenomena (HONNOREZ, 1969). The first zone stretches along the internal wall, the lip and the upper part of the outside north-ern flank of the crater called "fossa". The second concerns the Vulcano-Vulcanello Isthmus, at the place known as "Acqua Calda" and several spots on the beaches of "Porto di Levante".

The island is divided into four units which were formed in the following order (KEL-LER, 1967):
1) To the south, old Vulcano, characterized by basalts and trachytic andesites. These volcanic cones, broken down by erosion, are scored by erosion with their flanks sloping gently down to ten metres below sea level (at 600 metres from the shore).
2) To the northwest, the Lentia zone, formed by liparitic walls forming high cliffs, 40 to 50 metres to the sea.
3) To the north, the small peninsula of Vulcanello, made up of leucite basanites, with 10 to 15 metre cliffs along the Vulcano-Lipari channel.
4) To the northeast, Fossa, the still active part of the island, the products of which vary from trachytes and liparites. This part slopes gently toward the sea along

cliffs and shelves and is determined by the cohesion of the eroded materials (soft lava rock).

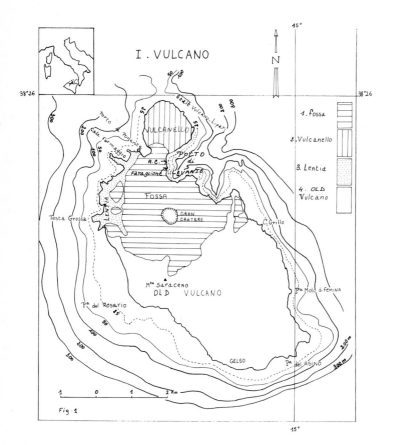

Fig. 1. Vulcano Island: Geographical position and petrographical areas
1 - Fossa: Trachites and liparites; 2 - Vulcanello: Leucite basanites;
3 - Lentia: Liparites; 4 - Old Vulcano: Basalts and trachitic andesites

I. Sampling and Dosing Methods (VALETTE, 1969)

a) Sediment sampling (Fig. 2)

In the course of three surveys, we carried out about fifty corings and about one hundred dredgings around the island. The samples were obtained by diving down to 70 meters in the coastal zones, or from boats for greater depths.

When diving, each sampling line included a coring in the deepest zone (30 to 70 m); then, in a line perpendicular to the coast, different dredgings (generally three) were taken at decreasing depths. Once cored and dredged, the places were marked by a float which was then located and its position marked on a chart; the methods of hand-coring used is that described by B. CHASSEFIERE (1968); plastic tubes from 4.5 to 9.9 cm in diametre are used. The "useful" length of core depended upon the thickness the granulometry and consistency of the sediment.

The tubes and cork bungs were carefully washed before immersion in order to avoid contamination; as soon as it was landed, the core was freed from its "orange peel" (stiff plastic) and both its ends were hermetically sealed. Then, as much as possible, the tubes were stored upright and in a cool place. The cores were partially opened by a circular saw; the final millimetre was then cut with a stainless steel scalpel. The dredgings were then collected by hand and placed immediately in a plastic bag.

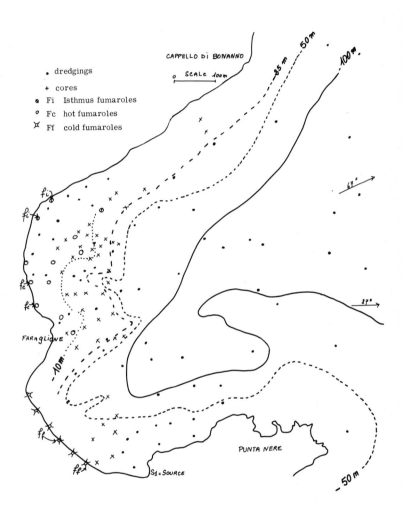

Fig. 2. Position of sampling points

As far as work from the boat was concerned, we concentrated mainly on dredgings. Core samples taken in heterometric sediments of Vulcano turned out to be very delicate and did not yield good results.

Two sorts of dredges were used: a) the stainless steel Genovese dredge shut by dropping a messenger, and b) a Berthois cone, which is a very simple instrument giving excellent results.

b) Methods of dosage

The crude samples were dried in an oven in a flow of nitrogen, then crushed in an agate mortar and sieved on a nylon mesh at seventy microns. Part of the sample was mechanically mixed to two parts of spectrographic buffer and 4.5 parts of graphite. The mixing was carried out in a fritted aluminium bowl with oscillating discs.

The dosages were done in a quantometre (A.R.L. Fica type 2-9000); the working of the quantometre in a continuous arc was chosen (MOAL, 1968). The program included 27 elements (5 major ones, and 22 other elements which can be dosed as trace elements). In this study, we have selected among the 22 elements those which showed values above the sensitivity threshold of the instruments, i.e. Mn, Cu, Ni, Co, Sr, Ba, V, Cr, Pb, Zn, B.

Three successive burnings were carried out for each sample. Each series of 5 samples analysed was framed by two reference samples, one white one giving the base value, or constant, and a synthetic sample containing the elements with variable and known contents. Once programmed, the major elements gave an idea of the matrix and in certain cases allowed correction.

Intensity correction permitted us to increase analytic precision and to lower the dosage limits. The average analytic precision obtained in quantitative analysis was from 5 to 15 %. The reproducibility factor varied between 5 and 10 % for most of the elements. The precision of measurements by this method was satisfactory; the results obtained for the analysis of the international and national geochemical standards were within the tolerated limits.

II. The Distribution of Elements in Marine Sediments

Figure 2 shows the positions of the different samplings in the Bay of Levant, but several badly positioned samplings have been left out.

Horizontal distribution in the Bay of Levant

From the results obtained by dredgings and the top part of the cores from the Bay of Levant, we constructed the distribution charts for 11 elements.

By concentration we refer to average values of the elements (GOLDSCHMIDT, 1954; BOSTROM, 1970; BOSTROM and FISCHER, 1971; AHRENS, 1968).

Manganese (Fig. 3):
The zones with a strong concentration of manganese are situated almost exclusively in the deep sector (between the axis, south and off shore part of the bay); there, the contents are greater than 1100 ppm (with a maximum of over 2000); the fumarolian zones, the spring and the rest of the sediments of the bay rarely contain more than 600 (on the average 660) ppm.

Copper (Fig. 4):
The curves of equal values in copper content clearly follow the Bay curves, above all in the south of the bay values reach 200 ppm in calm and deep zones and less than 66 near the shore. In places near Faraglione and opposite the place called "Acqua Calda", values over 70 ppm can be found. In contrast, the sediments situated around fumaroles on the North beach are poor with values under 50 ppm.

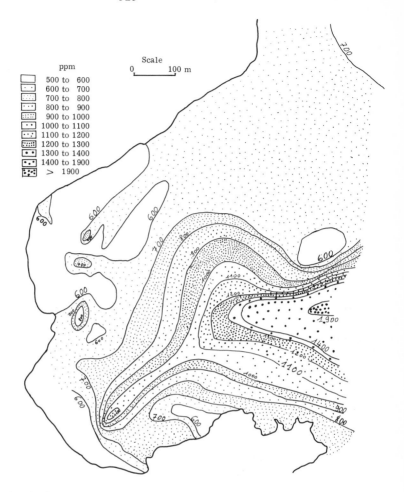

Fig. 3. Distribution chart of manganese

Nickel (Fig. 5):
The distribution chart for nickel shows concentrations greater than 22 ppm in the southeast of the bay and near the port. Hot and cold fumaroles are distinguished by zones of low content (less than 5 ppm) while the average value in the sediments of the bay is around 10 and 15 ppm. Near the spring there exists a slight increase in nickel content.

Cobalt (LUNDEGARDH, 1949):
Its distribution is very similar to that of nickel. Here too, one finds high concentrations (over 20 ppm) in the southern part of the bay, mainly in the axial and deep sector, and opposite the spring. As we remarked for nickel, this latter zone takes here values as important as in the centre of the bay. Lower contents suggest real dispersion[1] of cobalt in the sediments situated in the hot and cold fumaroles in the north and in the south.

---

1) The word dispersion is used here as meaning a flow of water which could dilute the content in trace elements.

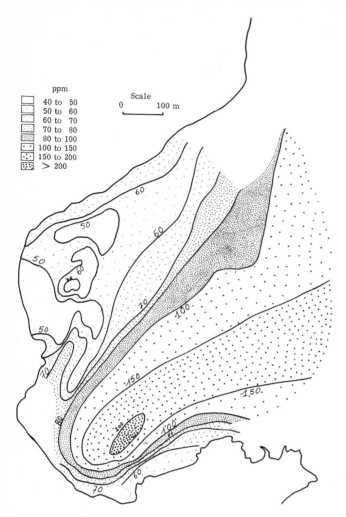

Fig. 4. Distribution chart of copper

Chromium (Fig. 6):
The chart for chromium shows the presence of extremely concentrated zones (contents over 90 ppm) in the spring sector and in the hot fumarole area (over 60 ppm). The average values for the bay are around 30 ppm and it seems that there is local dispersion by the cold fumaroles (less than 20 ppm). A small basin situated near the deep zone, on a straight line out from the Faraglione, is formed by sediments containing more than 40 ppm chromium. Yet again there is thus an extremely weak concentration over the deep zones.

Vanadium:
Its distribution is like that of chromium, and this element shows high contents (over 200 ppm) in the sector of the hot fumaroles and near Faraglione and the zone of the spring. The cold fumaroles are distinguished by low values (less than 100 ppm). A zone cutting across the north of the bay as far as the deep axial zone shows a concentration of over 140 ppm.

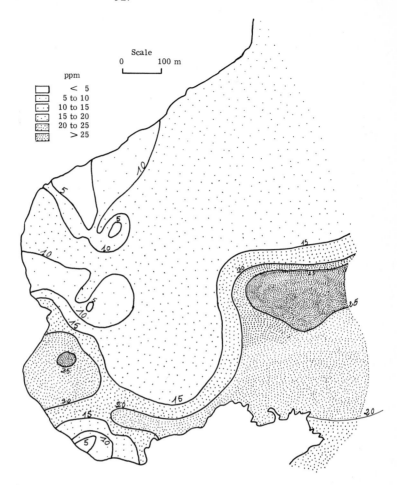

Fig. 5. Distribution chart of nickel

Strontium:
The distribution chart of this element shows the predominating influence of the hot fumaroles in the case of this element. The place called "Acqua Calda", north of Faraglione, is marked by contents over 1200 ppm. In contrast, the spring sector shows considerable poverty in strontium and there the values drop to less than 500 ppm. The coastal zones have values approaching 900 ppm while the deep zones contain between 600 or 700 ppm for this element.

Barium (Fig. 7):
Its distribution follows more or less that of strontium, but its variations are much less great. The coastal zones have average values greater than 700 ppm, whereas sediment of less than 600 ppm in barium are to be found in the deep zones. The hot fumarole areas show richer barium contents in the sediments, sometimes greater than 1000 ppm for this element. Here again the spring sector is characterized by a marked lowering in content in the sediment which contains less than 600 ppm of this element.

328

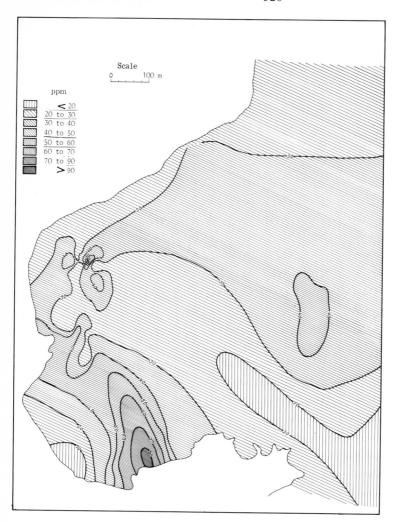

Fig. 6. Distribution chart of chromium

Lead:
The average contents of the Bay are situated between 20 and 30 ppm. The deep zone is a sector containing more than 30 ppm of lead, a content which is found along the north beach of the bay, near the Vulcano-Vulcanello isthmus. Zones with a high concentration nearly meet in the hot fumaroles sector where sediments with less than 20 ppm of lead separate them. In the cold fumaroles sector to the south, these contents go down to less than 10 ppm.

Zinc (Fig. 8):
Clear concentrations of zinc (over 90 ppm) are visible in the centre and to the south of the Bay. In this case as well, very high contents can be found near the Vulcano-Vulcanello isthmus (over 90 ppm). The hot and cold fumarole zones are generally marked by weak contents (less than 50 ppm). Locally some emissions seem, on the contrary, to concentrate zinc (over 70 ppm). This is what happens for the group of fumaroles situated off "Acqua Calda" and the "Faraglione".

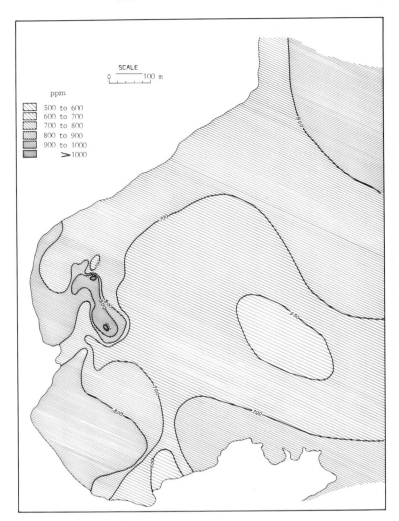

Fig. 7. Distribution chart of barium

Boron:
Its main disposition follows the bathymetry and the deep zones possess high contents in boron (over 70 ppm). Here again this disposition is modified locally by a high concentration towards the isthmus. According to their geographical sector, the hot fumaroles sometimes show a lowering, sometimes an increase in content (more than 50 ppm) of the sediments in boron. The sector of the cold fumaroles is on the contrary always very poor (less than 20 ppm).

The study of the distribution charts of Mn, Cu, Ni, Co has shown a clear relation between the concentration of these elements and bathymetry. The contents decrease from the deep zones towards the shore. Locally, in some places, this gradient is accentuated by contact with volcanic emissions. The fumarole sectors, in particular, are characterized by weak contents. We think that it is possible to speak of deep water concentrates in this case.

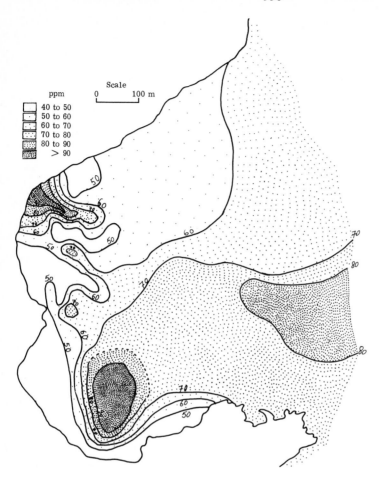

Scale

ppm   0   100 m

- 40 to 50
- 50 to 60
- 60 to 70
- 70 to 80
- 80 to 90
- > 90

Fig. 8. Distribution chart of zinc

The different distribution charts of V, Cr, Sr, Ba constitute a proof of the increase in the content of these elements in the sediments around the later volcanic phenomena. They are <u>fumarolian concentrates</u>. We can note that the curves of equal value content have very different profiles for the four elements. The hot fumaroles sector with its high contents in Cr, V, Sr, Ba contrasts sometimes with the spring, where we can note a concentration of Cr and V by a lower content of Sr and Ba. There are also for these 4 elements minor differences in the fumarole sector itself.

It is possible to deduce from this that the volcanic phenomena of the Bay of Levant do not play a constant role (influx, feed, deposition, environment). This last hypothesis could explain the contrast between the fumaroles and the spring; in the case of the spring, water at a high temperature and with a low pH follows the fissures of the rock and finally runs into the sea at surface level, while in the case of the fumaroles there is percolation (gas, water vapour bearing salts to a greater or lesser degree and at a high temperature and low pH). This proceeds through the soft sediment, thereby bringing about a better leaching of the materials and a better way of trapping certain trace elements. The distribution of Pb, Zn, and B shows a concentration of these elements both in the deep sector and the hot fumaroles sector. We shall call them <u>mixed concentrates</u>. We may think that the coastal zone rich in Pb, Zn, and B is continued in the sediments of Vulcano-Vulcanello isthmus, or that there exist here fumaroles of a special composition.

The study of the horizontal distribution of the traces shows that the most influencial factors are as follows: on the one hand bathymetry, and on the other, the existence of volcanic phenomena (springs and fumaroles). However, the bathymetric factor is difficult to dissociated from distance from the shore, from granulometry, and from the mineralogy of the sediment.

III. The Relation between Trace Elements and Granulometry

a) The granulometry of the Bay of Levant

Generally speaking, the sediments surrounding Vulcano are composed of several populations. Their grading is marked by, a) good and average sorting in terms of depth and hydrodynamic conditions, and, b) a better grading of the coarser part.

We constructed granulometric charts for the lutites (less than $40\,\mu$) and the sands. The detailed study of the Bay of Levant shows a regular zonation of lutite contents: the deep sectors are covered with sediments rich in fine particles (Fig. 9 - 10).

The granulometry chart for sands shows that the shore is lined by a band of varying width of coarse sediment except on the north beach. The isthmus linking Vulcanello to Vulcano is indeed marked by a sedimentation where medium and coarser sands are mixed.

On leaving the shore, we can see an intrication of the varied granulometric sediment-ary zones to a greater or lesser degree. To the north, the normal sedimentological succession is interrupted by an area of mixed elements. To the south of "Porto di Levante", one sector presents fine deposits on the calm and relatively deep bed of the bay. The extension of these fine materials is interrupted by a recurrence of coarse sediments brought from the north.

b) The relation of the grain size and the horizontal distribution of the trace elements

We have compared the horizontal distribution of the trace elements with the granulo-metric charts of the sediment of the Bay of Levant.

1. The comparative study of the geochemical charts and the distribution of the trace elements and of the lutites shows a positive correlation for Mn, Cu, and Pb (the quantity of lutites and elements grows at the same time). It is inverted for Sr and Ba (Fig. 9).

2. The comparison of the granulometry charts for the sands and the distribution of the traces is difficult to carry out (Fig. 10). In the Porto di Levante, the fine se-dimentation zones are rare and confined. They are, in general, rich in Mn, Cu, Zn and B. Chromium reaches average values in this area. More specifically in the centre and in the axis of the bay V and Pb are concentrated. On the coastal coarse and heterometrical zones, the role of granulometry is largely masked by the geographical position of the samples and, above all, by the interference of the fumaroles.

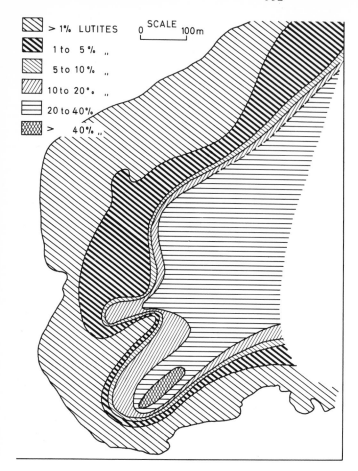

Fig. 9. Distribution chart of lutites

IV. The Relation between Trace Elements and Mineralogy

a) Mineralogical nature of the sediments (GOLDSCHMIDT, 1937)

The sieving of sediments taken around the island shows their weak content in lutites and their great richness in sand.

- The study of total sands and heavy minerals by means of smears and slides allowed us to recognize the following mineral species: quartz, potassic feldspars, plagio-clase, augite, diopside, peridots, hematite, magnetite, pyrite, and gypsum (DE FIORE, 1921; BERNAUER, 1935; HONNOREZ, 1969).

- The analysis carried out on the fraction of sediments below 40 μ shows that it is not made up of clay in the mineralogical sense of the term. Most of the time it is a case of a mixture of fine particles of glass and feldspar. The presence of clay minerals such as montmorillonite, interstratified montmorillonite and kaolinite, and kaolinite was however recognized.

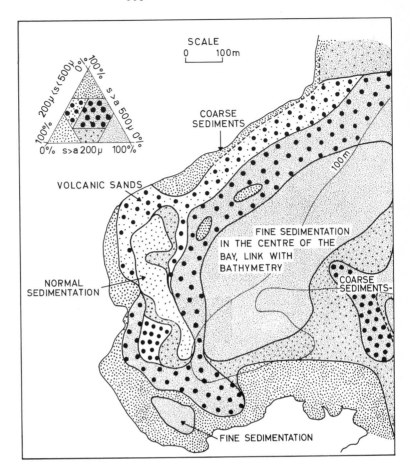

Fig. 10. Distribution of sands: granulometric chart

b) Distribution of minerals

The petrographic provinces on land are characterized mineralogically in the following manner:
- the old province (south of the Island) with olivine + augite + biotite + plagioclase,
- the province of Lentia (northwest) with augite + plagioclase + potassic feldspars,
- the province of Vulcanello (north) with olivine + augite + plagioclase + potassic feldspars + leucite,
- the province of Fossa (centre east) with augite + olivine rare + plagioclase + potassic feldspars.

The study of the distribution and the abundance of the different heavy minerals in the sea led us to establish four zones, the geographical position and the mineralogical nature of which link them directly to the land provinces.

- Zone Z1: olivine + augite + biotite; this zone stretches mainly from the Bay of Ponent to the west, as far as Capo Grillo in the east and corresponds to the old province of the south.

- Zone Z2: augite only; in very small quantities; it only subsists on the bed of the Cala Formaggio. This composition recalls that of the province of Lentia.

- Zone Z3: augite + olivine in abundance. This stretches in the north, from the Bay of Levant to the Bay of Ponent. It can be linked to the province of Vulcanello.

- Zone Z4: augite + olivine (rare). This zone lies in the Bay of Levant and recalls the province of La Fossa.

The boundaries between these different sectors are not always clearly defined.

c) The relations between mineralogy and trace elements

The Bay of Levant is the focal point of the different mineralogical zones Z2, Z3, Z4. Moreover the north of the Bay is marked by the presence of biotite (Fig. 11). Traditionally, most authors think that the essential carriers of trace elements are the minerals. These contain impurities (traces of different elements) by substitution in their lattice structure or in their microfissures (GONI, 1966; SMITH, 1966; TUREKIAN, 1963; TUREKIAN and WEDEPOHL, 1961).

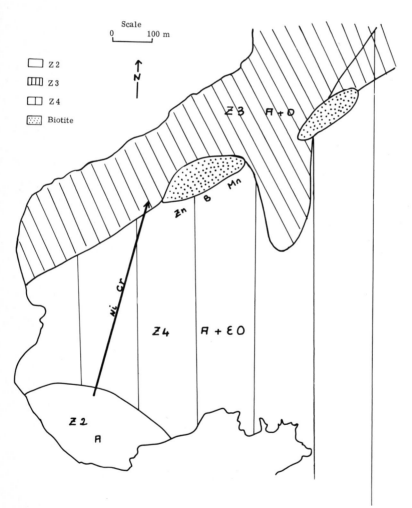

Fig. 11. Position of mineralogical sectors in the Bay of Levant
Z2: Zone with augite (A); Z3: Zone with augite (A) and abundant olivine (O);
Z4: Zone with augite (A) and rare olivine (ε O)

When we examine sediments taken from the mouth of the fumaroles through the microscope, we notice that minerals, especially leucite, olivine, and feldspar are changed (corrosion gulfs, transformations of optical properties) and this happens more often than in those which are submitted only to weathering erosion. If we take these results into account, the Bay of Levant, which is marked by the presence of different marine zones whose richness in olivine and biotite is variable, could present different content in elements (HEIER, 1962; SLAWSON and NACKOWSKI, 1959).

Indeed, olivine contains in addition to V, Mn, and Co mainly Cr and Ni (WEDEPOHL, 1963). It is therefore in these two elements that we should find variations in content in the direction of the arrow. That is what we found mainly in the south part of the bay. Elsewhere, the origin is more vague. The presence of biotite does not bring about any particular increase in content.

Moreover, some of these elements are fixed, absorbed or precipitated upon the minerals by fumarolian action. To the detrital mineral elements found in the "healthy" minerals must be added all the neogenesis minerals (gypsum, hydroscydes ...) which trap preferentially certain traces and in particular sulphides including pyrite, marcasite and chalcopyrite (FLEISCHER, 1955; KRAUSKOPF, 1967; WEDEPOHL, 1970).

Besides the sediments, a special place must be allotted to water: the water coming from the condensation of fumarole vapour; sea water covering the points of emissions (WOLGEMUTH and BROECKER, 1970).

Indeed, when an element such as manganese is concentrated in the deep zones, we do not know yet whether such a distribution corresponds to the absence of this element in the sediments and the water of the coastal zone or whether the Mn is nonexistent in the terrigenous fraction but present in the overlying water in soluble form; it would only precipitate in the sediment when the environmental conditions were suitable.

Conclusion

The analysis of the sediments of the Bay of Levant allowed us to follow the distribution of the different trace elements in detrital materials, and to see whether there exists a link between this distribution and granulometry, on one hand, and on the other, this distribution and mineralogy.

However, these relations are masked in the fumaroles zone by the direct action of the fumarolic modification of environment (temperature, pH, Eh), or by elements brought in. The hydrothermal activity in the "cleaning" of the sediments results in a mobilization of the Ba, Sr, V, and Cr contained in a detrital fraction. As for the elements Mn, Cu, Ni, Co, Pb, Zn and B, at the present time it is difficult to pass judgement.

However, this study allowed us to pose the problem of what fumaroles bring and the role they play. In order to reach a conclusion in a metallogenic sense, it will be necessary to carry on further research.

Acknowledgements

I should like to thank all the people who helped me by giving their advice in the writing of this paper. The paper was translated by A. Jappy.

References

AHRENS, L. H. : Origin and distribution of the elements. Pergamon Press (Oxford and New York), 1178 p. (1968).

BERGEAT, A. : Die Äolischen Inseln. Abh. bayer. Akad. Wiss. (München), Math. phys. Kl. XX, p. 192-202 (1899).

BERNAUER, F. : Rezente Erzbildung auf der Insel Vulcano. Teil I = Neues Jb. Miner Abh. (Stuttgart) 69, Beilagebände, p. 60-91. Teil II = Neues Jb. Miner. Abh. (Stuttgart) 75, Beilagebände, p. 54-71 (1935).

BOSTRÖM, K. : Submarine volcanism as a source for iron. Earth Planet. Sci. Lett. (Amsterdam), 9, p. 348-355 (1970).

BOSTRÖM, K. , FISHER, D. C. : Volcanogenic uranium and iron in Indian Ocean Sediments. Earth Planet. Sci. Lett. (Amsterdam), 11, p. 95-99 (1971).

CHASSEFIERE, B. : Sur la sédimentation et quelques aspects de l'hydrologie de l'étang THAU (Hérault). Thèse 3º cycle (Montpellier), 131 p. (1968).

DE FIORE, O. : Di un sulfuro di ferro delle fumarole sotto marine di Vulcano (Isola Eolie) formatosi nel 1916. Atti. Accad. naz. Lincei Rc. (Roma). Classe di scienze fisiche, matematiche e naturali. vol. 30, fasc. 3-4, ser. 5, sem. 2, 25 p. (1921).

FLEISCHER, H. : Minor elements in some sulfide minerals. Econ. geol. , 50th Anniv vol. , p. 970-1024 (1955).

GOLDSCHMIDT, V. M. : The principles of distribution of chemical elements in minerals and rocks. J. Chem. Soc. (London) March, p. 655-673 (1937).

—— Geochemistry. Clarendon Press (Oxford), 730 p. (1954).

GONI, J. : Contribution à l'étude de la localisation et de la distribution des éléments en trace dans les minéraux et les roches granitiques. Mém. B. R. G. M. (Orléan 45, p. 1-67 (1966).

HEIER, K. S. : Trace elements in felspars: a review. Norsk. geol. Tidsskr. (Kristiania, Bergen). Felspars volume no. 42, p. 415-454 (1962).

HONNOREZ, J. : La formation actuelle d'un gisement sous-marin de sulfures fumerolliens à Vulcano (mer Tyrrhénienne). Partie I: Les minéraux sulfurés des tufs immergés à faible profondeur. Mineral. Deposita (Berlin) 4, p. 114-131 (1969).

KELLER, J. : Personal cummunication (1967).

KRAUSKOPF, K. : Introduction ot geochemistry. McGraw-Hill (New York), 721 p. (1967).

LUNDEGARDH, P. : Aspects of the geochemistry of chromium, cobalt, nickel and zinc. Sver. geol. Unders. Afh. (Stockholm). Ser. c. , no. 513, p. 1-56 (1949).

MOAL, J. Y. : Le Quantomètre, pour les dosages des éléments en trace dans les matériaux naturels. Congrès de géochimie (Nancy) (1968).

SLAWSON, W. , NACKOWSKI, P. : Trace lead in potash feldspars associated with ore deposits. Econ. geol. (New Haven) 54, p. 1543-1555 (1959).

SMITH, J. V. : X-Ray emission microanalysis of common rock forming minerals. II: Olivine. J. Geol. (Chicago), 74, p. 1-16 (1966).

TUREKIAN, K. K. : Trace elements geochemistry. Trans. Am. geophys. Un. (Washington) 44, p. 526-532 (1963).

— Some aspects of the geochemistry of marine sediments. Chemical Oceanography. Academic Press (London and New York) t. 2, 1 vol, p. 81-126 (1965).

TUREKIAN, K. K. , WEDEPOHL, K. H. : Distribution of the elements in some major units of the earth's crust. Bull. geol. Soc. Am. 72, p. 175-191 (1961).

VALETTE, J. N. : Etude sédimentologique et géochimique des dépôts littoraux entourant l'île Vulcano (Sicile). Thèse 3º cycle (Paris), 175 p. (1969).

WEDEPOHL, K. H. : Die Nickel- und Chromgehalte von basaltischen Gesteinen und deren olivinführenden Einschlüssen. Neues Jb. Miner. Mh. (Stuttgart) 9-10, p. 237-242 (1963).

— Handbook of geochemistry. Springer (Berlin - Heidelberg) t. 2, part 2, p. 1-606 (1970).

WOLGEMUTH, K. , BROECKER, W. S. : Barium in the sea water. Earth Planet. Sci. Lett. (Amsterdam) 8, p. 372-378 (1970).

Address of the author:

Centre de Recherches de
Sédimentologie Marine
Centre Universitaire
66 Perpignan / France

Intergrowth and Crystallization Features in the Cambrian Mud Volcanoe of Decaturville, Missouri, U.S.A.

R. A. Zimmermann and G. C. Amstutz

Abstracts

Das Vorhandensein von Sulfiden und tonigen Substanzen in Sedimenten verzögert die Kristallisation während der Diagenese beträchtlich. Derartige noch weiche Sedimente können in benachbarte und höhere, bereits diagenetisch verfestigte Schichten intrudieren, wenn bei Bewegungen oder durch inhomogene Überlastung Spalten und Risse entstehen. Die vorliegende Arbeit beschreibt dafür ein Beispiel aus dem fossilen Schlammvulkan von Decaturville. Hier wurden wohl zum erstenmal auch Sulfide beobachtet, und die Arbeit konzentriert sich auf die Verwachsungsverhältnisse und die paragenetische Abfolge dieser Sulfide (Markasit, Pyrit, Zinkblende, Bleiglanz, Kupferkies). Die Identität mit dem Mississippi-Valley-Lagerstättentyp ist offenkundig.

The presence of sulfides and/or clay matter may considerably retard diagenetic crystallization. Semi-consolidated sediments of this kind may intrude adjacent and higher, more consolidated beds; movements or directed stresses resulting from unbalanced distribution of overlying sediment may cause joints and fissures to open up through which this soft sediment can intrude. The Decaturville mud volcanoe appears to have formed in this manner and it is probably the first such example with sulfides; the authors present here observations on the intergrowths and the paragenetic sequence of the various minerals which, in their order of decreasing abundance, are as follows: pyrite/marcasite, sphalerite, galena and chalcopyrite. The identity with the Mississippi Valley type of deposits is obvious.

- : -

Recent sedimentary mud volcanoes are known from many localities, especially from oil fields. Some are even reported to carry sulfides. Papers on fossil mud volcanoes are scarce and, to our knowledge, none of them contains sulfides. The present report may be the first account on sulfides in a fossil mud volcanoe and we thought, therefore, a detailed description might be of interest. Results of structural mapping in the Decaturville area are given in cross-section in Fig. 1.

The field aspects of the Decaturville mud volcanoe breccia were described in previous papers by AMSTUTZ (1965), AMSTUTZ and BUBENICEK (1967, Figs. 17 and 18), and AMSTUTZ and ZIMMERMANN (1965). Figures 2a and 2b clearly display the field aspects which, on a field trip to the area, were termed typical for mud volcanoe material by Professor GANSSER of Zürich, whose papers on recent mud volcanoes in Persia are well known. SNYDER and GERDEMANN (1965) pointed out that the Decaturville structure lies on a very long East-West fracture also running through other similar structures in the Middle West. AMSTUTZ (1965 a, 1965 b) has interpreted the details of the folded terrain within these structures as a product

a

Fig. 1. Outcrops of the sulfide breccia lie near the center of the Decaturville ring structure. The ring uplift (approximately 3.5 km in diameter) roughly corresponds to the circular tree belt; the structure of the northern part of the ring uplift is illustrated on the map and section. After ZIMMERMANN and AMSTUTZ (1965)

of tectonic diapir formation. The interpretation of this structure as an impact crater is made impossible by the following converging criteria:

1. the sequential tectonic events, described by the same authors (ZIMMERMANN and AMSTUTZ, 1965);

2. the sedimentary nature of large horizons of intraformational breccias inside and outside the structure and literally all over the Middle West;

3. the diagenetic features described in the present paper;

4. the existence of more than one (two or more) generations of brecciation (Fig. 3).

The Decaturville sulfide breccia contains variable amounts of pyrite, marcasite, sphalerite, galena and only specks of chalcopyrite, as illustrated in Fig. 4. Zoned pyrite investigated with the microprobe did not contain Ni, Co, As, or Cu. Of the whole rock rarely more than 5 % consist of sulfides. Of these, an average percent-

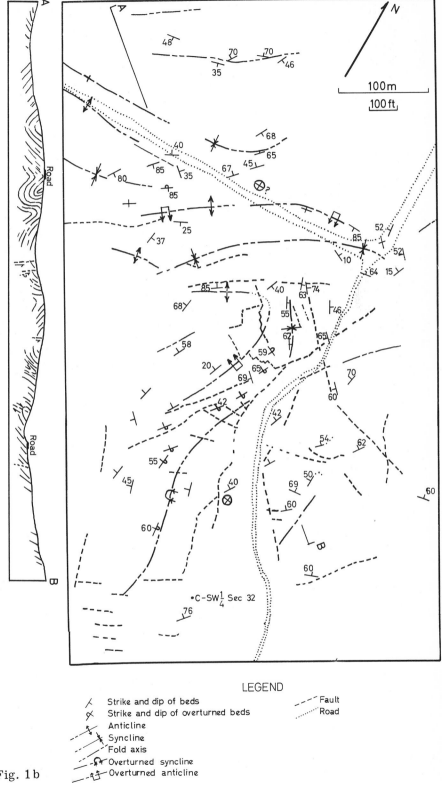

LEGEND

⊢ Strike and dip of beds
⊀ Strike and dip of overturned beds
   Anticline
   Syncline
   Fold axis
   Overturned syncline
   Overturned anticline

– ‒ Fault
······· Road

Fig. 1 b

Fig. 2a. Sandstone lense with flowage features engulfing limestone fragments (light grey) of the matrix breccia. Dark grey to black parts consist of sulfides

age would be 65 % pyrite and marcasite, 20 % sphalerite, and about 15 % of galena. Some streaks or bands contain up to 80 % sphalerite, whereas in others this sulfide is almost absent.

A partial intergrowth index was obtained by measuring the linear intergrowth of some 200 grains of sulfides. The index for sphalerite-galena is 9 %; the index for pyrite-marcasite with sphalerite is 91 %. Galena is very rarely intergrown with pyrite or marcasite (index below 1 %).

The average grain size of the sulfides in the matrix as determined from polished sections were determined as follows: pyrite and marcasite, $70\,\mu$; sphalerite, $60\,\mu$; galena, $50\,\mu$.

The earliest textures observed include the following classes and types:

1. Sulfide fragments showing intergrowths of pyrite, marcasite, calcite, sphalerite, and galena which are further classified according to the follwing texture types:

    (a) colloform pyrite-marcasite with banded sphalerite and zoned galena (or sphalerite) (Fig. 5);

    (b) crystals of sphalerite and calcite enclosed in later cement of pyrite (Fig. 6);

343

Fig. 2b. Intrusion veinlets of sulfide breccia in limestone. Base of sample is about 12 cm in length

    (c)   calcite with pyrite-marcasite with zonation of these, and with calcite commonly in inner parts of "nodules" (Fig. 7). This last type also occurs in texture classes (2) or (3), but then it is not fragmented. Along with this type (c) may be included skeletal and/or zoned rhombohedral intergrowths of calcite and pyrite (Fig. 8);

    (d)   the one fragment of colloform galena with zones of sphalerite might also be included here.

Type (a) and (b) were obviously crystallized in the sediment before brecciation and movement. Types (a) and (c) occur in fragments up to 4 cm in diameter. In the colloform type (a), pyrite grades into cocks-comb marcasite which in turn ends abruptly and is followed again by pyrite (or a band of sphalerite) deposited over the cocks-comb surface (Fig. 9). (The smallest galena grains lie at the bottom; larger ones in the middle, and largest galena grains at top of sphalerite bands; one therefore sees a galena-pyrite or galena-sphalerite intergrowth at the top of the sphalerite band; a galena-sphalerite intergrowth at the middle of the band; a galena-marcasite or galena-sphalerite intergrowth at the bottom of the sphalerite band.) The lowest pyrite band sometimes cements fragments of limestone (Fig. 3).

2. The next class of intergrowth consists of marcasite fragments cemented by pyrite.

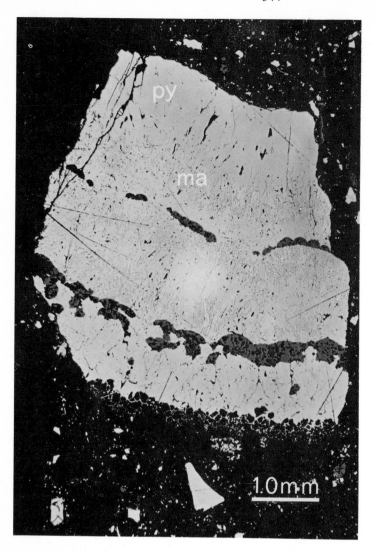

Fig. 3. Fragment of sulfides with bands of sphalerite, pyrite and marcasite deposited over carbonate breccia

3. The latest class of textures to form consists of the following two types:

   (a) small patches of "globular" pyrite-marcasite which, when observed under high magnification, are seen to consist of crystals in worm-shaped or even sub-colloform bunchy arrangement of pyrite and marcasite which encloses sulfide or carbonate breccia fragments (Fig. 10);

   (b) the same as (3) (a), but with pyrite-marcasite intergrown with calcite.

The occasional observation of two or perhaps more generations of brecciation points to different periods of mud volcanoe movements (including injections and perhaps extrusion; see Figs. 2a and 2b). Most of the large colloform fragments probably re-

Fig. 4. Matrix breccia consists of fragments of pyrite and marcasite (white), galena (white) (elongate rectangular fragment to the right of center of picture, at 2 o'clock), sphalerite (two large grey fragments in left half of picture) and limestone (dark grey)

present an important part of the pre-brecciation sulfide formation of the Decatur-ville sulfide breccia. However, some late pyrite-marcasite coatings on limestone fragments are also colloform. Much of the sphalerite occurs as independent grains or fragments, only intergrown with the country rock calcite. Small fragments of sphalerite may be either parts of sphalerite crystals, or parts of colloform sphale-rite. Big sphalerite fragments always possess a colloform texture (as shown in Figs. 3, 5 and 9).

Striking is the fact that most of the sulfide fragments contain one sulfide only and on-ly a minor portion is interlocked with other sulfides. This may indicate two things:

(a) that the original unbrecciated colloform bands of sulfides must have been thick-er than the sizes of the fragments now observed in polished and thin sections (< 2 mm size range);

(b) and that a large portion of the sulfide grains existed as isolated disseminated grains even before brecciation.

Though the breccia now occurs only partly stratiform and partly in irregular in-trusion structures with flowage textures (as seen in Fig. 2a), the nature of the country rock fragments and their intergrowth with the sulfides suggests that ori-ginally all of it was part of the normal stratabound Mississippi Valley-type mine-ralization of the area.

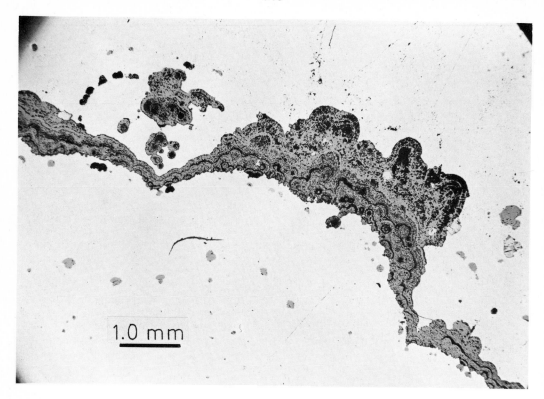

Fig. 5. Colloform pyrite-marcasite (white) and banded sphalerite (so-called schalenblende, grey). The colloform marcasite bands are made up of elongate grains lying at right angles to the margins of the bands. Galena (light grey) as small grains (extreme right of figure, intergrown with sphalerite and pyrite; at top of sphalerite band at right and left of picture; within the sphalerite band as polygonal or small irregularly shaped grains)

The age of the material also fits into this picture. In conclusion, the colloform banding and intergrowths with calcite and brecciated limestone indicate a typical formation. The brecciation together with the megascopic and microscopic flowage or "schlieren" textures and the small dykes or fissure fillings point to a mud volcanoe nature of the Decaturville breccia. As was mentioned at the beginning, this interpretation was first suggested on a field trip by Prof. A. GANSSER and we hope that the details presented here prove that this suggestion was correct.

Acknowledgements

Prof. P. Ramdohr made helpful observations regarding the ore microscopy for which the authors are grateful. They likewise are indebted to Mr. H.B. Hart of Decaturville for his kindness and permission to do field work on property of the Ozark Exploration Company. The financial assistance of the Deutsche Forschungsgemeinschaft to support field, travel and laboratory expenses is gratefully acknowledged.

Fig. 6. Crystals of calcite (dark grey) are earliest, followed by sphalerite (grey) and, finally, pyrite (white)

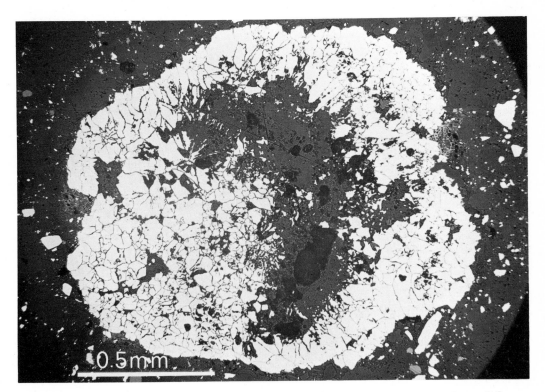

Fig. 7. Nodule with inner zone of calcite surrounded by marcasite with pyrite

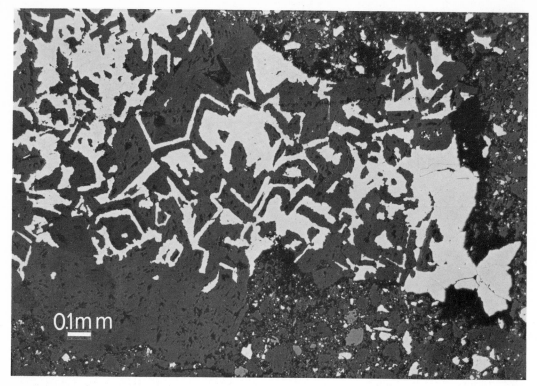

Fig. 8. Skeletal or zoned intergrowths of calcite and pyrite

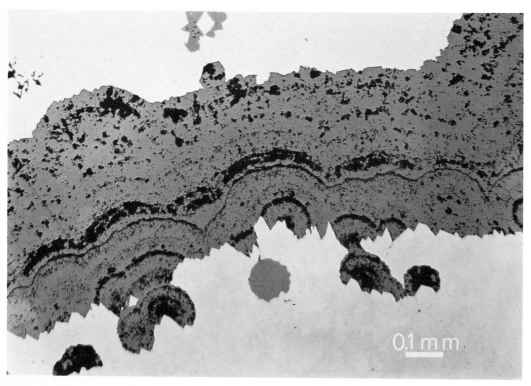

Fig. 9. The lower (concave) surface of the sphalerite (schalenblende) bands are bounded by underlying marcasite crystal faces

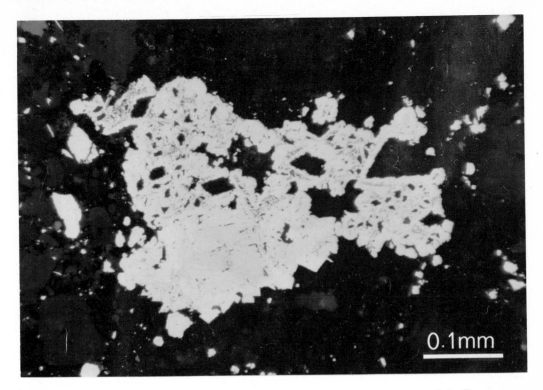

Fig. 10. Late pyrite often occurs in "globular" form; also in irregular shapes
similar to worm forms

Bibliography

AMSTUTZ, G. C.: Tectonic and petrographic observations on polygonal structures
    in Missouri. In: Geological problems in lunar research. Conference chm.:
    J. GREEN, Ed.: H. E. WHIPPLE. Annals of the New York Acad. Sci. 123,
    p. 876-894 (1965a).

— A morphological comparison of cone-in-cone structures and shatter cones.
    In: Geological problems in lunar research. Conference chm.: J. GREEN,
    Ed.: H. E. WHIPPLE. Annals of the New York Acad. Sci. 123, p. 1050-1056
    (1965b).

— BUBENICEK, L.. Diagenesis in sedimentary mineral deposits. In: Diagenesis
    in sediments, vol. 8. Ed.: G. LARSEN and G. V. CHILINGAR. Elsevier, New
    York, p. 417-475 (1967).

— ZIMMERMANN, R. A.: Decaturville sulfide breccia - a fossil mud volcanoe
    (Abstract). Geol. Soc. America Spec. Paper 87, p. 4 (1966).

SNYDER, F. G., GERDEMANN, P. E.: Explosive igneous activity along an Illinois-
    Missouri-Kansas axis. Amer. Jour. Sci. 263, p. 465-493 (1965).

ZIMMERMANN, R. A. , AMSTUTZ, G. C. : The polygonal structure at Decaturville, Missouri: new tectonic observations. In: Ries Colloquium, 25-26 June, 1965. N. Jb. Miner. Mh. , Nos. <u>9-11</u>, p. 288-307 (1965).

Address of the authors:

Mineralogisch-Petrographisches Institut
der Universität Heidelberg
69 Heidelberg / Germany

Springer-Verlag

Berlin · Heidelberg · New York

Recent Developments in Carbonate Sedimentology in Central Europe

Edited by **German Müller,**
Dr. rer. nat., o. Professor für
Mineralogie und Petrographie
an der Universität Heidelberg,
and **Gerald M. Friedman,**
Ph. D., Professor
of Geology, Rensselaer
Polytechnic Institute,
Troy, N.Y./USA

With 168 figures
VIII, 255 pages. 1968
Cloth DM 58,–; US $ 18.40

In the field of sedimentary research, ever increasing emphasis has been put on the investigation of carbonate and carbonate rocks during the past thirty years.

It is thus quite natural that in Central Europe – where classical carbonate investigations have already been carried out a hundred years ago – numerous scholars turned to the study of this type of sediment.

The present volume contains 30 papers summarizing the different subjects of a seminar held in Heidelberg in July 1967. Of course, these contributions represent only a small part of the work actually performed in the field of carbonate research in Central Europe. They give, however, a general survey of the work and the working methods employed in the different sectors of carbonate investigations.

Contents

Processes of Carbonate Formation and Diagenesis.
Microtexture and Microporosity of Carbonate Rocks.
Geochemistry of Carbonates and Carbonate Rocks.
Regional Carbonate Petrology:
Freshwater Carbonates and Carbonate Rocks.
Marine Carbonate Rocks.

■ **Prospectus on request**   **Applied Carbonate Petrology.**

Prices are subject to change without notice.

258 figures
XVI, 618 pages. 1972
Cloth DM 98,–;
US $ 31.10

Sand and Sandstone

By **Francis John Pettijohn,** Professor of Geology, The Johns Hopkins University, Baltimore, Md., U.S.A.;
Paul Edwin Potter, Professor of Geology, University of Cincinnati, Cincinnati, Ohio, U.S.A.;
and **Raymond Siever,** Professor of Geology, Harvard University, Cambridge, Mass., U.S.A.

Part one of "Sand and Sandstone" is an up-to-date summary of the compositional, textural and structural attributes of sandstones. Part two is an extended systematic treatment of all major species of sandstones including a treatment of the classification problem — a topic which has recently received much attention. This section of the book contains a glossary of rock names applied to sandstones. Of special interest is a related chapter on the volcaniclastic sands. Both chapters are profusely illustrated with photomicrographs. Part three emphasizes processes of sand generation: tracing sand back to its source areas, the transport of granular materials by fluid flow, soft-sediment deformation of sand, and the chemistry of diagenesis and related processes. The final section stresses the wider aspects of sand deposition, the geometrical form of sand bodies, environments of sand deposition and distribution of sandstones in space and time.
"Sand and Sandstone" is a comprehensive, though condensed, treatment of the salient facts about sandstones with emphasis on the principles governing the processes leading to the accumulation of sand.

 Springer-Verlag   Berlin Heidelberg New York
München · London · Paris · Tokyo · Sydney